AF517922

Commercial Fertilizers

Their Sources and Use

FIG. 1. Corn leaves showing nutrient deficiency symptoms. (*Bottom to top*) Nitrogen deficiency, phosphorus deficiency, potassium deficiency, probable magnesium deficiency, healthy leaf. (Courtesy, A. C. Richer, L. L. Huber, and B. L. Seem and the Pennsylvania Agricultural Experiment Station.)

Commercial Fertilizers

Their Sources and Use

— Fourth Edition —

by
Gilbert H. Collings

2017

Biotech Books

Delhi - 110 035

Gilbert, H. Collings (b. 1895–)
First Indian Impression 2017

ISBN 978-81-7622-065-1

Published by : **BIOTECH BOOKS**
1123/74, Deva Ram Park, Tri Nagar,
DELHI – 110 035
Phone: 27383999, (M) 09868812760
E-mail: biotechbooks@yahoo.co.in

Showroom : 4762-63/23, Ansari Road, Darya Ganj,
NEW DELHI - 110 002
Phone: 23245578

Printed at : **Chawla Offset Printers**
New Delhi – 110 052

PRINTED IN INDIA

This book

is dedicated to Professors

ROBERT J. DAVIDSON

At one time Chemist, South Carolina Agricultural Experiment Station, and later Dean, Scientific Department, Virginia Polytechnic Institute

and

CYRIL G. HOPKINS

Formerly Chief, Department of Agronomy, University of Illinois

and

JACOB G. LIPMAN

Formerly Dean, College of Agriculture of Rutgers University, and Director, New Jersey Agricultural Experiment Station

the investigators and teachers who were pioneers in the fields of agricultural chemistry, soil fertility, and soil microbiology, and at whose feet the author learned much that he in turn now attempts to pass on to students [illegible] is dedicated to Professors

ROBERT [illegible] DAVIDSON

[illegible] Chemist, South Carolina Agricultural Experiment Station, and [illegible] Department, Virginia Polytechnic Institute

[illegible]

CYRIL G. HOPKINS

Formerly Chief, Department of Agronomy, University of Illinois

JACOB G. LIPMAN

Formerly Dean, College of Agriculture of Rutgers University, and Director, New Jersey Agricultural Experiment Station

the investigators and leaders, who were pioneers

in the field of agricultural chemistry, soil

fertility, and soil microbiology, and to

whom in fact the author [illegible]

[illegible]

[illegible]

Preface to Fourth Edition

A new revised edition of "Commercial Fertilizers" is again required. The five war years that have transpired since the writing of the third revised edition of this book have been a period of great change. During this period research and experimentation in the fields of plant nutrition and fertilizer practice, while making some progress at some institutions, came to an almost complete stop at others. The fertilizer industry was forced by war conditions to make revolutionary adjustments and curtailments. Some fertilizer materials that were formerly found on the market disappeared from the market early during the war period and can now no longer be obtained. Whether these materials will again appear on the market is at present problematical.

During the war period the nitrogen fixation capacity of the nitrogen fixation plants of the United States was increased many times as a wartime necessity, and this fact will no doubt have an important bearing on the nitrogen consumption picture in this country in the immediate postwar years. Also, during the war period the American potash industry successfully expanded production to fill the wartime needs of this country for potash.

Regardless of wartime restrictions on publication, considerable new research information concerning plant nutrition, fertilizer practices, and fertilizer manufacture has been published during the last five years. From this wealth of new published information I have selected much that is suitable for the present edition of this book. Its incorporation in this edition has made necessary the modification of many statements given in the third edition and the rewriting of many paragraphs and sections of chapters. One new chapter on ammonium nitrate has been added for, since the writing of the third edition, ammonium nitrate has come on the market in large tonnages, and it bids fair to become a leader in its field.

Because production and consumption figures for recent years for some countries and states have not been published or are not yet reliable, and because I felt that many conditions as they existed during the war years were not normal, I have purposely used as illustrative material in many chapters the statistics, situations, or practices that prevailed in the normal years just prior to the outbreak of World War II.

It is my hope that this fourth revised edition will prove of even greater value than the three former editions for use in the classroom, for use on the farm, and for use in the fertilizer manufacturing plant.

GILBEART H. COLLINGS

Preface to First Edition

This book has been written in an attempt to satisfy a definite need in the agricultural colleges for a textbook on commercial fertilizers. Courses dealing with the subject already occupy a position of major importance in the curricula of the agricultural colleges in the southeastern states of the United States, and it now appears that they are destined to become of equal importance in the curricula of most of the agricultural colleges in the states east of the hundredth meridian, as well as in the Pacific Coast States.

The text is the outgrowth of the author's sixteen years of experience as an experiment station agronomist and teacher of soil, fertilizer, and crop courses at the Clemson Agricultural College. For a number of years the subject matter included herein has been presented to classes composed largely of juniors and seniors. It is assumed that the students who take this course have already had courses in Agronomy, Botany, Chemistry, Geology, and Soils, and, therefore, are familiar with the vocabularies of these subjects.

The manufacture and use of commercial fertilizers has progressed with the development of the science of plant nutrition. The fertilizer industry is about a hundred years old and, although it is passing through a widespread readjustment, it will, undoubtedly, continue to be one of our basic industries. Knowledge concerning plant nutrition has expanded rapidly during recent years and as a result many former concepts of plant nutrition have been greatly modified, thus making necessary important changes in fertilizer practices and recommendations.

Insofar as possible, it has been the aim of the author to give actual research findings or data, especially in the presentation of controversial subjects, in order that the student may make his own deductions and formulate his own conclusions.

The author is especially indebted to a large number of research workers in this and in other countries for much of the data, points of view, and illustrations that appear in the following pages. He is also indebted to many agronomists, chemists, and engineers who are connected with various fertilizer companies and associations. Each chapter has been read and constructive criticisms made by authorities connected with the organizations producing or marketing the products discussed. Thus, this book represents a composite of the opinions and conclusions of authorities within the fertilizer industry, as well as the conclusions of experiment station agronomists, especially in the fertilizer-using territory of the United States.

GILBEART H. COLLINGS

Contents

Commercial Fertilizers

FIG. 2. Thomas Green Clemson.

American farmer, chemist, engineer, author, and diplomat, whose bequest made possible the founding of Clemson Agricultural College, and whose efforts were influential in the founding of the Maryland Agricultural College.

Educated at the Sorbonne and the Royal School of Mines in Paris between 1828 and 1831, and later United States Chargé d'Affaires to Belgium (1844 to 1852), and Superintendent of Agricultural Affairs, United States Patent Office, 1860 to 1861, Clemson became personally acquainted with the fundamental chemical and plant research which led to the establishment in Europe of the practice of crop fertilization with chemical salts. His writings, especially those that appeared in the American Farmer between 1852 and 1861, probably did more to focus the attention of the American agriculturalist on the benefits to be derived from crop fertilization than did the writings of any other American of his time.

"One of the most interesting and important subjects to the agriculturalists is, of course, the means of keeping up, or restoring the fertility of his land; and that he may not work in the dark, it is essential that he should understand the nature and actions of soils, the functions of the plants, and the operation of fertilizers."—Rpt. Com. of Patents, p. 138, 1859.

"There can be no civilization without population, no population without food, and no food without phosphoric acid."—Rpt. Com. of Patents, p. 172, 1859.

1

Origin and Development of Use of Commercial Fertilizers

As shown by archeological research, the cultivation of crops was begun by man about 10,000 to 12,000 years ago. The development of this art by early man was, no doubt, soon followed by the application to the soil of animal manures and other materials for the purpose of increasing crop yields. It seems reasonable to believe that the first husbandman not only observed the influence of the droppings of animals on the growth of plants but also made practical use of this knowledge. Among the materials first utilized by man for fertilizers were animal manures, bones, wood ashes, wool wastes, guano, fish, chalk, and marl. The Celts (500 B.C.) and other European people are known to have used chalk or marl some hundreds of years before Christ. In America, the first white settlers found that the South American Indians had long been acquainted with the fertilizing value of guano, and that some of the North American Indians used fish for fertilizing purposes. Nevertheless, the extensive employment of fertilizing materials has become established only during the last 100 years.

Chemical salts which contain plant nutrients were not generally used as fertilizers by European and American farmers until during the last 85 years.

Some Early Discoveries Leading to Use of Chemical Fertilizers

The use of chemical fertilizers probably dates from 1665, when Sir Kenelm Digby wrote that he had increased the yield of crops by the application of saltpeter. Little was understood concerning the principles of fertilization, however, until 1804, when Nicholas Théodore de Saussure, of Switzerland, first directed the attention of the scientific world to the fact that the ash ingredients of plants were taken from the soil and that they were, in the main, essential for plant growth. About 50 years later, in 1855 and 1856, Justus von Liebig, a German chemist of the University of Giessen, laid great emphasis upon the necessity of supplying plants with phosphoric acid and potash. Liebig had proposed his now famous mineral theory in 1840, and had demonstrated the essential nature of potash in 1845. The most noted American chemist to support Liebig's mineral theory was Thomas Green Clemson, at one time Superintendent of Agricultural Affairs (1860–1861) of the United States Patent Office. Clemson wrote, in 1856, "The application of guano or any other manure, because of the presence of ammonia, is more than useless." The necessity for supplying plants with nitrogen was not recognized until 1857, when Lawes, Gilbert, and Pugh, of the Rothamsted Experiment Station, England, definitely established the essential nature of this element.

Materials Used as Fertilizers During Nineteenth Century

Ground bone was one of the first materials to be sold extensively for use as fertilizer. Bones are known to have been used for fertilizing purposes in England as early as the seventeenth century, and a commercial trade in bones was thoroughly established by the middle of the nineteenth century. Ground bone was first used in the United States in 1825.

Peruvian guano was first imported into the United States in 1832 and into England in about 1820. Although the deposits of Peruvian guano were discovered by the German traveler and botanist Humboldt in 1802, the fertilizing value of guano was not generally recognized until about 1840, and its extensive exportation did not take place until 1843. It was about this time that the Peruvian government began its exploitation of the deposits. The deposits of guano were consumed rapidly, for in 1875 it was estimated that there were only about 2,000,000 tons left in the deposits, whereas in 1840 the deposits had been estimated to contain at least 12,000,000 tons. The annual exportations of Peruvian guano showed a steady decrease from 1875 to the end of the century. Although the annual American consumption of Peruvian guano necessarily diminished with the exhaustion of the supply, it is well to emphasize the fact that the extensive employment of Peruvian guano as manure had served a very useful function in introducing to many American farmers the new practice of crop fertilization.

Exploitation of the Peruvian (later Chilean) deposits of sodium nitrate was begun with British capital in about 1830, and during that year the first small importation of Peruvian sodium nitrate was made into the

United States at Norfolk, Virginia. Importation of Peruvian sodium nitrate into Europe was begun about 1831. In 1850 nearly 240,000 metric tons of sodium nitrate were shipped out of Chile. During the remainder of the nineteenth century the use of sodium nitrate rapidly increased in popular favor among farmers in both Europe and America.

By-product ammonium salts were first produced in England in 1840, but the development of the by-product coke oven did not take place in the United States until 1893. By-product ammonium sulfate appeared on the American market a short time later.

FIG. 3. Sir John B. Lawes, Founder of the fertilizer industry. (Courtesy, Rothamsted Experiment Station, England.)

The German potash salts were first used for fertilizing purposes about 1860, and the exploitation of the German deposits of potash was begun in 1861. The first German potash was imported into the United States in 1869–70. German potash is said to have been first used by an American agricultural experiment station in 1872. During that year it was tested on the farm of the Pennsylvania State College. By the end of the nineteenth century potash salts had become thoroughly established in the fertilizer market.

The mining of phosphate rock was begun in the United States in 1868 and in North Africa in about 1899. Basic slag was first produced for fertilizer purposes in about 1880. It was first used as fertilizer in England in 1884 and thereafter gradually came into common use in Europe.

In 1839–1840 Baron Justus von Liebig demonstrated that the fertilizing value of bone could be increased by treatment with sulfuric acid—an idea that had been suggested by Escher in 1835. In 1842 Sir John B. Lawes took out a patent protecting a similar treatment of ground phosphate rock with sulfuric acid, and in 1843 he began the manufacture of what he called "super-phosphate" at Deptford, England and made a commercial success of the enterprise. A patent covering a similar procedure was taken out also in Scotland and Ireland by Dr. James Murray in 1842 but he failed to make a commercial success of his undertaking. A second plant was built by Lawes in 1857. In 1854 Edward Packard erected the first complete superphosphate factory, that is, one that included sulfuric acid works. In 1872 Lawes disposed of his fertilizer plants for about $1,500,000. It is from this period that the commercial fertilizer industry of today had its beginning. Sir John B. Lawes generally is considered the founder of the industry, although there are some who would confer this honor upon Baron Justus von Liebig. Superphosphates, which probably were made from bones, were first manufactured and sold in the United States in 1852.

The first mixed fertilizers manufactured in the United States for commercial purposes were made in Baltimore in 1849 by Dr. P. S. Chappell and William Davison. During this same year fertilizer factories were erected also in Germany. By 1854 a few hundred tons of mixed fertilizers were being manufactured annually in this country, largely in Richmond, Virginia, and Charleston, South Carolina. Fertilizer factories began operations in a small way in Philadelphia and Baltimore in 1855. It has been estimated that in 1856, 20,000 tons of commercial mixed fertilizers were manufactured in the United States, but it is interesting to note that during that year 60,000 tons of Peruvian and Mexican guanos were imported and consumed. In 1860 there were 7 fertilizer plants in operation in the United States. By 1880 the annual production of mixed fertilizers had increased to about 1,150,000 tons and by 1900 to about 2,000,000 tons.

The fertilizer materials available on the market in 1900 are shown in the accompanying table.

Table 1

FERTILIZER MATERIALS AVAILABLE FOR USE IN 1900

Nitrogen Sources		*Phosphoric Acid Sources*	*Potash Sources*
Inorganic	*Organic*		
Ammonium sulfate	Castor pomace	Basic slag	Kainite
Sodium nitrate	Cottonseed meal	Precipitated phosphate	Manure salts
	Dried blood	Superphosphate (then called acid phosphate)	Potassium chloride
	Fish scrap		Potassium sulfate
	Rough ammoniates		Sulfate of potash-magnesia
	Tankage		

Founding of the Rothamsted Experiment Station

The world-renowned Rothamsted Experiment Station was founded by John B. Lawes, later Sir John B. Lawes, on his estate at Harpenden, Hertfordshire, England. He began his plat tests in 1837 and in 1843 associated with him Henry Gilbert, later Sir Henry Gilbert. In 1889 he endowed a trust, known as the Lawes' Agricultural Trust, with about $500,000 to carry on the work of the station. This sum of money was part of a fortune which he had amassed from the manufacture of superphosphate. Lawes died in 1900, and his collaborator, Gilbert, in 1901. Sir A. D. Hall succeeded Gilbert as director of the experiment station in 1902, and Hall in turn was succeeded in 1912 by Sir E. J. Russell, who retired in 1943. Dr. W. G. Ogg was recently appointed Director of this station. The station's activities have dealt largely with soil and fertilizer problems. In 1893 all experimental results which had been secured since this experiment station began operation were published in six volumes known as the Rothamsted Memoirs. Additional volumes have been published at intervals since that time. In 1905 the results of the station's work were published by Director Hall in the "Book of the Rothamsted Experiments." Annual reports have been issued since 1908. Extensive contributions to the subject of plant nutrition have been made by this, the oldest of all agricultural experiment stations. The fertilizer industry of the world will always be indebted to the Rothamsted Experiment Station and to the men who have directed its affairs.

Founding of the American Agricultural Experiment Stations

A testing garden established by the Lords Proprietors on the Ashley River in South Carolina in 1669–70 appears to be the first agricultural experimental farm to be established in America for the purpose of improving agriculture.

In 1887 the Hatch Act was passed by the United States Congress. This act called for the establishment of an agricultural experiment station in every state in the Union. Prior to this time there were in existence 19 state-supported agricultural experiment stations, but sooner or later these were absorbed by the stations founded as a result of the Hatch Act.

Collectively, the state agricultural experiment stations have conducted much research dealing with plant nutrition, and soil and fertilizer problems. In fact, the American fertilizer industry has always depended largely upon these experiment stations to determine the best methods for using the industry's products. Only a few fertilizer manufacturers maintain research organizations.

Founding of the International Institute of Agriculture

The International Institute of Agriculture was founded in 1905 as the result of a conference called by the King of Italy at the suggestion of Mr. David Lubin of California. The purpose of this organization is to collect, compile, and disseminate statistics dealing with agriculture. The organization began to function in 1909 and has achieved noteworthy progress.

The Institute publishes several reviews which cover a wide range of agricultural subjects. Information concerning the utilization of fertilizers in agriculture is published in a monthly known as the "International Review of Agriculture." It also publishes an "International Yearbook of Agricultural Statistics," in which may be found statistics representing the production, consumption, and trade in fertilizer materials in all the leading countries of the world. It has been the custom, every few years, to publish a statistical review that deals with the world trade position of the fertilizer industry. Although, as pointed out by Clark and Sherman (1945), this publication has its limitations in a survey of plant food production and consumption for a particular country, it is nevertheless the most complete and accurate source of fertilizer statistics on a world-wide basis.

The International Institute of Agriculture is supported financially by

FIG. 4. Rothamsted House. The home of Sir John B. Lawes on his estate near Harpenden, Hertfordshire, England. (Courtesy, Rothamsted Experiment Station, England.)

most of the leading governments of the world. The fertilizer statistics compiled by this organization are of great value to the fertilizer industry.

Founding of the International Society of Soil Science

The International Society of Soil Science was founded in Rome, Italy, on May 19, 1924. The first president of the society was the late Dr. Jacob G. Lipman, Dean of Agriculture at Rutgers University in New Jersey. The purpose of the society is the encouragement and promotion of soil science in general. This is accomplished: first, through the publication of a review in which appear the research findings of the leading soil men of the world; second, through the establishment of a central office for soil science literature and documentation at the International Institute of Agriculture in Rome; and third, through the organization of congresses and conferences, and the formation of sections and commissions, to study various soil problems. The accomplishments of this society have proved of distinct value to the world's fertilizer industry, as well as to agriculture and science in general.

Expansion of Fertilizer Industry During Twentieth Century

The fertilizer industry, both in this country and in Europe, continued to expand during the latter part of the nineteenth and the early part of the twentieth centuries. In the United States the expansion was delayed somewhat because of the availability of enormous acreages of virgin land. In time, however, as these virgin soils were put under cultivation and as their fertility became depleted through cropping and erosion, the fertilizer industry made a substantial growth. Roughly, the adoption by farmers of the United States of the practice of using commercial fertilizers has progressed somewhat proportionally with the depletion of farm lands from the Atlantic seaboard westward. Today there are several states whose expenditures for fertilizers amounts to more than 10 per cent of their total cash farm income.

In the decade prior to World War I the only marked innovation that was made in the fertilizer industry was the development by German chemists of economically feasible methods for fixing atmospheric nitrogen. Following World War I there appeared on the American market many new synthetic nitrogenous fertilizers and also new fertilizer materials containing nitrogen and phosphoric acid and nitrogen, phosphoric acid, and potash. These materials were of both foreign and domestic manufacture.

FIG. 5. Monticello, the home of Thomas Jefferson on his plantation near the University of Virginia. Concerning this farm Jefferson wrote, "We now plow horizontally following the curvature of the hills and hollows on dead level, however curved the lines may be. Every furrow thus acts as a reservoir to receive and retain the water; scarcely an ounce of soil is now carried away. . . . In point of beauty nothing can excell that of our waving lines and rows winding along the face of our hills and valleys." (Courtesy, American Potash Institute.)

During the first third of the century the exportation of phosphate rock from the United States to Europe gradually declined, due largely to the increased production and exportation of phosphate rock from the North African deposits. Also during this period superphosphate of a high phosphoric acid content replaced a considerable tonnage of superphosphate of lower analysis.

The present American potash industry had its beginning during the first World War and for a decade following the war it passed through many vicissitudes. The domestic production of potash for 1931, which was derived entirely from distillery waste and the brines of Searles Lake, California, amounted to 63,880 tons. The discovery of potash in the Permian basin of the Southwest in 1912, and the successful mining which began at Carlsbad, New Mexico, in 1931 has succeeded in freeing this country from dependence upon foreign sources. The American production of potash during 1945–1946 was about 890,000 tons.

Table 2

FERTILIZER MATERIALS DISTRIBUTED BY THE UNITED STATES GOVERNMENT THROUGH THE TENNESSEE VALLEY AUTHORITY IN 1942 TO 1943

	(short tons)
Mixed fertilizer (mostly 0–14–14)	150,262
Normal superphosphate	1,217,143
Concentrated (double) superphosphate	22,872
Basic slag	39,545
Raw rock phosphate	32,020
Other phosphates (mostly calcium metaphosphate)	8,037
Manure salts	2,002
Gypsum	1,061
Borax	1,271
Total	1,474,213

Table 3

PRINCIPAL FERTILIZER MATERIALS SOLD AS SUCH BY COMMERCIAL AGENCIES IN 1942 TO 1943*

Material	*Continental U. S. (short tons)*	*Total U. S. (short tons)*
Normal superphosphate	802,605	803,056
Nitrate of soda	753,724	766,285
Miscellaneous phosphates†	175,973	178,428
Sulfate of ammonia	140,362	172,065
Liming materials	113,539	113,618
Cyanamid	59,183	59,183
Dried manures	57,450	57,450
Concentrated superphosphate	53,323	53,323
Muriate of potash	49,797	52,855
Miscellaneous potash materials‡	41,360	44,931
Miscellaneous materials§	29,800	29,800
Cottonseed meal	28,217	28,223
Other organics¶	25,397	25,397
Other chemical nitrogenous‖	22,299	22,330
Kainit and manure salts	22,225	22,225
Bone meal	17,509	17,510
Animal tankage	3,838	3,838
Totals	2,396,601	2,450,517

* Estimated to be 95 per cent complete.

† Phosphate rock, basic slag, precipitated bone, ammonium phosphate, base goods, etc.

‡ Sulfate of potash, nitrate of soda-potash, wood ashes, tobacco stems, etc.

§ Materials reported as miscellaneous, peat, borax, copper sulfate, manganese sulfate, gypsum, iron sulfate, aluminum sulfate, etc.

¶ Castor pomace, process tankage, fish scrap, soybean meal, guano, peanut meal, sewage sludge, cocoa hulls, compost, etc.

‖ Uramon, Cal-nitro, ammonium nitrate, ammonia, etc.

A very significant development that took place about 1930 in the fertilizer industry was the ammoniation of superphosphate. This development allows the practical use, by the farmer, of cheap forms of fixed atmospheric nitrogen. In 1943 ammonium nitrate came on the market in large amounts. Its economic importance is just beginning to be understood.

The distribution of fertilizers by government agencies was started in 1935 as part of the "Test Farm Demonstration" program of the Tennessee Valley Authority. The fertilizer distributed has been largely superphosphate. The farmer receives the phosphate without cost other than paying transportation charges and assuming certain obligations.

The principal fertilizer materials sold directly to consumers in the United States during the 1942–1943 season is given in Table 3.

The present world consumption of the principal chemical fertilizer nitrogen products is, in order of magnitude, about as follows: sodium nitrate, ammonium sulfate, cyanamide, cottonseed meal, calcium nitrate, ammonium phosphate, ammonia (anhydrous), ammonium nitrate, and urea.

Table 4

GROWTH OF THE FERTILIZER INDUSTRY SINCE 1859

Year	*Number of Establishments*	*Capital Invested*	*Value of Products*
1859	47	$ 466,000	$ 891,000
1869	126	4,914,000	5,815,000
1879	364	17,913,660	23,650,795
1889	390	40,594,168	39,180,844
1899	422	60,695,753	44,657,385
1909	550	121,537,451	103,960,213
1919	600	311,633,259	281,143,587
1929	638	..	232,510,936
1939	764	300,000,000	185,684,328
1944*	1100	325,000,000	476,000,000

* Estimated.

At the present time there are between 800 and 900 companies in the United States engaged in the manufacture, distribution, and sale of fertilizers. These companies operate in excess of 1,100 plants. American farmers spend about $476,000,000 annually for fertilizers and approximately one-third of this expenditure comes out of the pockets of the cotton farmers. Table 4 shows figures compiled by the Bureau of Census, United States Department of Commerce, illustrating the growth by 10-

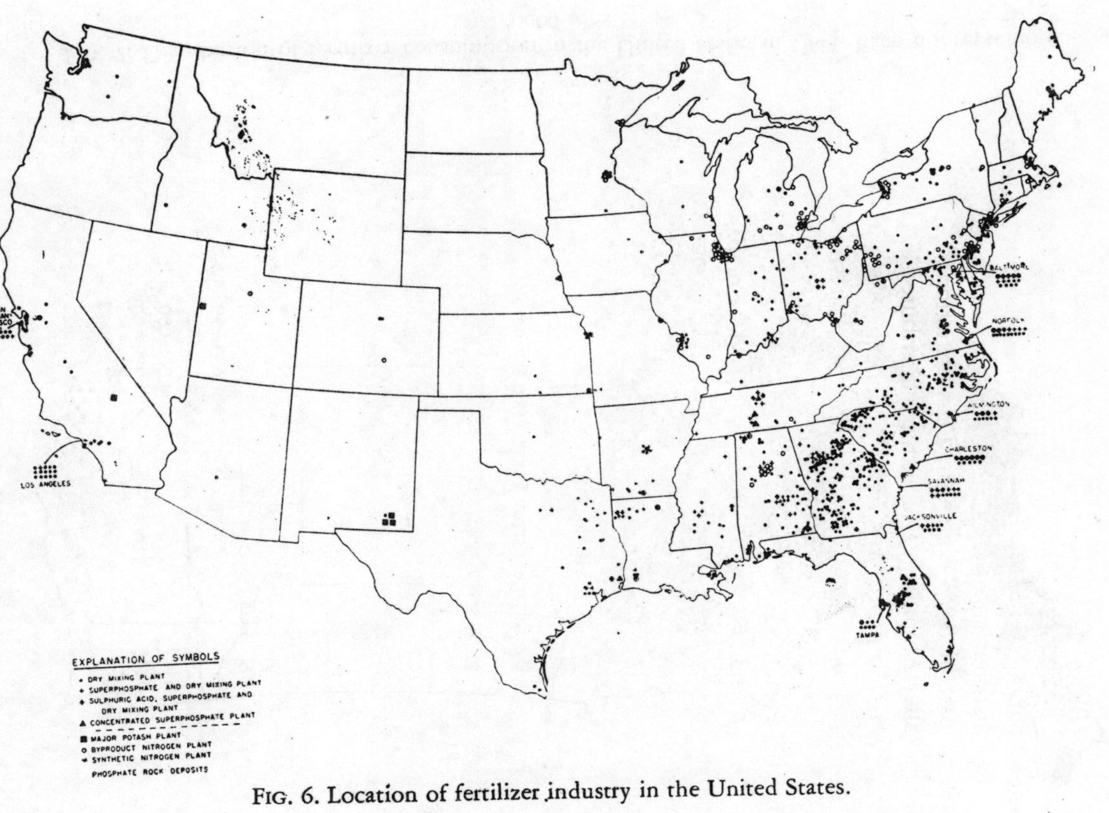

FIG. 6. Location of fertilizer industry in the United States.

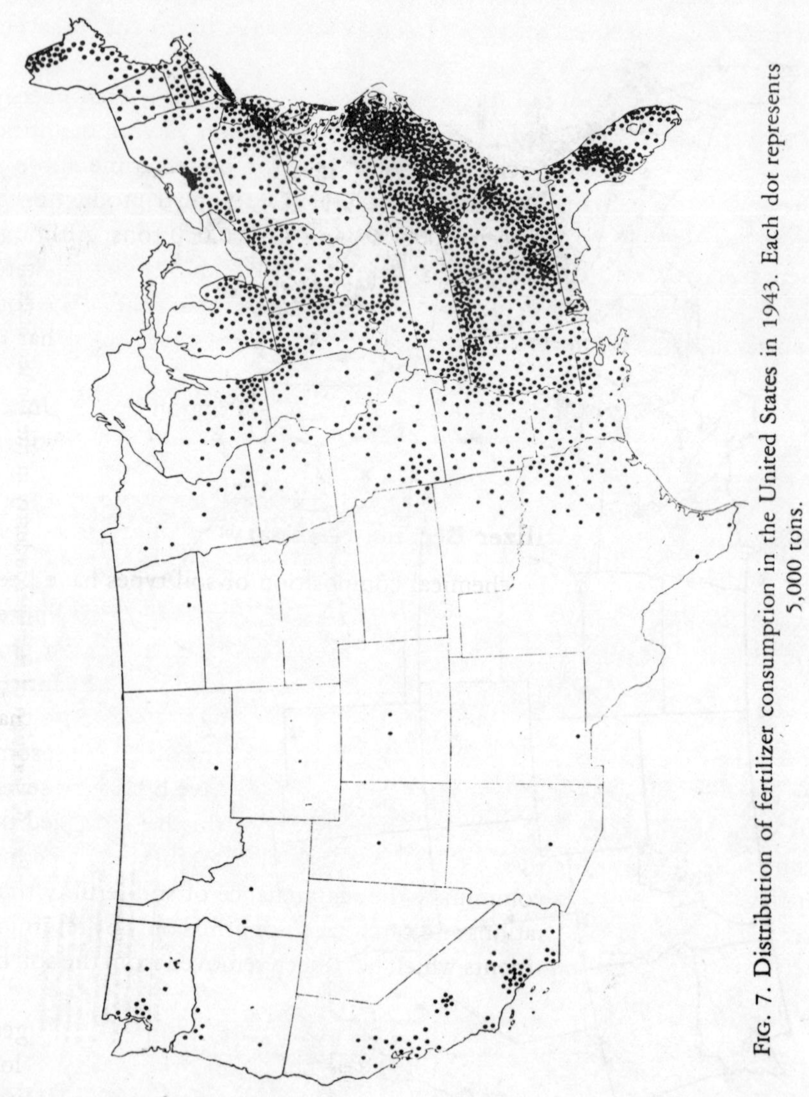

FIG. 7. Distribution of fertilizer consumption in the United States in 1943. Each dot represents 5,000 tons.

year periods that has taken place in the fertilizer industry since 1859. Capital invested in fertilizer establishments increased from around $466,000 in 1859 to about $476,000,000 in 1944. These establishments annually average jobs to the extent of 20,000 and a peak employment of several times this number.

Our domestic plant capacity for the production of finished fertilizers—that is, those compounded plant foods which contain varying quantities of nitrogen, phosphoric acid, and potash—at the present time approximates 12,000,000 to 14,000,000 tons a year. Our annual production in recent years has been between 8,000,000 and 12,000,000 tons. Although between 1900 and 1945 the consumption of fertilizers in the United States increased about 26 per cent, the plant food content of the 1945 consumption of mixed fertilizers was about 39 per cent more than that of 1900.

In Table 5 is given the average fertilizer consumption of the United States, by states, since 1910, as estimated by the National Fertilizer Association.

Why the Use of Fertilizer Became Necessary

Wide variations in the chemical composition of soil types have been noted by many investigators. Almost every soil type contains limited amounts of one or more of the nutrients necessary for optimum crop growth. As cropping and leaching remove from the soil large quantities of plant nutrients few soils can be cropped economically for more than a few decades continuously without requiring supplementation with some essential plant nutrient. In addition, erosion losses have been very severe in many parts of the country and in some sections this has hastened the time when the use of fertilizers became necessary. In establishing a permanent agriculture in any community the maintenance of soil fertility transcends all other considerations and can be accomplished only by returning to the soil those plant nutrients which have been removed from the soil by cropping, leaching, and erosion.

It has been estimated by Baker (1931) that 250,000 tons of nitrogen, 350,000 tons of phosphoric acid, and 3,500,000 tons of potash are lost annually in solution in the drainage water from the soils of the United States. The loss of nutrients by erosion, especially from the soils of the southern and southeastern states, amounts to many times that lost in drainage water. The seriousness of erosion has been recognized from the earliest days of the Republic. In 1817 Thomas Jefferson wrote minutely

Table 5

CONSUMPTION OF FERTILIZER IN TONS IN THE UNITED STATES FOR SELECTED YEARS, 1910 TO 1945

State	1910–1914 Average	1920–1924 Average	1930–1934 Average	1938	1940	1942	1944	1945*
NORTH ATLANTIC	982,608	1,073,934	978,443	1,177,234	1,217,851	1,252,775	1,544,375	1,684,127
Maine	113,760	168,275	173,930	140,000	140,000	143,000	215,000	229,348
New Hampshire	15,050	15,800	11,800	16,500	12,500	16,609	16,000	19,451
Vermont	13,040	17,200	13,568	20,067	14,585	14,311	14,054	19,236
Massachusetts	48,140	62,819	62,121	69,175	64,998	70,580	75,863	76,964
Rhode Island	7,300	8,860	8,201	12,200	11,500	12,326	14,000	15,000
Connecticut	54,200	71,000	57,200	56,274	61,753	64,913	69,000	83,755
New York	257,580	246,000	238,600	332,881	369,911	362,013	478,000	530,893
New Jersey	142,415	163,011	139,482	171,722	183,567	211,023	234,458	245,173
Pennsylvania	331,123	320,969	273,541	358,415	359,037	358,000	428,000	464,307
EAST NORTH CENTRAL	401 592	622,969	544,700	760,462	898,949	1,118,953	1,681,874	1,973,164
Ohio	165,652	298,058	243,673	324,228	363,320	427,679	576,011	623,263
Indiana	177,654	204,004	143,462	220,967	255,059	334,593	466,850	548,295
Illinois	6,200	14,991	25,097	36,132	49,753	72,744	185,565	257,014
Michigan	48,786	92,116	100,603	132,702	166,564	187,517	265,332	305,564
Wisconsin	3,300	13,800	31,865	46,433	64,253	96,420	188,116	239,028
WEST NORTH CENTRAL	54,036	75,182	73,936	115,918	142,380	138,298	318,248	391,242
Minnesota	1,800	6,000	11,606	13,778	18,627	25,135	52,527	68,546
Iowa	..	3,700	13,319	11,146	13,745	25,361	74,846	117,777
Missouri	47,326	58,596	43,972	40,301	87,577	70,937	146,504	160,024
North Dakota	..	160	696	650	1,800	2,350	3,050	3,300
South Dakota	..	145	120	150	500	700	300	325
Nebraska	..	470	400	1,794	2,200	2,400	2,500	3,125
Kansas	4,910	6,111	3,823	18,099	17,931	11,415	38,521	38,145

SOUTH ATLANTIC	3,697,374	3,445,593	3,104,299	3,752,617	3,724,120	4,135,711	4,849,256	5,266,329
Delaware	36,600	42,541	35,192	36,281	35,500	37,600	44,790	49,000
Maryland	163,670	155,006	145,471	166,408	160,315	182,720	207,485	218,144
District of Columbia	..	..	..	1,700	1,800	2,000	1,500	1,500
Virginia	385,353	429,785	350,391	405,179	393,069	420,668	515,578	594,829
West Virginia	30,680	36,468	41,658	53,500	40,000	47,665	46,890	53,753
North Carolina	750,252	1,012,353	941,559	1,104,788	1,076,730	1,243,012	1,369,751	1,512,013
South Carolina	984,826	752,361	589,583	660,963	685,310	698,704	799,083	822,875
Georgia	1,146,284	683,013	587,541	768,323	762,725	852,765	1,039,679	1,105,564
Florida	199,709	334,066	412,904	555.475	568,671	650,577	824,500	908,651
SOUTH CENTRAL	951,730	892,898	978,645	1,401,568	1,549,158	1,728,375	2,227,833	2,306,391
Kentucky	69,200	82,400	79,039	109,968	123,102	146,420	246,119	282,134
Tennessee	75,739	94,731	103,061	128,291	147,311	174,298	271,545	277,090
Alabama	476 829	346,444	383,100	528,850	575,900	574,150	720,000	722,500
Mississippi	131,235	149,871	190,378	325,836	319,508	377,082	417,624	403,557
Arkansas	51,370	62,451	60,170	67,800	101,000	140,458	131,083	126,650
Louisiana	88,880	90,278	91,427	148,542	156,775	173,241	216,245	239,560
Oklahoma	1,500	3,220	4,720	8,005	7,363	8,752	20,708	25,072
Texas	56,977	63,503	66,750	84,276	118,199	133,974	204,509	229,828
WESTERN	48,514	84,787	169,594	281,001	298,505	409,003	580,108	678,137
Montana	..	..	450	5,000	4,500	4,500	7,300	8,000
Idaho	..	340	1,220	9,015	7,000	9,600	20,000	30,000
Wyoming	..	..	460	1,500	2,100	2,000	2,500	3,000
Colorado	..	230	1,715	4,393	5,557	8,264	10,000	15,379
New Mexico	20	780	1,065	2,240	2,406	4,000	7,002	8,000
Arizona	..	425	1,980	6,500	6,902	10,000	21,600	29,453
Utah	..	300	1,760	2,000	3,050	2,500	7,000	6,026
Nevada	..	28	148	500	500	600	400	750
Washington	1,300	5,400	15,600	27,000	30,000	44,999	50,000	56,504
Oregon	3,960	6,900	10,400	14,500	19,500	21,895	36,000	37,434
California	43,234	70,384	134,796	208,353	216,990	301,145	418,306	483,591
* United States	6 135,854	6,195,363	5,849,617	7,488,800	7,830,963	8,783,115	11,201,694	12,299,390

* Preliminary.

on just how horizontal or contour plowing was done in Albemarle County, Virginia, and why. It is well to recall that nearly half of the cultivated land in the United States is in clean cultivation, and is in a region where the rains are of a torrential nature. Fippin (1945) has reported the annual loss by erosion per acre of some plant nutrients from the row crop acreage of the Tennessee River system to be 84.6 pounds of calcium, 97.9 pounds of magnesium, 212.2 pounds of potassium, 13.0 pounds of phosphorus—all expressed as oxides—and 7.1 pounds of nitrogen.

Bennett and Lowdermilk (1938) have estimated that 90,000,000 tons of phosphorus, potassium, nitrogen, calcium, and magnesium are contained in the soil eroded from the United States annually. This is 60 times the plant food found in the commercial fertilizer produced annually at that time in the United States.

Plant Nutrients Lost by Crop Removal

According to the late Dr. Jacob G. Lipman (1930), Director of the New Jersey Agricultural Experiment Station, cultivated crops removed from the soils of the United States in 1927, a representative year, approximately 32.5 pounds of nitrogen, 12.5 pounds of phosphoric acid, 30.75 pounds of potash, and 13 pounds of lime per acre. He concluded that after due allowance is made for the nitrogen returned to the land in animal manures, for that fixed by bacteria, and that brought down by rain and other atmospheric precipitation, there was still a net loss of 3,000,000 to 4,000,000 tons of nitrogen annually from our 300,000,000 acres of arable land. Lipman also estimated that the nitrogen in the commercial fertilizers

FIG. 8. Influence of complete fertilizer on the growth of cotton. (*Left*) Unfertilized. (*Right*) Fertilized. (Courtesy, Mississippi Agricultural Experiment Station.)

Table 6

ANNUAL BALANCE OF PLANT NUTRIENTS IN SOILS OF THE UNITED STATES FOR 1930

	Nitrogen (*tons*)	*Phosphorus* (*tons*)	*Potassium* (*tons*)	*Calcium* (*tons*)	*Magnesium* (*tons*)	*Sulfur* (*tons*)
Losses (harvested crops, grazing, erosion, leaching)	22,899,046	4,221,302	50,108,560	68,185,730	24,557,881	12,043,911
Additions (fertilizer and liming materials, manure and bedding, rainfall, irrigation waters, seeds, nitrogen fixed)	16,253,862	1,447,835	5,151,076	12,561,673	4,040,813	9,029,690
Net annual loss	6,645,184	2,773,467	44,957,484	55,624,057	20,517,068	3,014,221

replaced less than a tenth of that lost by crop removal alone. His estimates indicate also a net loss through crop removal of nearly 2,000,000 tons of phosphoric acid, of which less than half was returned in commercial fertilizers, and a net loss of nearly 5,000,000 tons of potash from this cause, of which only one-fifteenth was returned in fertilizers. Again in 1936 Lipman and Conybeare estimated the net annual loss of nitrogen, phosphorus, potassium, calcium, magnesium, and sulfur from the soils of the United States. These estimates of nutrient losses for the year 1930 are shown in the table on page 17.

Accompanying the recent increase in the population of the world there has been a somewhat corresponding increase in the amount of produce taken from the soil for direct or indirect human consumption, and in turn a corresponding increase in the loss of plant nutrients from the soil. It is well to recall that the population of the world in 1800 has been estimated at 800 millions. During the next hundred years, that is from 1800 to 1900, the population increased to 1730 millions. At the beginning of World War II the world population was estimated at 2,115 millions. It seems unreasonable to expect that such a rate of increase can be projected into the future for any great length of time without a somewhat corresponding increase in the consumption of commercial fertilizers.

Relationship of Fertilizer Industry to Other Industries

The fertilizer industry has always been closely related to many other industries. It consumes many by-products of the packing plant. It also furnishes an outlet for basic slag from the steel industry, potash from the cement industry, ammonia from the industries engaged in the destructive distillation of coal and petroleum, sulfur from the copper industry, and meal from the cottonseed oil industry. The successes and failures of the fertilizer industry inevitably must influence the welfare of many other industries.

Use of Fertilizer in Foreign Countries Compared with that in America

During normal years farmers in some of the European countries use much more fertilizer per acre than American farmers, yet American soils cannot be said to be on the whole inferior or superior to those of Europe. The Soil Improvement Committee of the National Fertilizer Association has estimated that in 1934, a representative prewar year, the consumption of the plant nutrients, nitrogen, phosphoric acid, and potash, per acre of cultivated land was about 14 pounds in the United States, whereas it was about 18 pounds in France, 57 pounds in Germany, 84 pounds in Belgium, and 106 pounds in Holland. It is a well-known fact that the average crop yield per acre in Europe is very much larger than is the average yield per acre in the United States. This is certainly due in part to the use by European farmers of larger quantities of plant nutrients per acre.

In Table 7 is given, by countries, the world average consumption for the years 1937, 1938, and 1939, of the three plant nutrients, nitrogen, phosphoric acid, and potash, as compiled from data secured by Clark and Sherman (1946), of the Bureau of Plant Industry, Soils, and Agricultural Engineering, Agricultural Research Administration. These are the latest figures available for individual countries. It will be observed that Germany leads all other countries in the total consumption of these nutrients and in the consumption of nitrogen and potash. The United States ranks second in total consumption of plant nutrients and in the consumption of nitrogen and potash, but it ranks first in the consumption of phosphoric acid. Three countries, the United States, Germany, and France, with about

FIG. 9. The most famous agricultural field in the world—Broadbalk field of the Rothamsted Experiment Station, England. This field has grown wheat every year since 1843. Each plat receives a different combination of fertilizers. (Courtesy, *American Hortigraphs and Agronomic Review.*) (*Royal Air Force Official Copyright Reserved.*)

Table 7

World Apparent Consumption of Nitrogen, Phosphoric Acid, and Potash by Countries, Average 1937, 1938, and 1939

Country	*Nitrogen (N), tons*	*Phosphoric Acid (P_2O_5), tons*	*Potash (K_2O), tons*	*Total, tons*
Germany	571,384	564,117	950,507	2,086,068
United States	334,603	675,223	353,424	1,363,250
France	154,770	353,274	262,365	770,408
Soviet Union	139,892	386,731	148,250	674,873
Japan	252,824	232,909	101,758	587,492
Italy	116,229	219,486	14,503	350,214
Netherlands	62,081	100,858	167,447	330,386
United Kingdom	50,790	180,472	75,711	306,972
Spain	103,727	165,565	28,113	297,405
Australia	12,550	249,595	8,495	270,640
Belgium	52,181	81,087	64,612	197,880
Poland	26,109	32,872	81,340	140,322
Denmark	40,542	67,163	32,250	139,955
Sweden	27,690	50,917	47,776	126,383
Korea	93,690	21,017	1,500	116,207
New Zealand	4,276	101,903	6,749	112,928
Czechoslovakia	24,852	48,646	27,868	101,366
Egypt	73,733	8,955	293	82,981
Canada	14,738	33,953	22,608	71,299
Union of South Africa	9,112	41,383	2,989	53,485
Finland	7,866	28,741	15,731	52,339
Formosa	34,180	13,970	2,707	50,827
Portugal	15,928	32,865	1,029	49,823
Peru	19,392	16,408	9,850	45,650
Eire	6,765	26,717	7,486	40,969
Switzerland	3,911	24,219	8,098	36,228
Norway	7,191	14,369	15,182	36,742
Latvia	3,767	22,923	7,797	34,488
Dutch East Indies	18,864	11,413	2,133	32,410
Puerto Rico	14,446	4,626	12,375	31,447
China	27,035	480	220	27,735
Austria	5,184	12,832	9,059	27,075
Hawaii	13,612	32	10,597	24,241
Lithuania	1,319	18,085	3,868	23,272
Ceylon	8,468	9,119	4,461	22,048
British India	16,288	1,612	1,788	19,688
Algeria	3,145	12,134	4,238	19,517
Greece	5,211	7,376	3,619	16,205
Esthonia	911	10,396	2,747	14,054
Hungary	2,333	10,371	1,100	13,804
Canary Islands	4,521	2,653	3,200	10,374
Philippines	6,649	108	200	6,957
Yugoslavia	1,668	4,397	474	6,539
Mexico	2,802	..	..	2,802
Brazil	1,938			1,938
All others	30,947	25,764	14,745	86,823
World total	2,431,114	3,942,136	2,541,261	8,914,511

one-tenth of the world population, consume collectively 47 per cent of all the plant nutrients consumed in the world.

Agricultural Value of Fertilizers Applied During This Century

During the last 10 years fertilizer consumption in the United States has increased about 115 per cent, fertilizer nitrogen consumption 127 per cent, fertilizer phosphoric acid consumption 146 per cent, and fertilizer potash consumption 144 per cent. No one will ever know to what extent commercial fertilizers have aided in the production of larger yields of crops during this century in both Europe and America. The late Dr. Cyril G. Hopkins, of the Illinois Agricultural Experiment Station, addressed a questionnaire to some of the leading agricultural authorities in European countries asking for an estimate as to what part of the increase in crop yields which had been noted in these countries in recent years might be due to the use of commercial fertilizers. France and Germany each credited the use of fertilizers with 50 per cent of the increased production per acre. The Netherlands stated that fertilizer was the largest factor, and this was also the opinion of the Rothamsted Experiment Station, England.

Where Fertilizers Are Used in the United States

Approximately one-half of the fertilizer consumed in this country is applied in the South Atlantic and Gulf States, and it is largely used in the production of cotton. The New England and Middle Atlantic States are also consuming large quantities of fertilizer, especially in the production of potatoes and other truck crops. In the states north of the Ohio River and in some states west of the Mississippi River and in the Pacific Coast region the consumption of commercial fertilizers is increasing rapidly, and especially is this true of those brands of complete fertilizers that carry a high percentage of phosphoric acid. Nitrogenous fertilizers will probably never be used extensively in the Great Plains region. At the present time the application of nitrogen to the soils of this area seldom effects much increase in yields of grain, except in unusual seasons when water sufficient to maintain excess growth is available to the crop. Most of the carriers of the rarer elements sold in recent years have been consumed in Florida.

In Table 8 is given the estimated acreage of land fertilized in the United States in 1942 (Smalley and Engle, 1943).

In 1945 nearly 60 per cent of the fertilizer manufactured in this country was used in the South where the states that lead in the use of fertilizer are,

Table 8

ESTIMATED ACREAGE OF LAND FERTILIZED IN THE UNITED STATES IN 1942

State	*Total Crop Acreage*	*Acres Fertilized*	*Per Cent Fertilized*
Maine	1,231,600	333,100	27
New Hampshire	379,300	124,500	33
Vermont	1,026,900	279,500	27
Massachusetts	437,600	347,800	79
Connecticut	369,600	237,900	64
Rhode Island	50,000	40,900	82
New England	3,495,000	1,363,700	39
New York	6,574,200	2,482,000	38
New Jersey	793,200	593,400	75
Pennsylvania	5,853,400	2,816,300	48
Delaware	375,900	199,600	53
Maryland	1,620,800	944,200	58
Virginia	3,851,400	2,831,000	74
West Virginia	1,422,800	738,100	52
Middle Atlantic	20,491,700	10,604,600	52
Ohio	10,359,500	4,823,000	47
Indiana	10,433,400	4,759,300	46
Michigan	7,922,700	2,892,400	37
Kentucky	5,496,500	3,300,000	60
East North Central	34,212,100	15,774,700	46
Illinois	19,236,600	1,073,400	6
Wisconsin	9,991,400	1,886,500	19
Minnesota	18,568,500	783,700	4
Iowa	21,530,300	688,900	3
Missouri	12,121,100	1,884,000	16
West North Central	81,447,900	6,316,500	8
North Carolina	6,457,200	5,517,000	85
South Carolina	4,897,300	3,532,000	72
Georgia	8,423,000	6,317,000	75
Florida	1,569,700	1,367,000	87
South Eastern	21,347,200	16,733,000	78

Table 8—(Continued)

State	Total Crop Acreage	Acres Fertilized	Per Cent Fertilized
Alabama	6,748,300	4,271,500	63
Mississippi	7,126,300	3,542,000	50
Tennessee	6,566,700	2,436,700	37
Louisiana	4,123,500	1,731,300	42
Arkansas	6,659,000	1,835,700	28
South Central	31,223,800	13,817,200	44
North Dakota	17,952,200	47,000	.3
South Dakota	15,148,200	14,000	.1
Nebraska	19,320,400	41,000	.2
Kansas	21,996,800	414,300	2
Oklahoma	12,763,600	287,000	2
Texas	26,512,600	1,457,000	5
Great Plains	113,693,800	2,260,300	2
Montana	6,932,700	81,500	1
Idaho	3,083,700	164,400	5
Wyoming	1,674,200	40,000	2
Nevada	463,600	12,000	3
Utah	1,123,000	80,300	7
Colorado	6,003,300	158,700	3
Arizona	789,300	52,800	7
New Mexico	1,692,100	40,300	2
Mountain	21,761,900	630,000	3
Washington	3,697,500	314,500	9
Oregon	2,591,500	143,700	6
California	6,210,800	1,299,800	21
Pacific Coast	12,499,800	1,758,000	14
Grand total, U.S.	340,173,200	69,258,000	20

in the order named, North Carolina, Georgia, and South Carolina. The leading states in the North are Ohio, Pennsylvania, and New York.

Distribution of Fertilizer Consumption by Crops Fertilized

Over half of the fertilizer used in the United States in 1938, a representative prewar year, was applied to the three crops: corn, cotton, and wheat. The corn crop now consumes a larger tonnage of fertilizer than the cotton crop. Formerly the cotton crop was the greatest consumer of fertilizer but recent curtailment of crop acreages by the government has brought about a shift in the percentage consumption of fertilizer on several crops.

In Table 9 are given the quantities of fertilizer used on certain crops in 1938.

Table 9

QUANTITIES OF FERTILIZER USED ON CERTAIN CROPS IN THE UNITED STATES IN 1942

Crop	Tons of Fertilizer	Per Cent of Total
Corn	2,203,900	22.02
Cotton	1,460,623	14.59
Vegetables	894,050	8.93
Wheat	780,300	7.79
Potatoes	708,650	7.08
Hay	704,850	7.04
Pasture	583,850	7.04
Fruits and nuts	667,100	6.66
Tobacco	536,400	5.34
Oats	496,200	4.95
Peanuts	183,600	1.83
Sweet potatoes	148,700	1.48
Barley*	119,300	1.19
Soybeans	67,700	.67
Sugar beets	40,200	.40
Sugar cane	32,500	.32
Rice	25,200	.25
Buckwheat	13,200	.13
Dry beans	14,000	.13
Cowpeas	13,200	.13
Rye	11,700	.11
Other crops	302,333	3.02
Total	10,007,556	100.00

* Includes mixed grains.

FIG. 10. Influence of complete fertilizer on the growth of corn. (*Left*) No fertilizer. (*Right*) 400 pounds of a complete fertilizer. (Courtesy, North Carolina Agricultural Experiment Station.)

Fertilizers Controlled by Governmental Monopoly

It is interesting to note that until after World War I, of the three essential plant nutrients—nitrogen, phosphoric acid, and potash—phosphoric acid was the only one that was subject to the fundamental economic laws of an open market regulated by supply and demand. Foreign monopoly controlled to a great extent the price of nitrogen and potash. In recent years concern has been felt over the marked tendency of the government of the United States to edge into the fertilizer business. This tendency should be carefully watched and studied by all students of commercial fertilizers. Calvin Coolidge, when president of the United States, spoke very wisely on this subject when he said: "When the Government once enters a business it must occupy the field alone. No one can compete against it. The result is a paralyzing monopoly."

The fertilizer industry in the United States has been built around our natural deposits of phosphate rock. In Germany the industry originally centered on potash, but at the beginning of World War II it centered largely on the production of fixed atmospheric nitrogen.

Unlike most American industries, the fertilizer industry of the United States enjoys no tariff protection, so that during normal years if American fertilizer prices get out of line with world prices, imports will follow.

The national possession of sources of fertilizer materials by any country makes for safety for that country, while their absence tends to create uneasiness.

World Production and Reserves of Fertilizer Minerals

The world production of fertilizer minerals, just prior to World War II, was about 18,000,000 metric tons. In Table 10 are given the world reserves of some fertilizer minerals as estimated by the U.S. Geological Survey. These reserves will last the world only about a thousand years at the rate of consumption of these materials during the years just prior to World War II.

Table 10

WORLD RESERVES OF THE FERTILIZER MINERALS

	(*Metric Tons**)
Phosphates	16,879,000,000
Potash salts	2,338,000,000
Nitrates	245,300,000
Sulfur	56,000,000
Pyrites	183,000,000
Totals	19,701,300,000

* A metric ton is equal to 1,000 kilograms, or 2204.6 pounds.

Are We Approaching the Saturation Point for Fertilizers?

The South Atlantic States lead all other sections of the United States in the consumption of commercial fertilizers. The farmers of these states seldom undertake the production of a crop without first applying a commercial fertilizer. Yet in these states the saturation point for the use of fertilizers is not a serious factor at the present time. A few years ago Pridmore, of the National Fertilizer Association, estimated that if the farmers of the Southeast had followed the recommendations of their state experiment stations during a recent year they would have used 6,496,750 tons of fertilizer under their cotton crop alone. During the year in question the total fertilizer tonnage for the entire South amounted to only 2,296,366 tons. If such a condition exists in the Southeast, where the practice of using commercial fertilizers is so thoroughly intrenched, the saturation point for fertilizers in other sections of the country can be of only minor moment for some time to come.

In 1945 The National Planning Association made an estimate of increases in fertilizer consumption that could be made to increase crop yields if disregard be made of economic conditions. This estimate is given in Table 11.

Table 11 shows that fertilizer consumption in this country might profitably be doubled.

Table 11

ESTIMATE OF FERTILIZER THAT COULD BE USED TO INCREASE CROP YIELDS

Crop	Fertilizer Used 1000 tons 1942	Increased Fertilizer for High Production	
		Per Cent	1000 tons
Cotton	1,461	50	703
Corn	2,204	100	2,204
Wheat	780	100	480
Potatoes	709	0	0
Tobacco	536	0	0
Vegetables	894	0	0
Oats, barley, and rye	627	50	313
Sweet potatoes	149	0	0
Fruits	667	0	0
Hay	705	500	3,525
Pasture	584	500	2,920
Other crops	692	50	346
Total	10,008		10,818

Recent Development in Manufacture of Concentrated Fertilizers

In recent years much effort has been expended in an attempt to manufacture and to develop a market for fertilizer materials that carry 30 per cent or more of plant nutrients. These materials have been designated as concentrated fertilizers. Concentrated fertilizers have the distinct advantage over lower analysis fertilizers in that through their use a saving can be made in transportation and handling charges per unit of plant food. At the present time the farmer's reluctance to purchase higher analysis materials than he has been accustomed to use is delaying the extensive manufacture of these fertilizers. During the period of World War II the regulations of the War Production Board and other temporary governmental agencies restricted the consumption of low analysis materials and forced an adjustment of the fertilizer industry toward materials of higher analysis.

It appears now that in the future the manufacture of mixed commercial fertilizers will be even more strictly in the hands of the chemists than it has been in the past. This may tend to center production in fewer hands than is the case today.

2

Source, Production, and Use of Sodium Nitrate

Sodium nitrate has been used for a longer time and is better known by farmers who use fertilizers than any other inorganic nitrogenous fertilizer. For about a century sodium nitrate was obtained almost entirely from natural deposits in Chile. In recent years large quantities have been produced from synthetic ammonia and sodium carbonate.

World Deposits of Natural Nitrates

Deposits of the nitrates of aluminum, ammonium, barium, calcium, copper, iron, magnesium, potassium and sodium have been found in various localities of the earth. Only the deposits of sodium and the potassium nitrates are commercially important. Probably because of the ready loss by leaching of sodium nitrate in humid regions, only comparatively few deposits have been found. Small deposits are known to exist in Egypt, the Argentine Republic, Columbia, Protectorate of Southwest Africa, Mexico, and the United States, but the largest and most important deposit is found in Chile in the desert of Atacama and Tarapaca. Mansfield and Boardman (1932) have reported the analyses of a number of samples of nitrates that have been collected from small deposits in various localities in the United States.

Incrustations of calcium nitrate on the soil are often found in localities where the temperature and moisture conditions have been such that a rapid accumulation of nitrates could result. These deposits, which have been produced by the evaporation of soil water, occur as a silky efflorescence on the surface of the ground, but seldom are they found in quantities sufficient to be mined for fertilizer purposes. All the calcium nitrate found on the fertilizer market today is of synthetic origin.

Potassium nitrate is known to exist in small deposits in Bengal, Rhodesia, Russia, Canada, and the United States, and rather recently a deposit was found in Central Australia. These deposits are of little economic importance. A small amount of potassium nitrate occurs in the Chilean deposits of sodium nitrate and is now being recovered as a by-product of the refining of the crude sodium nitrate.

Chilean Sodium Nitrate Deposits

The existence of the Chilean deposits was first mentioned in 1809 when Taddes Haenke, a German living in Bolivia, wrote of the use of natural nitrate as a fertilizer. Nitrate from this deposit is said to have been mined first by the Spaniards in 1813, and first exported from Peru (to whom the deposits then belonged) in 1830. The desert in which the nitrate is found is a part of the plateau of Tarapaca, which is located in the northern part of Chile. The plateau is about 450 miles long north and south, and 15 to 90 miles wide east and west, and has an elevation of from 3,000 to 9,000 feet. It is shut in on both sides by mountain ranges: on the east by the Andes, and on the west by the coast hills. It seldom rains on this plateau, and when it does the precipitation is so light that the water soon evaporates. Probably at one time the area was humid and very productive, as is indicated by the buried fossils of trees.

The nitrate deposits generally occur on the eastern slopes of the coast mountain range and according to Bain and Mulliken (1924), lie at elevations ranging from 4,000 to 7,000 feet. They are exceedingly patchy and vary in thickness from a fraction of an inch to 30 feet, and average about 3 to 4 feet. The deposits are located 15 to 50 miles from the coast.

The Chilean reserves of sodium nitrate have been estimated by Mansfield (1926) to contain 245,300,000 metric tons, an amount sufficient to last about a hundred years at the rate of production the year the estimate was made. The reserves are probably much larger than these figures indicate, however, as considerable promising territory remains unexplored.

Origin of Chilean Sodium Nitrate Deposits

The origin of the Chilean sodium nitrate deposits remains today practically as much a mystery as when the deposits were first discovered. Many theories have been advanced to account for their origin. Nitrogen fixation by nonsymbiotic soil microörganisms and electrical fixation of atmospheric nitrogen in the atmosphere have been suggested, but neither explanation accounts for the presence of the iodine that is found in the deposit. For many years the seaweed theory of origin was accepted by many. According to this theory it is assumed that when the plateau was raised above the sea large quantities of marine plant and animal refuse were put in a position where decay and nitrification would take place, and that the present deposits of nitrate resulted from this decay. The exponents of this theory overlook the tremendous quantity of organic materials that would have been required to form so large a deposit of nitrate. This theory does give an explanation, however, for the presence of iodine in the nitrate deposit and of phosphorus in the overlying rocks. Both iodine and phosphorus could have been derived from the seaweed. Nevertheless, it does not explain the absence of bromine in the deposit, for bromine also is a constituent of seaweed.

Another theory that has received some support would account for the deposits as accumulations resulting from nitrification of animal manures which accumulated in the surrounding highlands from great herds of vicunas and llamas. It is supposed that the nitrates were washed down from the highlands and accumulated through evaporation of water in the present valley. This theory does not explain how the deposits of common salt or iodine originated. Large phosphate deposits which would probably have resulted from such decompositions have not been found. Still another somewhat similar theory assumes that the nitrate has been formed from the nitrogen of guano deposits that may have accumulated on the shores of a great soda lake.

The most commonly accepted theory, although it is subject to criticisms similar to those of the theories already discussed, is that suggested by Singewald and Miller (1916), that Chilean sodium nitrate had its origin

FIG. 11. Morrow plats—America's oldest soil experimental plats, University of Illinois. Begun in 1879 by Dean George E. Morrow. (Courtesy, Illinois Agricultural Experiment Station.)

in the evaporation of nitrate-carrying waters which have reached the plateau after leaching through the bedded sedimentary rocks and gravels of surrounding highlands.

Nature of Chilean Sodium Nitrate Deposits

Covering the surface of the Tarapaca plateau is a deposit of a mixture of sand and gypsum carrying a small amount of phosphate. This layer is called *costra*. Just beneath the costra lies a layer of conglomerate rock made up largely of gravel and clay. This layer is called *congelo* and contains a high percentage of common salt. Underneath these two layers lies the deposit carrying the nitrate called *caliche*. Strictly speaking, the word caliche has been applied to the product incorrectly. The word is derived from the Latin word *calx* and if used properly should mean a lime or calcareous deposit rather than a deposit of sodium salt. The caliche is very impure. Besides much sand and earthy materials, the deposit contains varying percentages of sodium, magnesium, and calcium chlorides and sulfates, together with smaller quantities of iodates and borates. The percentage of sodium nitrate in the deposit may vary from 5 to 70 per cent. Today large quantities of 30 per cent material and above are rather rare, but formerly material containing as high as 50 per cent and above was worked. With the best modern methods of mining and purification 8 per cent material can be worked economically (Curtis, 1932), but deposits containing less than 15 per cent of sodium nitrate were not mined profitably until rather recently. Underlying the caliche is a deposit called *coba*, which is composed largely of clay with sea fossils scattered through it.

The combined thickness of the costra and the congelo overburden may vary from 3 to 10 feet and the deposits of caliche from a few inches to 10 feet. The average thickness of the cliche is about 3 feet. The beds that are now being worked may be located anywhere from the surface of the deposit to 24 feet below the surface.

Chemical Composition of Caliche

Caliche is made up largely of a matrix of insoluble earthy material, varying in size from a clay to pebbles, which is cemented together with a mixture of soluble salts. These salts are made up largely of sodium nitrate, sodium chloride and sodium sulfate. Other salts such as the iodates, borates, and perchlorates of potassium, calcium, and magnesium may also be present. Some of the potassium is present in the form of the mineral polyhalite. The content of sodium nitrate in the caliche now being worked varies on the average from about 12 to 25 per cent. The average range of composition of the caliche usually mined is given in the following table.

Table 12

AVERAGE RANGE OF COMPOSITION OF REPRESENTATIVE SAMPLES OF CALICHE

	Per Cent
Sodium nitrate ($NaNO_3$)	12 –25
Potassium nitrate (KNO_3)	.2 – 3
Sodium chloride (NaCl)	8 –25
Sodium sulfate (Na_2SO_4)	2 –12
Calcium sulfate ($CaSO_4$)	2 – 6
Magnesium sulfate ($MgSO_4$)	0 – 3
Sodium biborate ($Na_2B_2O_4$)	1 – 3
Sodium iodate ($NaIO_3$)	.05– .1
Sodium perchlorate ($NaClO_4$)	.1 – .5
Insoluble matter	23 –70

Methods Used in Mining of Chilean Sodium Nitrate

While a number of methods have been used at one time or another for the recovery of Chilean nitrate from the natural deposits, there are only two in use at the present time. In the older of these the miners first remove with hand shovels any soft material that may overlie the caliche at a given location. Special drillers then sink pits through the caliche layer, the work being carried on wholly by means of iron bars which are wielded by hand. The bottoms of these holes are enlarged and filled with low-grade blasting powder which is made locally. These charges are then exploded, and the caliche, together with the overburden, is broken up. The

FIG. 12. Caliche being loaded into cars by electric shovels. (Courtesy, Chilean Nitrate Educational Bureau.)

FIG. 13. Partial view of Chilean sodium nitrate plant located 40 miles east of the port of Tocopilla, Chile. (Courtesy, Chilean Nitrate Educational Bureau.)

miners hand sort from the costra and congelo the larger pieces of caliche that are of sufficiently high grade for treatment in the plant. These pieces are then loaded into small railway cars, either directly or after a short haul by mule cart to a gathering point, and then shipped by rail to the plant for treatment. It has been estimated that when this method is used the industry recovers and markets only about half a ton of sodium nitrate for every ton of sodium nitrate in the unbroken caliche.

In recent years a superior method of mining has been developed. The general procedure, where the deposits are sufficiently extensive and the grade high enough to permit regular working, is to open the ground by trenching. These trenches are carried forward laterally across the ground to be mined by the use of drag-lines, scrapers or steam shovels. The caliche is then drilled with compressed air-driven jack-hammers, blasted by high-grade blasting powders, sorted, and the broken pieces, which may be as large as six feet in length, loaded into cars by large electric shovels, and hauled to the plant. The waste rock is thrown back into the trench and the caliche conducted to the crushers over temporary narrow-gauge railway lines.

Methods Used in Refining Caliche

Two methods of refining caliche are now being employed: the Shanks or old process; and the Guggenheim or new process. The real expansion of the Chilean nitrate industry came with the introduction of the Shanks process by J. T. Humberstone about 1884.

Shanks Process of Refining Caliche

At the plant the caliche is crushed to about three-inch size and placed in small steel tanks which hold 60 to 100 tons of material and which are equipped with steam coils. A solution of the various caliche salts is then run onto the caliche and the charge is boiled. As a result the nitrate is dissolved from the caliche, making a hot solution with a high nitrate content. This solution is then run into settling tanks to be clarified. The caliche, from which most of the nitrate has been removed, is washed with weaker solutions and the remaining material is finally discharged as waste. The strong solution, after it is clarified in the settling tanks, is run into shallow pans and allowed to stand in the open air to cool. In the cooling process the nitrate crystallizes out. The nitrate is removed from the pans, allowed to drain, and is then stored in large piles where in time it becomes sufficiently dry to be crushed, after which it is packed in jute sacks and shipped by rail to the nearest port. The ports generally handling the product are Iquique, Tocopilla, Antofagasta, and Taltal. At the seaport it is loaded into lighters and delivered alongside ships for foreign shipment. The solution from the cooling pans is returned to the boiling tanks and is used to treat more caliche as it is received from the crushers.

The Shanks process of refining proved fairly economical so long as a supply of high-grade caliche was procurable and little competition was being met from air-fixed nitrogen. But as the higher-grade deposits gradually became exhausted, the plants were forced to handle lower grades until they are now treating caliches of about 15 to 20 per cent sodium nitrate, with resulting increased costs. Competition with synthetic nitrogen fertilizers will in all probability force the abandonment of the Shanks process with its crude accompanying method of mining. At the time this is written the importations of the Shanks product into the United States are greatly restricted.

Guggenheim Process of Refining Caliche

In 1926 a new process of refining caliche was developed by the firm of Guggenheim Brothers of New York City. This process, known as the Guggenheim process, has caused modern mechanical methods to be applied to the Chilean nitrate industry, and in this way the cost of production has been lowered and the utilization of much lower grade caliches has become possible. This new process, together with its mechanical mining, is now in use in only a few plants, but it would seem that the old methods must be replaced gradually by the new.

Under the new process the caliche is crushed to a size less than five-eighths inch in diameter. It is then loaded into large concrete tanks, each holding approximately 7,500 tons, and there treated with warm solutions which have received their warmth from waste heat that becomes available in the plant. These solutions are then cooled to about freezing point by mechanical refrigeration in large cooling tanks, during which the nitrate crystallizes out. The crystallized nitrate resulting therefrom is separated from the liquid by centrifugal driers which discharge the nitrate with a moisture content of about 2.5 per cent. The solutions are then rewarmed and returned to the dissolving tanks, where the same cycle of operations is repeated.

The nitrate from the centrifugal driers is sent to a graining plant where it is melted in large iron pots and then sprayed in the molten form into an enclosed chamber. On cooling, the drops solidify into small shotlike pellets. The product thus obtained, having lost all its moisture in the fusion operation, is packed in waterproof, paper-lined, burlap bags which protect it to such an extent that it is delivered to the consumer with a moisture

FIG. 14. Loading railway car with Chilean sodium nitrate. (Courtesy, Chilean Nitrate Educational Bureau, Inc.)

content of less than 0.5 per cent, and with a sodium nitrate content of about 99 per cent. This refined product, having been largely freed of its impurities, approaches the composition of synthetic sodium nitrate.

The outstanding features of the Guggenheim process are: first, the ability to extract sodium nitrate efficiently from caliche of lower grade than is considered economical for use in the Shanks process; second, the conservation of all heat wastes in the operation; and third, the employment of about one-fourth the number of men necessary for similar production under the old process. Thus, it is possible to introduce mechanical methods for handling the caliche on a larger scale than heretofore, the available reserves of nitrate are greatly expanded, the output per unit of fuel consumed is materially increased and, incidentally, a purer product is delivered. It is claimed that these newer methods of mining and refining give promise of doubling the life of the Chilean deposits and that 90 per cent of the available nitrates can be recovered, as compared with about 55 per cent recoverable under the Shanks process.

Organization of Chilean Nitrate Industry

From the beginning of the fertilizer industry until 1914, the Chilean sodium nitrate industry practically controlled the world market for nitrates. This was due to the fact that Chile had a natural monopoly of the world's supply of nitrates, a fact that was fully appreciated by Chileans. This monopoly allowed the Chilean government to levy an export tax on nitrate that yielded an income sufficient to carry, at different times, from about one-sixth to two-thirds of the total cost of the government. Until the Chilean government secured a half interest in the industry in 1930, the tax amounted to about $12.30 per metric ton of product, a figure that, according to Mayer (1926), had been maintained since 1880. It has been estimated by Ernst (1921) that this tax supported the Chilean government during the 50 years preceding 1921 to the extent of $900,000,000. He has also estimated (1927) that the consumers of the United States paid in 1925, a representative year, about $15,000,000 to the support of the Chilean government. Because of its monopoly the Chilean nitrate industry, prior to 1914, was able to fix prices for sodium nitrate that have allowed a profit even for the most inefficient producers.

During World War I and afterward many countries began the commercial production of synthetic nitrates and other nitrogenous fertilizers. Competition from these sources, together with tariff and other import barriers, and the world depression, forced a reorganization of the Chilean nitrate industry in 1930. The present arrangement, under which nitrate sales are centralized and the Chilean Government receives a fixed percentage (25 per cent) of the profits of the industry instead of the export tax, was put into effect in 1933. Thirteen companies now produce Chilean sodium nitrate and three of these account for approximately 87 per cent of the nitrate produced.

Following the reorganization of the Chilean sodium nitrate industry, world consumption of Chilean sodium nitrate increased steadily and showed a gain of about 150 per cent during the five-year period ending with 1936–1937. In 1939–1940, according to figures of the United States Department of Commerce, Chile exported 747,700 tons of nitrate to the United States. This importation included both sodium nitrate and sodium potassium nitrate. The imports of Chilean sodium nitrate into the United States since 1916, as given by Merz and Fletcher (1940), are shown in the following table.

Table 13

IMPORTS OF CHILEAN SODIUM NITRATE SINCE 1916

Year	*Short Tons*	*Year*	*Short Tons*
1916	1,364,463	1926	1,024,014
1917	1,728,399	1927	838,636
1918	2,066,615	1928	1,156,860
1919	456,354	1929	1,042,113
1920	1,480,519	1930	636,825
1921	413,474	1931	616,687
1922	607,560	1932	56,482
1923	998,680	1933	137,610
1924	1,105,001	1934	328,750
1925	1,245,693	1935	437,635

Synthetic Sodium Nitrate

During the period between the two world wars the successful production of nitric acid from atmospheric nitrogen allowed the production of sodium nitrate at a price which would allow synthetic sodium nitrate to compete on the market with Chilean sodium nitrate. The first shipment of synthetic sodium nitrate was made in 1929. Briefly, the most common method of manufacture of synthetic sodium nitrate is accomplished as follows: Gaseous nitrogen is combined with hydrogen, obtained from coal gas, to form ammonia. The ammonia thus formed is oxidized by suitable catalysts to nitrogen oxides. The nitrogen oxides are then absorbed

FIG. 15. Storage piles of Arcadian sodium nitrate at the Hopewell, Virginia Plant. (Courtesy, The Barrett Division, Allied Chemical and Dye Corporation.)

in a solution of sodium carbonate that has been obtained from sodium chloride by the well-known Solvay process. The resulting solution of sodium nitrate is then partly evaporated and the solid material allowed to crystallize. In 1936 another method of sodium nitrate manufacture was perfected which consisted of treating sodium chloride with nitric acid. The synthetic sodium nitrate now found on the American market is marketed under the brand name of Arcadian. It is manufactured at Hopewell, Virginia.

In Norway synthetic nitrate is made from calcium nitrate. Calcium nitrate, according to Merz (1937), is passed through a sodium zeolite to obtain a sodium nitrate solution, the sodium zeolite being simultaneously converted into calcium zeolite. The calcium zeolite is reconverted into the sodium zeolite for reuse by passing sea water through it.

By-product Sodium Nitrate

During the World War II period, by-product sodium nitrate was produced at Holston, Tennessee. This sodium nitrate contained only 14 per cent nitrogen. In 1944 15,000 tons were produced but its production ceased with the war.

Physical Characteristics of Sodium Nitrate

Two brands of Chilean sodium nitrate are now found on the American market—original Old Style and Champion. The Old Style brand is the sodium nitrate produced by the Shanks process. It may vary in color from a pink through yellow to a gray or violet. The yellow color is caused by the presence of manganese nitrate, and the violet color is due largely to the presence of manganese chloride. The Old Style or Standard brand of sodium nitrate is crystalline and somewhat deliquescent but owing to an improved mechanical condition and bagging it has less tendency to become lumpy than was formerly the case. It is alkaline to litmus.

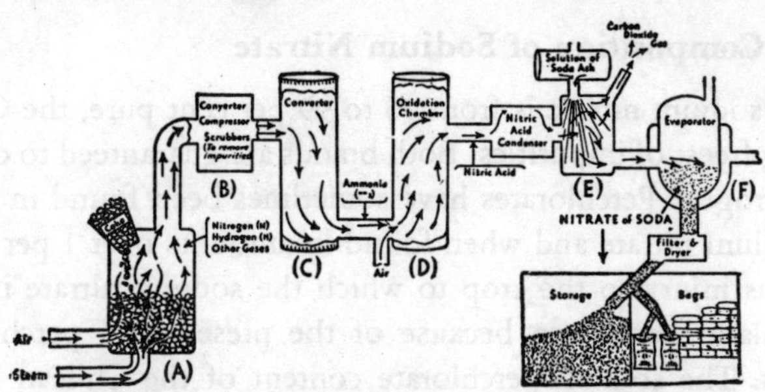

FIG. 16. Principal steps in the fixation of atmospheric nitrogen as sodium nitrate. (Courtesy, The Barrett Division, Allied Chemical and Dye Corporation.)

Explanation of diagram: In (A) air, containing the nitrogen, and steam are blown through glowing coke, producing a number of gases including nitrogen and hydrogen; (B) indicates converters and compressors, and "scrubbers" to remove impurities from the hydrogen and nitrogen gases; in converter (C) the hydrogen and nitrogen are combined, under pressure and heat, to make ammonia gas which passes into another converter (D) where it is burned with air (oxidized) to make nitric acid; when nitric acid reacts, as in mixer (E), with a solution of soda ash it forms a neutral or slighly alkaline solution of sodium nitrate and liberates carbon dioxide gas; the nitrate solution is evaporated in (F); then the nitrate is filtered and dried, and is conveyed to storage or put in moisture-resisting bags for shipment.

The new process or Champion brand of Chilean sodium nitrate is produced by the Guggenheim method and appears on the market in a granulated form made up of shotlike pellets. It is very white in color. Because of its granulated form and low moisture content it has a better mechanical condition than the old process product. Nevertheless, many farmers prefer the Old Style sodium nitrate. The greater part of the Chilean nitrate consumed in this country is now shipped in bulk from Chile and bagged at the ports in the United States. Both the old and new process sodium nitrate are now marketed in 100- and 200-pound bags.

The synthetic Arcadian (American) sodium nitrate is produced in the form of cubical crystals with rounded edges, about the size of No. 6 bird shot, and is guaranteed to carry 16 per cent of nitrogen. It is low in moisture content and is in a good mechanical condition for distribution, but has a slightly greater tendency to "blow" than the Champion Chilean brand when broadcast during windy weather.

Chemical Composition of Sodium Nitrate

Chilean sodium nitrate is from 96 to 99 per cent pure, the Champion brand being freest of impurities. Both brands are guaranteed to contain 16 per cent nitrogen. Perchlorates have sometimes been found in Old-Style Chilean sodium nitrate and when found in amounts over 1 per cent may cause serious injury to the crop to which the sodium nitrate is applied. Little complaint has arisen because of the presence of perchlorates in recent years. The sodium perchlorate content of the Chilean brands of sodium nitrate found on the market today may vary from a trace to 0.94 per cent, the greatest amount being found in the Shanks process product.

As impurities, Chilean sodium nitrate carries small amounts of potassium, boron, magnesium, iodine, calcium, chlorine, and sulfur, and

Table 14

Composition of Representative Samples of Chilean Sodium Nitrate

	Old-Style (*per cent*)	*Champion* (*per cent*)
Moisture at 130° C.	0.940	0.220
Insoluble matter	0.060	0.120
Iron oxide and alumina	*None*	*None*
Sodium chloride (NaCl)	0.638	0.470
Potassium sulfate (K_2SO_4)	0.111	0.279
Sodium chlorate ($NaClO_3$) including perchlorate	0.105	0.041
Sodium borate ($Na_2B_4O_7$)	0.096	0.038
Sodium iodate ($NaIO_3$) including bromate	0.030	0.064
Calcium nitrate ($Ca(NO_3)_2$)	0.044	0.073
Magnesium nitrate ($Mg(NO_3)_2$)	0.200	0.280
Potassium nitrate (KNO_3)	1.782	0.170
Sodium nitrate ($NaNO_3$)	95.994	98.245
	100.00	100.00
Potash	0.890	0.230
Nitrogen (Devarda)	16.100	16.250
Nitrogen (by difference)	16.108	16.271

spectrographic analyses show traces of twenty-two other elements. It should be mentioned, however, that spectrographic analyses of most crude fertilizer materials show traces of a large number of elements. Shive (1936) has secured results that seemed to show that the entire boron and manganese requirements of some plants might be satisfied by the amounts of these elements present as impurities in normal applications of Old Style Chilean sodium nitrate when the plants are grown in culture solutions and washed sand cultures.

The composition of representative samples of Old Style and Champion brands of Chilean sodium nitrate, as reported to the author by the Chilean Nitrate of Soda Educational Bureau in 1938, is given in Table 14. The composition of a representative sample of Arcadian sodium nitrate, as reported to the author by the Barrett Company in 1946, is given in Table 15.

Sodium nitrate is commonly referred to by the southern farmer as "soda." Chemically speaking, the term soda may refer either to the chemical compounds $Na_2CO_3.10H_2O$ or $NaHCO_3$.

Table 15

Typical Analysis of Arcadian, the American Nitrate of Soda

Sodium nitrate	99.32% (not less than 99.5% dry basis)
Total nitrogen	16.37%
Insoluble	*Trace*
Moisture	0.31% (not more than 0.50%)
Sodium chloride and other solubles	0.37% (not more than 0.50% NaCl)
Alkalinity as Na_2O	0.01% (less than)
Sodium nitrite	0.05% (" ")
Sodium sulfate	0.20% (" ")
Calcium oxide	0.10% (" ")
Magnesium oxide	0.10% (" ")
Fe_2O_3 and AL_2O_3	0.01% (" ")

The presence of iodine—about 1.5 pounds per ton—has led American and European dealers of Chilean sodium nitrate to claim that sodium nitrate when used as a fertilizer for food crops has some virtue as a preventive of goiter if these crops are consumed by man. No data are available which would substantiate such claims, but the theory contains elements of plausibility. McHargue, Roy, and Pelphrey (1930) have secured data showing that moderate applications of sodium nitrate increased the iodine content of red clover more than sixteen times. Also, McHargue, Young, and Calfee (1935) later reported that crude Chilean nitrate, raw rock phosphate, and limestone rocks may contain enough iodine to influence the iodine content of crops when they are applied to crops in adequate amounts.

Availability of Sodium Nitrate

It is generally conceded by agronomists that the nitrogen of fertilizers that is readily soluble in water is available to crops. Sodium nitrate is readily soluble in water and because of this it may diffuse rapidly in the soil. It is therefore immediately available for plant absorption following application and solution, but because of this and the nature of the salt it is also subject to loss by leaching. Most field crops, after they reach an age at which they begin to synthesize their own food, appear to utilize most of their nitrogen to best advantage in the nitrate form, and some crops, such as the sugar beet, are particularly responsive to nitrates.

Influence of Sodium Nitrate on Physical Condition of Soil

As sodium compounds have the power to deflocculate soil colloids, the continuous application of sodium nitrate in large quantities to some soils, particularly to those that are high in colloidal matter, will result in the production of a poor physical condition. This influence of sodium nitrate on the physical condition of the soil was early noted by Hall (1904) at the Rothamsted Experiment Station in England, and by B. E. Brown (1909) and more recently by Aldrich *et al.* (1945) in this country. Aldrich, whose work was done in California, found that if gypsum is applied with the sodium nitrate the percolation of water is not so poor as without it. Also, in 1945 Fireman *et al.* showed that of 51 western soils studied the permeability of over one-half of these soils was seriously reduced by the application of sodium nitrate and that the average reduction in permeability rates was 41 per cent. Schollenberger and Dreilbelbis (1930) and others pointed out that in humid regions this influence on soil structure is rarely brought about by the application of sodium nitrate in quantities as generally applied. Similar results but less pronounced may be secured from the use of potassium nitrate when applied under similar conditions.

Influence of Sodium Nitrate on Soil Acidity

The influence of sodium nitrate on soil acidity is noted often in experimental work. Sodium nitrate is a neutral salt but it is physiologically basic and therefore it has the power to reduce, to some extent, the acidity of the soil, and may thus be employed with additional advantage on acid soils if some form of lime is not being used. Its influence in this respect is even greater than that of calcium or potassium nitrate, but not as great as that of calcium cyanamide. Theoretically, 100 pounds of sodium nitrate has a neutralizing value equivalent to about 60 pounds of ground limestone. Skinner and Beattie (1917) applied sodium nitrate at the rate of 150 pounds per acre to one-half of each of 14 plats. After five years of cropping the lime requirement of the top six inches of the soil of each half plat was determined. Their results showed that with two exceptions the lime requirement of the soil was materially reduced through the use of sodium nitrate.

Gardner (1917) determined and reported the lime requirement in 1910 and 1915 of the surface seven inches of soil of all plats of the famous Pennsylvania experiments that have received nitrogen as sodium nitrate. The fertilizer treatments have been in operation since 1883. His results

Table 16

INFLUENCE OF SODIUM NITRATE ON THE LIME REQUIREMENT OF THE SOIL (JORDAN PLATS, PENNSYLVANIA EXPERIMENT STATION)

Nitrogen Applied as Sodium Nitrate Every Two Years	*Lime Requirement in Pounds of $CaCO_3$ per Acre*	
	1910	*1915*
24 lbs.	843	1983
48 lbs.	730	1941
72 lbs.	750	1898

are given in Table 16. They also show that an application of sodium nitrate will reduce the lime requirements of acid soils. Incidentally, these data also show the rapidity with which a soil tends to become acid after the development of acidity has begun. As might be expected, the quantities of sodium nitrate applied in these experiments have not greatly retarded the development of soil acidity.

Although sodium nitrate may neutralize a small amount of soil acidity it should not be applied with the idea of correcting the soil's need for lime. The ineffectiveness of such a procedure is demonstrated by the Pennsylvania experiments. It is well to keep in mind that the influence of sodium, potassium and calcium nitrates on soil reaction is just opposite to that produced by the use of ammonium sulfate and other ammoniacal salts. Applications of ammonium sulfate to the soil result in increased acidity. This is due in part to the fact that some of the ammonium ions are absorbed by plants and the excess sulfate ions are left in the soil, and in part to the fact that some of the ammonium ions are converted to nitric acid which, in turn, neutralizes some of the soil calcium. Some farmers have been known to follow the practice of alternating the use of sodium nitrate and ammonium sulfate but this practice will not in itself maintain the reaction of the soil or prevent a drift to greater acidity.

Influence of Sodium Nitrate on Liberation of Potash

An idea current among some fertilizer men, that has persisted for a quarter of a century, is to the effect that the sodium which is left in the soil after the nitrate ion of sodium nitrate has been absorbed by plants will give rise to a substitution of bases in the soil which will result in again making fixed potash available for plant use. This influence of sodium nitrate is of doubtful value, for the standard oxidation reduction potentials of potassium and sodium are such as to lead one to believe that the quantity so set free would be small. Sodium may, however, supplement potassium as a plant nutrient. This influence has been noted by many investigators, including those at the Rothamsted Experiment Station, England.

Use of Sodium Nitrate in Arid and Semiarid Regions

Because of the deflocculating effect of sodium on soil colloids, sodium nitrate probably should not be employed under arid or semiarid conditions as the main source of nitrogen if the soil contains a high content of sodium carbonate. Under these conditions nitrogen carriers which are acid-forming fertilizers, such as ammonium sulfate, would appear to be preferable to sodium nitrate.

There is only a little experimental evidence available that would indicate the best practice to follow in nitrogen fertilization in arid regions. It is known that nitrates are often preferred for crops growing under conditions of low rainfall, for under these conditions ammonification and the oxidation of ammonia to nitrites and then to nitrates is slow. Since relatively high concentrations of the carbonates and chlorides of sodium and potassium rather effectively precipitate calcium from the soil solution, the uniformly favorable crop response obtained from applications of calcium nitrate to alkaline soils may be ascribed in part to the nitrate form of nitrogen and in part to the soluble calcium combined with the strong nitrate anion.

Influence of Sodium Nitrate on Conservation of Soil Lime

The application of sodium nitrate to the soil may aid in conserving calcium and magnesium in the soil. An application of sodium nitrate will reduce the quantity of calcium and magnesium salts and increase the quantity of sodium salts that normally pass out of the soil in the drainage water. Bases are leached from the soil only in combination with strong anions so that the quantity of bases removed from the soil in the drainage water is correlated with the amount of strong anions arising in the soil. Also, the basic elements removed from the soil in drainage water depend upon the nature and quantity of the basic elements present in the soil colloidal complex. In most soils in humid climates anions combine largely with calcium and magnesium when the soil is neutral, and with iron, aluminum, and manganese when the soil is distinctly acid; whereas, in arid and semiarid climates anions combine largely with sodium and potassium. When sodium nitrate is applied to humid soils the sodium not absorbed by plants neutralizes soil acidity and in this way helps to conserve the calcium and magnesium supply of the soil. At the Rothamsted Experiment Station, England, it was determined by Voelcker (1874) that con-

FIG. 17. Bagging Arcadian sodium nitrate at Hopewell, Virginia. (Courtesy. The Barrett Division, Allied Chemical and Dye Corporation.)

tinued fertilization with sodium nitrate on Broadbalk field reduced the annual loss of lime. Similar results have been secured by many investigators in this country.

Visible Influence of Sodium Nitrate on Plant Growth

Nitrogen is the first limiting element of plant growth in many of the soils of the eastern and southern sections of the United States. For this reason the application of any readily available nitrogenous fertilizer often materially increases crop growth. This effect following an application of sodium nitrate is often noted by the layman, with the result that in the past sodium nitrate has been looked upon by many as a stimulant. This is an erroneous conception and should be discredited.

How Sodium Nitrate Should Be Applied

Because sodium nitrate is readily and entirely soluble in water, and because the nitrate ion is not absorbed by soil colloids, it should be applied for early utilization by the crop following application. This principle is applicable to all nitrate fertilizers. The loss of nitrates in drainage water is especially severe when nitrates are applied to sands, and particularly sands with open subsoils, from which they may be leached easily. Larger quantities of sodium nitrate may be applied to clays than to sands, but even in such cases the application of quantities greater than that which the crop is in position to utilize within a reasonable length of time should be avoided. When large amounts are to be used it will be found best to apply sodium nitrate in two or more applications. Such a practice will minimize the loss of nitrogen by leaching, and at the same time will avoid injuring plant roots by applying excessive quantities of soluble salt at one time. The first application may be broadcast or applied in the hill or drill before planting, and the second and later applications may be applied as side-dressings or top-dressings. Top-dressings of sodium nitrate often appear to be more effective during cool than during warm seasons.

Exhaustive Effect of Sodium Nitrate on Soil

Many farmers and soil investigators have called attention to the apparent exhaustive effect upon the fertility of the soil of the continuous use of sodium nitrate when sodium nitrate was the only fertilizer used. This effect may also be observed where any other readily available nitrogenous fertilizer is used in the same way. The results noted arise from the improper use of fertilizer in general. The increase in plant growth induced by the application of the sodium nitrate results in the absorption from the soil by crops of large quantities of phosphoric acid and potash. In time, this increased removal of phosphoric acid and potash results in a marked deficiency of these nutrients, and a diminished crop yield results. Such a condition can be avoided by employing a more rational system of fertilization whereby an application of other nutrients, which are likely to become deficient in the soil, is made in addition to the sodium nitrate.

Efficiency of Sodium Nitrate

In general, comparative tests with nitrogeneous fertilizers have shown that sodium nitrate is as efficient for most crops as any other carrier of nitrogen. Sodium nitrate often produces the highest yield of crops in comparative nitrogen fertilizer tests conducted on acid soils. This is probably due in part to the presence of the sodium ion in addition to the nitrate ion, and to the fact that the acidity of the soil is reduced or at least not increased.

In the following table are given the results of a nine-year test of the relative crop-producing power of sodium nitrate as compared with that of ammonium nitrate and ammonium sulfate when applied to cotton at the rate of 30 pounds of nitrogen per acre. These results are representative of the data secured from many similar tests that have been conducted by the various state experiment stations.

Table 17

RELATIVE CROP-PRODUCING VALUE OF SODIUM NITRATE, AMMONIUM NITRATE, AND AMMONIUM SULFATE APPLIED TO COTTON AT THE STONEVILLE, MISSISSIPPI, EXPERIMENT STATION, 1921–1929

Source of Nitrogen	*Nine-year Average*		
	Pounds Seed Cotton per Acre		*Per Cent Increase*
	Yield	*Increase*	
Check—no fertilizer	1503.85	..	..
Sodium nitrate	1950.51	446.66	29.70
Ammonium nitrate	1918.33	414.48	27.56
Ammonium sulfate	1858.70	354.85	23.59

In well-drained soils non-nitrate forms of nitrogenous fertilizers are converted largely to the nitrate form before they are extensively assimilated by general field crops. This conversion may result in a loss to the plant of a part of the nitrogen, and it would therefore necessarily lower the efficiency of those fertilizers. Twenty-one year results secured at the New Jersey Experiment Station by Lipman and Blair (1918) have shown that, under the conditions of the experiment, crops recovered more nitrogen from sodium nitrate than from ammonium sulfate, dried blood or green manure.

Relative Agricultural Value of Brands of Sodium Nitrate

A great deal of propaganda has been carried on in recent years to convince the farmer of the superior value of a given brand of sodium nitrate over that of others. Much of this propaganda has been fallacious and has often led the farmer to erroneous conclusions.

Comparative tests of different brands of sodium nitrate found on the American market have been made by a number of American agricultural

Table 18

Comparative Yields of Seed Cotton with Three Brands of Sodium Nitrate

State	*Station*	*How Used*	*Years*	*Yields Seed Cotton*		
				Arcadian	*Champion*	*Old Style*
Arkansas	Marianna	Under	5	240.4*	x	214.9*
Texas	College Sta.	Under	3	261†	x	245†
Louisiana	Baton Rouge	Under	6	1510	x	1472
	St. Joseph	Under	7	2068	x	2068
	St. Joseph	Side	4	2103	x	2094
	Calhoun	Under	8	1193	x	1183
Mississippi	Stoneville	Under	7	1354	x	1353
	Stoneville	Under	9	1092	1051	1102
	Stoneville	Under	3	1771	1789	x
	Stoneville	Under	7	1583	1514	x
	Stoneville	Under	7	920	913	x
	Stoneville	Under	3	1247	1241	x
Georgia	Main	Under	4	1034	x	1013
South Carolina	Clemson	Side	5	1136	1183	1134
	Sandhill	Side	5	763	782	785
	Pee Dee	Side	6	1891	1888	1883

* Increases over checks, seed cotton.
† Yield—lint cotton.

experiment stations but few have reported results which would indicate that one brand of sodium nitrate is materially superior to another. In Table 18 are given results of some comparative tests conducted in six states. This data has been compiled from the results reported by Gray (1938), Stewart (1936), Haddon (1936), Dowell (1937), Kuykendall (1937, 1938), Bledsoe (1937), and Cooper *et al.* (1937).

Residual Effects of Sodium Nitrate

Although sodium nitrate, like all nitrates, may readily be leached from the soil, many farmers have noted that applications of sodium nitrate frequently show a prolonged residual effect, sometimes observable several seasons after application. This has been apparent often on pasture land and in orchards. The effect noted is probably due largely to a residual influence of the increased growth of the root systems of the crops fertilized although it might be due in part to a bacterial fixation. Much nitrogen is consumed by microörganisms, particularly in soils that contain large amounts of high-carbonate organic matter.

Influences of Sodium Nitrate on Soil Flora

The number of soil microörganisms and their kinds and activities depend, among other things, upon the kinds and amounts of food or energy-yielding material present in the soil. Soil organisms are largely plant organisms and they feed upon nutrients applied in fertilizers as do the higher plants. For this reason, an application of sodium nitrate or any other readily available nutrient may, and often does, bring about a marked increase in the numbers and activities of some of the soil microörganisms.

Quantities of Sodium Nitrate Applied to Crops

The amounts of sodium nitrate applied per acre to farm crops vary widely in the various crop-producing areas of the United States. Usually, 30 to 500 pounds per acre are applied to general field crops. Truck crops, and especially those truck crops that are grown primarily for their foliage, often receive very heavy applications—1000 pounds or more—of sodium nitrate per acre.

Although the legumes have the power to get nitrogen indirectly from the atmosphere, experimental work has shown that the application of small amounts of sodium nitrate or other readily available nitrogenous fertilizer either under the crop or as a top-dressing, or both, is often a profitable practice to follow in the production of some leguminous crops.

Florists often recommend that a solution consisting of 1 ounce of sodium nitrate to 2 gallons of water be used for fertilizing flowers in greenhouses. Sodium nitrate is also used advantageously in starter solutions and in irrigation water that is to be applied to truck crops. The sodium nitrate is used most advantageously at a rate not exceeding 50 pounds to each 7000 to 8000 gallons of irrigation water.

3

Manufacture and Use of Ammonium Sulfate

The consumption of ammonium sulfate for fertilizer purposes has now exceeded that of sodium nitrate, which for years was the world's principal carrier of nitrogen. Ammonium sulfate is the source of the greater part of the nitrogen found in the mixed commercial fertilizers produced in the United States and accounts for about half of the nitrogen consumed in fertilizers by the world.

Source of Ammonium Sulfate

Most of the ammonium sulfate used in this country is a by-product of the destructive distillation of coal, bituminous shales and bones. Other by-products of the same process are illuminating gas, shale oils, and bone char. Bituminous coal suitable for destructive distillation varies in its nitrogen content from about 0.8 to 1.5 per cent, of which only about 25 per cent can be recovered when the coal is heated in closed retorts. When the coal is heated in open fires the nitrogen escapes into the atmosphere. The heating of coal to high temperatures results in many of its complex compounds being broken down, and in a part of them escaping as gas. If the heating is accomplished under conditions where the air is excluded, the process is known as destructive distillation.

Small amounts of ammonium sulfate are being made in the United States from nitrogen fixed by the cyanamide process and small quantities of synthetic ammonium sulfate were imported from Germany and Japan in the years prior to World War II. A promising method of manufacturing ammonium sulfate from the sulfur and nitrogen originating in coal is under experimentation in Europe.

Methods Used in Coking of Coal

In this country three methods are used for the coking of coal: (1) The coking of coal alone in what is known as the beehive coke oven. The

FIG. 18. Coke ovens, South Chicago, Illinois. This type of plant supplies most of the ammonium sulfate used as fertilizer in the United States. (Courtesy, The Koppers Company.)

volatile part of the coal, instead of being caught and held as a valuable by-product, is burned and furnishes the heat necessary for the coking. Because of the tremendous waste associated with the coking of coal by means of the beehive oven, it is only a matter of time until this oven will be replaced entirely by ovens of the recovery type. During World War I ammonium sulfate sold for $110.00 a ton and this helped considerably in the change-over from beehive ovens. (2) The coking of coal in the coal retort. This retort is specially designed to supply illuminating gas. The principal by-products of this process are ammonia, coke, and tar. (3) The coking of coal in the by-product coke oven. This oven is primarily designed for the production of metallurgical coke. The by-products of this process are ammonia, tar, and illuminating gas. Today, approximately 97 per cent of the coke manufactured in the United States is made in by-product ovens.

Recovery of Ammonia from Coal Gas

The gas that volatilizes as coal is coked is a mixture made up largely of ammonia, carbon dioxide, hydrogen, methane, water vapor and finely divided particles of tar. Small amounts of many other substances, such as ethylene, naphthalene, and other hydrocarbons, may be present also. At one time the ammonia was recovered by cooling the gas and bubbling it through water. The ammonia gas was absorbed by the water and formed an ammonia or gas liquor. Enough sulfuric acid was added to the ammonia liquor to neutralize the ammonia and crystallize out the ammonium sulfate salt. This was a crude process and the ammonium sulfate produced by this method contained a high percentage of impurities. For this reason, this method was largely replaced by a process in which the ammonia liquor is distilled before the ammonia is neutralized with sulfuric acid. This is accomplished by treating the ammonia liquor with an alkali, usually lime, which is added to set free the fixed ammonia. The ammonia is then driven off by means of steam. This mixture of ammonia and steam is then led into tanks of sulfuric acid where the ammonium sulfate is precipitated. The salt is then removed by long-handled copper ladles and allowed to drain. Sometimes it is washed with weak ammonia water for the purpose of removing any free acid still adhering to the product. It is then further dried in a centrifuge, or sometimes by exposing the material to a current of hot air, after which it is bagged for shipment.

In recent years modern coke oven plants have adopted a still more efficient method of ammonia recovery in which the oven gases pass directly into a weak 60° (Baumé) solution of sulfuric acid where they are deprived of their ammonia. This is known as the direct method and is now used by practically all of the by-product producers in this country. In general, the process consists of passing the gases from the ovens through coolers and a tar extractor, and then to a saturator containing dilute sulfuric acid. A certain amount of gas liquor is obtained in the coolers, but this liquor is usually distilled and the ammonia passed into the saturator with the oven gases.

Because of the unprofitableness of ammonia recovery less than half of the coal-gas plants in the United States make any attempt to recover the ammonia. In 1940 our ammonium sulfate production of about 670,000 tons was produced in about 108 coke works located in 18 states.

At this point it is of interest to note that should phosphoric acid become a cheap industrial agent, the manufacture of ammonium phosphate may replace the manufacture of ammonium sulfate at the coke oven plants.

By-products Secured from Coal Distillation

It has been estimated that a net ton of coking coal yields about 30 pounds of ammonium sulfate. It also yields about 1440 pounds of coke, 9.0 gallons of tar, 2.5 gallons of crude benzol, and 10,000 cubic feet of gas.

Synthetic Ammonium Sulfate

In 1923 synthetic ammonium sulfate from the German nitrogen fixation plants came into the world market in large quantities. It is now manufactured in limited quantities in many other countries. Ammonium sulfate is not, as might be supposed, manufactured to any great extent by treating ammonia with sulfuric acid, although this can be done. The method of manufacture followed is known as the gypsum process. By this method ammonia water is conducted into a tank containing pulverized gypsum suspended in water. At the same time carbon dioxide is bubbled through the mixture under 5 atmospheres pressure. Ammonium carbonate is formed first, which in turn reacts with the calcium sulfate to form ammonium sulfate. The ammonium sulfate remains in solution and is recovered by evaporation after the resultant calcium carbonate has been filtered off.

Synthetic ammonium sulfate that is manufactured in the United States is sold under the trade name Arcadian. Arcadian ammonium sulfate differs from the ordinary commercial grade ammonium sulfate in that it has been neutralized, dried, and then screened to remove lumps.

Characteristics of Ammonium Sulfate

Commercial ammonium sulfate varies in color from a white through yellow to a gray salt. Occasionally various tints of brown, blue, yellow or gray may be present. These tints are due to impurities such as ferric ferrocyanide and arsenious sulfide, or to traces of tar or its derivatives.

The by-product ammonium sulfate now found on the market contains practically no free acid. The better grades have a free sulfuric acid content of less than 0.15 per cent. Some of the commercial grades might slightly exceed this figure, but the free acid content is seldom over 0.5 per cent. This was not always so. At one time free acid adhering to the crystals of ammonium sulfate caused rotting of shipping bags and the formation of lumps and hard cakes. This condition has been overcome by more thorough washing of the salt and in some cases by neutralizing the free acid with ammonia. It has been shown that large crystals are to be preferred to small ones as the tendency to cake is thus reduced.

Pure ammonium sulfate contains 21.2 per cent nitrogen and 27.5 per cent sulfur. The commercial product is sold for fertilizer under a guaranteed analysis of 20 per cent nitrogen. It is a fine, crystalline salt, almost moisture-free and requires no grinding before use. It does not absorb moisture if it is stored in a reasonably dry place. At ordinary temperatures the solubility of ammonium sulfate in water is practically equal to that of sodium nitrate. Tidmore and Williamson (1932) and many other investigators have shown that the nitrogen of ammonium sulfate may be retained by the soil in a form which does not leach as readily as sodium nitrate.

Presence of Thiocyanates in By-product Ammonium Sulfate

Thiocyanates, such as NH_4CNS, may also be produced in the destructive distillation of coal and may be present in by-product ammonium sul-

FIG. 19. Unfertilized and fertilized barley. (Courtesy, Missouri Agricultural Experiment Station.)

fate. Thiocyanates are toxic to plants, but their presence in ammonium sulfate may be determined easily by making a solution of the salt and then adding a little ferric chloride. If the cyanates are present a reddish coloration of the solution is produced. The toxic ingredient is seldom found in domestic ammonium sulfate, but formerly it was often present in quantities sufficient to produce injury to plants, being found especially in ammonium sulfate of European origin.

Mixing Superphosphate with Ammonium Sulfate

When ammonium sulfate is added to a fertilizer mixture containing superphosphate, and the mixture is allowed to stand for a time, the mixture is likely to become hard owing to a chemical reaction that may take place between the ammonium sulfate and the monocalcium phosphate, as follows:

$$(NH_4)_2SO_4 + CaH_4(PO_4)_2.H_2O + H_2O = CaSO_4.2H_2O + 2NH_4H_2PO_4$$

For this reason, when ammonium sulfate is used in the preparation of complete fertilizers containing superphosphate, sufficient time must be allowed for the curing or completion of this chemical reaction. Once the mixture has been cured there is no secondary caking. The gypsum which is formed in the reaction has 2 molecules of water of crystallization, and as only 1 molecule is supplied by the monocalcium phosphate the other molecule of water is taken from the free moisture in the mixture. This moisture absorption is very advantageous in preparing a dry fertilizer mixture. The addition of woody or fibrous organic material to the mixture previous to curing also reduces the extent of caking.

Influence of Ammonium Sulfate on Soil Acidity and Toxicity

Ammonium sulfate, like other ammoniacal fertilizer materials, is physiologically acid and, as shown by Barnette and Hester (1930) and Pierre (1931), and many other investigators, produces, when added to soil, an increase in the hydrogen-ion concentration of the soil solution. When the acidity of the soil is greatly increased many beneficial soil processes are interfered with. The rate of nitrification in particular is decreased, for nitrification does not proceed extensively in the soil when the pH of the soil solution is below 6.0. That the increase in soil acidity resulting from the use of ammonium sulfate can easily be corrected through the use of lime or other basic materials, has been shown by Hall and Gimmingham (1907), at the Rothamsted Experiment Station, Noll, Gardner, and Irvin

(1931), at the Pennsylvania Experiment Station, and investigators at many other experiment stations.

Results of much research have shown also that the continuous use of ammonium sulfate without lime may result in the production in the soil of compounds that are toxic to crops. The free acid produced when ammonium sulfate is added to the soil in humid regions probably first reacts with the calcium and magnesium compounds of the soil if such are present. When these compounds are absent it reacts with the iron, aluminum, and manganese compounds of the soil. The iron, aluminum, and manganese compounds which are thus rendered soluble, appear to be responsible for a large portion of the toxic effects so often noted where ammonium sulfate has been applied to the same soil for a number of years. If a quantity of lime sufficient to neutralize the acidity produced by ammonium sulfate is present in the soil to which the ammonium sulfate is applied the toxic iron, aluminum, and manganese compounds will not be formed.

The application of ammonium sulfate has been known also to produce chlorine toxicity in crops when applied with chlorine-carrying fertilizer materials.

Influence of Ammonium Sulfate on Lime Requirement of Soil

According to White and Holben (1925), 40-year results secured from the Jordan plats located on a limestone soil at the Pennsylvania Experiment Station, showed that the increase in the lime requirement of plats receiving ammonium sulfate above that of the lime requirement of plats receiving sodium nitrate was equal to about 67 pounds of calcium carbonate for each 100 pounds of ammonium sulfate applied. Hartwell and Damon (1927), of the Rhode Island Experiment station, concluded that the theoretical requirement of about 7 pounds of calcium carbonate to 1 pound of nitrogen supplied as ammonium sulfate should be the basis of liming to maintain land fertilized with ammonium sulfate at the same reaction as if receiving sodium nitrate.

Allison and Cook (1917) found that for every 100 pounds of ammonium sulfate applied to a soil an increase in soil acidity occurred equivalent to that which could be neutralized by 143 pounds of calcium carbonate. Theoretically, this is what might be expected. In general, for most acid soils the proper addition of lime appears to be about 1.5 to 2.0 tons of ground limestone per ton of ammonium sulfate applied. Many agronomists now recommend that a sufficient amount of limestone be mixed with the ammonium sulfate to make a physiologically neutral mixture before the ammonium sulfate is applied. Such a practice is unnecessary if the pH of the soil is being maintained at an optimum value by liming as often as necessary.

Application of Ammonium Sulfate

Ammonium sulfate may be put under the crop or applied as a top or side dressing. Like all ammoniacal compounds, ammonium sulfate is resistant to leaching and because of this may under some conditions be more suitable than nitrates for use at planting time.

Ammonium sulfate usually is applied to crops in about the same manner and in about the same amounts, on an equivalent nitrogen basis, as sodium nitrate. Seventy-five pounds of ammonium sulfate is equivalent to about 100 pounds of sodium nitrate. From 50 to 250 pounds per acre of ammonium sulfate is the most common application. Recently very good results have been reported from plowing under ammonium sulfate when the land is broken, and rice growers of California have successfully applied ammonium sulfate by plane.

Loss of Ammonia Following Application of Ammonium Sulfate

Following an application, the ammonia of ammonium sulfate is readily fixed by the colloidal complex of acid soils. It appears, however, that alkaline soils of low base holding capacity cannot absorb and hold ammonia, for Jewett (1942) found that considerable quantities of ammonia were lost in drainage water following an application of ammonium sulfate to the alkaline soils of the Sudan.

Influence of Ammonium Sulfate on Growth of Various Crops

Preferably, unless lime is applied, ammonium sulfate should not be applied to most humid soils as the only source of nitrogen continuously for a long term of years, nor should it be applied in large quantities to non-acid-tolerant crops. On the other hand, ammonium sulfate can be applied successfully to soils upon which are to be grown cranberries, blueberries watermelons, and rhododendrons, or other plants that thrive well on acid soils. In some areas potato growers favor ammonium sulfate because by maintaining an acid condition of the soil it aids in controlling potato scab. The author has observed that the repeated application of ammonium sulfate to the pasture land of Park Field of the Rothamsted Experiment Station, England, has resulted in the dominance of acid-tolerant grasses.

Is Nitrification of Ammonium Sulfate Necessary?

For many years it was generally thought by agronomists, but without conclusive supporting evidence, that all field crops use nitrates exclusively as their source of nitrogen, and that all non-nitrate nitrogenous fertilizers must be converted into the nitrate form before they could be assimilated by plants. Later, however, some nitrogen in ammoniacal form was shown to be necessary for optimum growth of potatoes, buckwheat, and other crops, especially during the seedling stage of development.

That seedlings absorb ammonia as such, and often in quantities greater than they absorb nitrates, has been shown by Shive (1926), of the New Jersey Experiment Station. Shive found that as seedlings become older the absorption ratio of nitrogen as ammonia to nitrogen as nitrate becomes smaller until finally it is less than unity. In general, it might be said that if the reaction of the soil solution is slightly acid, neutral, or alkaline young plants will absorb ammoniacal nitrogen rapidly and by preference. On the other hand, if the soil solution is strongly acid seedlings seem to favor the nitrate form of nitrogen.

Bizzell (1937) has shown from data secured from the Cornell lysimeters, that the nitrogen composition of plants receiving their nitrogen from sodium nitrate was about the same as those plants receiving their nitrogen from ammonium sulfate.

It has been estabiished fairly conclusively that for crops such as those tolerant to acid soil conditions the ammoniacal form of nitrogen is effective. It also appears that the ammoniacal forms would be superior for most crops if the masking effect of the resulting acidity could be overcome. According to Russell (1932) plants which are relatively rich in carbohydrates, such as barley, maize, and pumpkins, readily utilize ammonia, whereas peas and vetches which contain less carbohydrate, utilize ammonia only when calcium carbonate is present. Plants that grow well under conditions of poor soil aeration and in soils that are low in soluble potassium, sodium, and calcium appear to tolerate part or all of their nitrogen in the ammoniacal form. Stewart, Thomas and Horner (1925) showed that pineapples are able to utilize all their nitrogen in the form of ammoniacal compounds, and Kelly (1911), of the Hawaii Experiment Station, demonstrated that rice grown on submerged soils absorbs nitrogen in the ammoniacal form and that this nutrient should be applied to rice as an ammonium salt, or as organic nitrogen. Kelly has shown also that when nitrates are applied to rice on submerged soils they are partly reduced to nitrites, and that if nitrites accumulate in the soil to any considerable extent injury to the rice is likely to result. This last observation has been confirmed by Kapp (1933).

Nitrate nitrogen appears to be superior for plants that require short growing periods, such as the sugar beet, and especially is this true when the fertilizer is to be applied to alkaline soils. In fact, as has been pointed out by Pierre and Bower (1943), a relatively high pH of the soil is most favorable for the absorption of ammonium ions by plants.

Allison (1931) concluded that higher plants may utilize more ammoniacal nitrogen than is generally supposed.

Use of Ammonium Sulfate in Growing Turf for Lawn or Golf Courses

In early years the Rhode Island Experiment Station and other experiment stations in the northeastern and middle Atlantic states reported that a fine, firm, and springy turf could be secured and kept free from clover and weeds only on soils that are fairly acid. According to their observations, the acid-tolerant grasses, such as bent, fescue, and redtop, produce excellent greens, especially when fertilized with ammonium sulfate or other acid-forming ammoniacal fertilizers. Later, Sprague and Evaul

FIG. 20. Part of the famous Jordan Soil Fertility Plats at Pennsylvania State College—oldest extensive soil fertility experiments in America. These experiments were started in 1881. There are 144 plats. (Courtesy, Pennsylvania State College.)

(1930), of the New Jersey Experiment Station, reported that they were able to secure a vigorous growth of bent grass on limed soils. They suggested that the attempt to control weeds by making the soil strongly acid is unwise, for certain weeds are as tolerant of acid conditions as is grass.

In those sections of the country where Kentucky bluegrass and Dutch clover grow well, and are desired, it is generally advised that the pH of the soil be maintained at approximately 6.5 if ammonium sulfate is used. Otherwise, the use of a nitrogenous fertilizer which is physiologically basic would appear to be preferable.

Relative Value of Sodium Nitrate and Ammonium Sulfate

Long-time results reported by Russell (1919) from the Rothamsted Experiment Station indicate that under the conditions of the experiment, sodium nitrate produced slightly larger yields of barley and wheat than did ammonium sulfate. These results are given in Table 19. The barley data were secured on Hoos field and represent an average of 60 years, 1852 to 1911. The wheat data were secured on Broadbalk field and represent an average of 20 years, 1878 to 1898. Phosphoric acid and potash were applied to all plats. Russell (1932) thinks that although ammonium salts appear to be inferior to nitrate salts in many experiments, the reverse should be true where the masking effect of acid residues is overcome.

Table 19

RELATIVE VALUE OF SODIUM NITRATE AND AMMONIUM SULFATE FOR BARLEY AND WHEAT ON THE HOOS AND BROADBALK FIELDS, RESPECTIVELY, OF THE ROTHAMSTED EXPERIMENT STATION, ENGLAND

Hoos Field		*Broadbalk Field*	
60-year average yield of barley per acre		*20-year average yield of wheat per acre*	
Carrier of Nitrogen	*Bu.*	*Carrier of Nitrogen*	*Bu.*
No nitrogen	19.7	No nitrogen	14.4
43 lbs. as sodium nitrate	42.7	86 lbs. as sodium nitrate	31.4
43 lbs. as ammonium sulfate	41.5	86 lbs. as ammonium sulfate	28.8

Many studies have been conducted by the various state experiment stations in this country to determine the relative crop-producing value of sodium nitrate and ammonium sulfate. A large number of these studies have been of short time duration and many are of local importance only. As might be expected, variable results have been secured because of the

FIG. 21. Nitrogen retained from percolating solutions of sodium nitrate and ammonium sulfate. The percolating solution tested drips from the larger pot at the top upon the column of three pots each containing dry soil deficient in nitrogen and planted to milo. (Courtesy, Dr. John P. Conrad and the California Agricultural Experiment Station.)

variations in soil, climate, and cropping systems employed. The hydrogen-ion concentration of the soil solution appears to be the major factor which has determined the results of many of these experiments. Harvey (1925), after a careful study of the data of many experiments, reached the conclusion that by adjusting the hydrogen-ion concentration of a soil in respect to the preferences of a given crop one could show an apparent superiority of ammonium sulfate over sodium nitrate and vice versa. He believes that through failure to consider all the factors influencing the tests, many experiments have led to erroneous conclusions regarding the relative merits of these two materials.

Results secured by Garner and Brown (1919), of the Maryland Experiment Station, are given in Table 20 as an example of many short-time studies. These investigators attempted to determine the relative value of ammonium sulfate and other nitrogenous fertilizers, when applied to tobacco at rates equivalent to about 24 pounds of nitrogen per acre. Their experiments covered a period of six years. In the New Jersey cylinder experiments Lipman, Blair and Prince (1931) have reported a 15 per cent higher average recovery of nitrogen by crops from sodium nitrate than from ammonium sulfate.

Table 20

RELATIVE VALUE OF VARIOUS CARRIERS OF NITROGEN FOR TOBACCO

Carrier of Nitrogen	*Average Yield in lbs. per Acre*	*Average Gross Value per Acre*	*Average Value of Crop Less Cost of Fertilizer*	*Average Value of Tobacco per 100 lbs.*
No nitrogen..........	868	$102.71	$ 97.54	$11.83
Sodium nitrate........	1195	162.85	143.95	13.63
Ammonium sulfate....	1228	168.43	159.73	13.71
Tankage.............	1102	138.26	127.83	12.55
Cottonseed meal.......	1104	157.23	144.95	14.24

Among the most often cited of the long-time experiments that have been made in this country are those of the Jordan plats of the Pennsylvania Experiment Station. The relative crop-producing values of sodium nitrate and ammonium sulfate during a period of 55 years, as reported from this station by Noll, Gardner and Irvin (1931), and by Noll from unpublished data, for three rates of application are indicated in Table 21. These results would lead one to believe that ammonium sulfate is a better source of nitrogen than sodium nitrate for field crops provided the soil is maintained at a soil reaction which is optimum for the crops grown.

White (1925), of the Pennsylvania Experiment Station, in summing up the comparative values of ammonium sulfate and sodium nitrate, as interpreted from the data obtained from the Pennsylvania plat experiments for the first 40 years, concludes as follows:

"We must conclude that data obtained from the 40-year plots as to yields from sodium nitrate and ammonium sulphate do not indicate the relative value of these two fertilizers when properly used. The experiments, however, are accurate and give very valuable information, possibly more valuable because of the fact that they develop so clearly the improper use of ammonium sulphate. The grower, however, should use these results to point the way to profitable use of ammonium sulphate and not condemn the product which like a fine piece of machinery does not give the best results under misuse."

Table 21

A Comparison of Total Yields in Pounds per Acre of All Crops in a Rotation on Complete Commercial Fertilizer Plats Receiving Nitrogen as Sodium Nitrate and Ammonium Sulfate

	24 lbs. of Nitrogen		*48 lbs. of Nitrogen*		*72 lbs. of Nitrogen*	
	Sodium Nitrate	*Ammonium Sulfate*	*Sodium Nitrate*	*Ammonium Sulfate*	*Sodium Nitrate*	*Ammonium Sulfate*
1882–1889........	18,084	18,431	18,512	20,034	18,577	19,441
1890–1897........	19,012	18,198	19,413	19,035	19,901	18,130
1898–1905........	16,679	16,015	17,661	15,950	17,964	15,368
1906–1913........	18,475	16,432	19,276	14,892	19,456	12,716
1914–1921........	17,856	15,211	18,297	11,876	18,868	9,272
Average 1882–1921........	18,021	16,857	18,632	16,357	18,953	14,985
Unlimed 1922–1937........	15,114	9,338	15,284	5,792	15,869	5,063
Limed 1922–1937........	17,532	19,394	18,465	20,013	18,715	19,227

In the final analysis, the relative crop producing values of sodium nitrate and ammonium sulfate, like the relative crop producing values of any nitric source of nitrogen as compared with any ammonic source of nitrogen, is dependent largely upon two factors: the acidity producing powers of the products compared, and their relative susceptibility to loss by leaching before they can be utilized by the crop fertilized.

4

Manufacture and Use of Ammonium Nitrate

Within the last three years ammonium nitrate has come on the market in quantities that would suggest that sodium nitrate and ammonium sulfate will meet increasingly serious competition from this new source of nitrogen in the years immediately ahead.

Source of Ammonium Nitrate

Ammonium nitrate is manufactured by passing ammonia gas into nitric acid, and it is now available from private and government plants of the United States and Canada. It is manufactured for use in the production of explosives and for fertilizer purposes. Very little of the domestic output of ammonium nitrate was utilized in commercial fertilizers before 1935. It was during that year that one of the American nitrogen producers put on the market a solution of ammonium nitrate and ammonia for use in ammoniating superphosphate. While many of the ordnance ammonia plants that produced anhydrous ammonia during the World War II period have now been closed, others have been converted to the production of ammonium nitrate. Among the latter are the Jayhawk Ordnance, Baxter Springs, Kansas; the Ozark Ordnance, El Dorado, Arkansas, and the Buckeye Ordnance, South Point, Ohio. Large stocks of ammonium nitrate were disposed of for fertilizer both in this country and in Europe following World Wars I and II. At the beginning of World War II, the Tennessee Valley Authority of Wilson Dam, Alabama, constructed a new synthetic ammonia plant, modernized one of the old nitrate plants remaining from World War I, and began the production of ammonium nitrate in August, 1942. In 1943 much of this output was made available to agriculture. This plant has a capacity of about 150 tons of ammonium nitrate a day. Canada has three ammonium nitrate plants with a capacity of 120,000 tons of nitrogen a year.

Chemical Composition of Ammonium Nitrate

Pure ammonium nitrate contains about 35 per cent of nitrogen but the commercial product as sold contains 32.5 and 33.5 per cent of nitrogen. One-half of the nitrogen in ammonium nitrate is in the ammoniacal form and one-half in the nitrate form. Because it is entirely soluble in water when used as a fertilizer, ammonium nitrate leaves no residue in the soil. Because of its composition and solubility it is perhaps the quickest acting nitrogenous fertilizer known.

Physical Characteristics of Ammonium Nitrate

Ammonium nitrate crystals or particles are white in color. As produced for the fertilizer trade the particles may be single crystals or they may be aggregates of crystals produced either by graining or spraying. The grained ammonium nitrate is about twice as porous as the monocrystalline material and the sprayed material about five times as porous. Ammonium nitrate is hygroscopic under humid conditions and cakes so badly that the untreated salt cannot be applied with the ordinary fertilizer drill. Moisture is the principal factor causing the material to cake. It is very soluble being in

FIG. 22. Pasture demonstration near Madison, Wisconsin. (*Left*) 175 pounds per acre of ammonium nitrate. (*Right*) Unfertilized. (Courtesy, Prof. C. J. Chapman and the Wisconsin Agricultural Experiment Station.)

fact, more than twice as soluble as sodium nitrate. Its solubility as reported by Ross *et al.* (1943) is given in the following table.

Table 22

SOLUBILITY OF NITROGEN FERTILIZER MATERIALS IN WATER

Temp. (degrees F.)	*Grams Dissolved by 100 g. of Water*			
	Ammonium Nitrate	*Sodium Nitrate*	*Ammonium Sulfate*	*Urea*
0.......	118.3	73.0	70.1	66.7
68.......	187.4	87.6	75.4	103.3
104.......	280.2	102.0	81.2	166.7
140.......	418.1	121.7	87.3	250.9
176.......	635.3	148.1	94.2	400.0

Ross *et al.* (1946) have emphasized the fact that ammonium nitrate differs from other fertilizer materials in that it is capable of undergoing several crystalline modifications and that these changes are accompanied by measurable volume and thermal changes. Ammonium nitrate is cubic above 125.2° C.; tetragonal between 125.2° and 84.2°; orthorhombic between 84.2° and 32.3°; orthorhombic pseudotetragonal between 32.3° and −18°; and tetragonal below −18°.

When exposed to the atmosphere ammonium nitrate by itself is regarded as a hazardous material in that it will support combustion and ignites and detonates, although not easily, under certain conditions that are not fully or adequately understood.

Production of Commercial Ammonium Nitrate

The chief objection to pure ammonium nitrate crystals or granules is their tendency, depending upon humidity and temperature, to absorb water from the atmosphere and to harden or cake. Until recent years ammonium nitrate crystals, because of their hygroscopic character, were considered hopeless as a fertilizer. Unfortunately, the first ammonium nitrate to come on the market was not intended for fertilizer use and was in poor physical condition and difficult to use in mixed fertilizers or for direct application. Now, however, the picture has entirely changed. In order to overcome its objectional characteristic, processes have been devised for conditioning the ammonium nitrate and for coating the granules with waxes and dusts to ensure against absorption of water. Much progress has been made in this respect since 1943, so that the

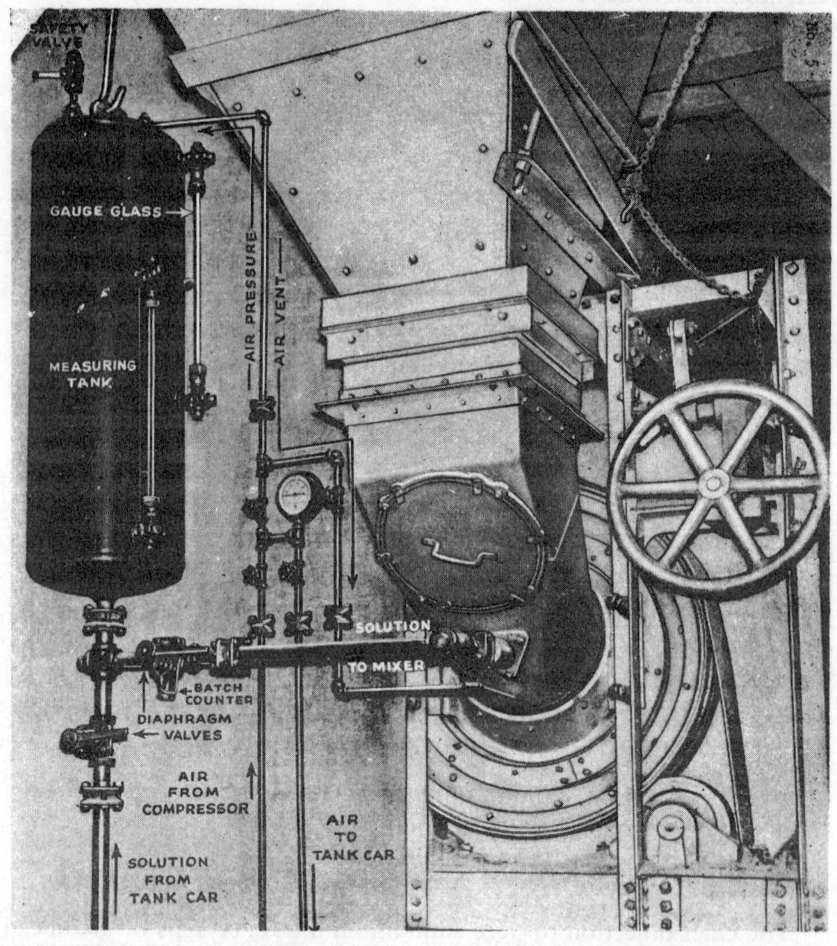

FIG. 23. An arrangement for measuring and introducing Barrett's ammoniating solutions into rotary-type fertilizer mixers. (Courtesy, The Barrett Division, Allied Chemical and Dye Corporation.)

products on the market today are workable and acceptable to producers of mixed fertilizers and to farmers. When grained and mixed with 3 to 5 per cent of a suitable conditioning agent, such as tricalcium phosphate, Kieselguhr, plaster of Paris, or kaolin and other forms of clay, the tendency of ammonium nitrate to take up moisture is greatly reduced. Ammonium nitrate that is to be used in very humid regions is further treated with 0.5 to 1 per cent of a water repellent, such as mixtures of paraffin with petrolatum or asphaltum and of petrolatum and paraffin with asphaltum or rosin. When thus conditioned and packed in waterproof bags, burlap, or multiwall-paper bags lined with asphalt, ammonium nitrate will remain in good condition for many months. In addition, it has been found that the production of large crystals or granules has aided somewhat in the production of a better physical condition. The TVA product now on the market contains about 35 per cent of plus 20-mesh, 50 per cent of 20–40-mesh, and 15 per cent of minus 40-mesh granules. The Canadian material on the American market has large granules; about 80 per cent will not pass through a 20-mesh sieve and less than 5 per cent is smaller than 40-mesh. These improvements which lower the absorption of water are such that ammonium nitrate can now be put on the fertilizer market in about as good a physical condition as that of sodium nitrate and ammonium sulfate.

Ammonium Nitrate Mixtures

Ammonium nitrate dissolved in aqueous ammonia is now used in many mixing plants that ammoniate their superphosphate. Its use in this way results in an excellent mechanical condition of the mixture. Also, in this form nitrate nitrogen is made available for use in commercial mixed fertilizers at a cost competing with that of ammonium salts. The use of nitrogen solutions is discussed in another chapter.

Solid ammonium nitrate may be satisfactorily mixed with many other fertilizer materials if reasonable consideration is given to (1) physical constants and chemical reactions, (2) a proper neutralization of the superphosphates, and (3) the use of a minimum amount of incompatible compounds.

In Germany, following World War I, ammonium nitrate was mixed with gypsum to the extent of 60 per cent, and excellent results were reported from the use of this mixture both in Europe and in this country. In Europe ammonium nitrate was also commonly sold in mixtures with potash salts or ammonium sulfate. It is now used in the manufacture of ammonium sulfate nitrate, potassium ammonium nitrate, ammonium nitrate-calcium carbonate and similar materials.

When ammonium nitrate is mixed with ammonium sulfate, pulverized limestone, superphosphate, and most other materials used in the manufacture of complete fertilizers, it loses somewhat its dangerous quality of detonation although a small amount of ammonia may be lost. For example, ammonium nitrate mixed with ammonium sulfate in equimolecular proportions, and ammonium nitrate mixed with 80- to 100-mesh pulverized limestone in the proportion of 60 to 35 have been found entirely non-hazardous by the Underwriters Laboratories, Inc.

With the exception of calcium cyanamide, ammonium nitrate is the most soluble compound in fertilizer mixtures. This allows thorough blending with other materials.

Hardesty and Parker (1943) have studied the influence of ammonium nitrate when used as a component of mixed fertilizer. They conclude that mixed fertilizer formulas calling for the use of ammonium nitrate should include 25 to 50 pounds of some active basic material per 1000 pounds of superphosphate to insure neutralization of the superphosphate. For this purpose they suggest hydrated lime, powdered calcium cyanamide, ammonia and cement flue dust. They also suggest that fertilizers containing ammonium nitrate be formulated with as low a moisture content as possible and that wet dolomite, wet fillers and conditioning agents be avoided. This would also apply to all incompatible compounds. The most incompatible compounds are sodium nitrate, potassium nitrate, sodium chloride, small amounts of magnesium chloride that are often found in some low-grade potash salts, and especially urea. Dry urea and ammonium nitrate should never be used in the same mixture. Sodium and potassium nitrates reduce the amount of ammonium nitrate that can be used without hygroscopic effects, sodium chloride with ammonium nitrate forms sodium nitrate as a reaction product, and magnesium chloride forms by reaction magnesium nitrate, a very hygroscopic compound.

Smith (1944) states that ammonium nitrate that has become lumpy should be broken up, preferably with wooden mauls or stompers on a non-inflammable floor. Also, that sulfur and organic dusts should be kept away from ammonium nitrate prior to batching or mixing.

In England, Germany, Norway, and the United States, manufacturers have combined ammonium nitrate variously with precipitated calcium carbonate, pulverized chalk, high calcium limestone, and dolomite. These mixtures, usually containing 16 to 20.5 per cent of nitrogen and 16 to 50 per cent of lime materials, have been put on the market under the trade name of Nitro-Chalk, Cal-Nitro, A-N-L and others as trademarked by the respective manufacturers and sellers. They are referred to as ammonium limestone materials. Just prior to World War II, 70,000 tons of these materials were imported annually from Norway. The A-N-L and Cal-Nitro are American products and have been sold rather extensively in this country in recent years. They contain 10.25 per cent nitrate nitrogen, 10.25 per cent ammonia nitrogen, 9.0 per cent CaO and 7.0 per cent MgO.

The color of the ammonium nitrate limestone materials is not important since it results from the color of the lime material, or from the siliceous dust or other powder used as a protective coating on the particles, or from pigments added as a mark of identification.

Fertilizing Value of Ammonium Nitrate

Ammonium nitrate has been used to a considerable extent in fertilizers for the past 15 years, and has proved to be excellent because of its solubility and composition. It is readily available to plants as soon as it dissolves in the soil solution. In addition, because it contains nitrogen in both the nitrate and ammoniacal forms, it is ideally suited for grass crops and other crops that prefer nitrogen in both forms. It is more resistant to leaching than sodium nitrate but less so than ammonium sulfate. Its properties are somewhat similar to those of a 50-50 mixture of sodium nitrate and ammonium sulfate.

Table 23

YIELD OF SEED COTTON WITH DIFFERENT SOURCES OF NITROGEN IN COMPLETE FERTILIZERS* IN FIELD TESTS ON PROMINENT SOIL TYPES
(pounds per acre)

Location and Soil	*Duration*	*Ammonium Nitrate*	*Ammonium Sulfate*	*Ammonium Chloride*	*Urea*	*Sodium Nitrate*
NORTH CAROLINA						
Marlboro sandy loam	1926–27	1428	1573	1428	1457	1620
Marlboro fine sandy loam	1925	1403	1288	1472	1311	1495
Greenville sandy loam	1926–27	2015	2040	1415	1905	1695
Portsmouth sandy loam	1921–22	1172	1290	1017	1025	1270
Dunbar fine sandy loam	1920	930	1052	1152	820	834
Coxville sandy loam	1920	700	692	786	640	642
Cecil fine sandy loam	1924–25	957	993	905	950	1051
Appling sandy loam	1934	1163	1075	1088	1263	1313
Davidson clay	1926–27	1357	1418	1360	1430	1450
SOUTH CAROLINA						
Norfolk fine sandy loam	1921–25	1054	1008	924	1047	1091
Norfolk sandy loam	1924	1140	1120	1080	1020	1100
Marlboro sandy loam	1921–23	1447	1337	1327	1340	1330
Average of all tests		1231	1240	1163	1184	1241

* Source of materials in fertilizer: P_2O_5 from superphosphate, K_2O from potassium sulfate, no lime in fertilizer mixture.

Soils tend to become acid following the continuous use of ammonium nitrate because the soil becomes impoverished of its bases as a result of leaching and the removal of larger crop yields following the application of the fertilizer.

Ammonium nitrate may be applied before the crop is planted or as side- and top- applications when the crop is young, provided the crop foliage is not wet. In the Mississippi Delta and on the grasslands of the Northwest, fall applications of ammonium nitrate have been found substantially equal to spring applications. Fitts (1945), of the Nebraska Experiment Station, secured better distribution of the ammonium nitrate by applying it in irrigation water, and almost as good crop response as that obtained from side-dressing.

Skinner (1944) has reported the results of 23 cotton tests conducted in North and South Carolina with different sources of nitrogen which included ammonium nitrate. His results are given in the table on p. 71.

5

Manufacture and Use of Synthetic Nitrogenous Fertilizers

Synthetic nitrogen was first produced commercially in the United States in 1922. Production rapidly increased so that since 1930 more nitrogen has been fixed annually from the atmosphere than has been made available for fertilizer use from any other source. According to O'Brien (1937), just before the outbreak of World War II, about 63 per cent of the world's supply of nitrogen came from the air. The capacity of the United States to produce synthetic nitrogen in the last 20 years is shown in Table 24.

Table 24

CAPACITY OF THE UNITED STATES TO PRODUCE SYNTHETIC NITROGEN IN RECENT YEARS

Year	*Tons of Nitrogen*
1925	15,300
1930	195,500
1935	341,350
1940	380,300
1944	1,186,100*

* 431,100 tons in private plants; 735,000 tons in government plants.

During World War II the government erected many new nitrogen fixation plants, and ten of these were in operation at the end of the war. Also, during this period three new plants were erected in Canada.

There are three nitrogen fixing processes with which the fertilizer industry is primarily concerned. These are the arc, the cyanamide, and the direct synthetic ammonia processes. The primary products of these processes are dilute nitric acid, crude calcium cyanamide, and anhydrous or aqueous ammonia, respectively. The nitric acid and aqueous ammonia must be combined with other chemicals before they can be used for fertilizing crops.

Arc Process

The arc process is the simplest process of nitrogen fixation and was the first to be developed commercially. Under the direction of C. S. Bradley and R. Lovejoy, the Atmospheric Products Company, in the period 1902 to 1904, made an unsuccessful attempt to establish the arc process on a commercial scale at Niagara Falls, N. Y. In 1903, however, a successful experimental plant was set up by two Norwegians, Prof. Christian Birkeland and Engineer Samuel Eyde, at Ankerlökken, Norway. In this country a plant was built in 1913 at Nitrolee, S. C., but was soon abandoned. Little progress has been made since World War I in increasing the world's nitrogen fixation capacity by this process.

The basic reaction of the arc process involves the union of nitrogen and oxygen gases by means of an electric arc. The process is similar to that reaction which is known to take place between nitrogen and oxygen in the atmosphere and which has been estimated to be responsible for about 8 pounds of fixed nitrogen per acre that is brought down annually by rain. Burns *et al.* (1922) estimated that 100,000,000 tons of nitrogen are fixed annually in the world's atmosphere through the lightning discharge of electrical storms. In the arc process the reaction is secured by blowing air through a flaming electric arc, which has a temperature of about 3500° C. Only about 2 per cent of the air thus heated is caused to combine. The nitric oxide produced is then converted into nitric acid by the action of water and atmospheric oxygen. The reactions involved are about as follows:

$$N_2 + O_2 = 2NO$$
$$2NO + O_2 = 2NO_2$$
$$3NO_2 + H_2O = 2HNO_3 + NO$$

The principal objections to the arc process are the extremely large power requirement and the high capital cost involved. This process requires about five times as much electricity to fix a ton of nitrogen as is required by the cyanamide process. In view of the development of other processes for nitrogen fixation, these two objections to the arc process eliminate it as a possible large producer of fixed nitrogen in this country.

FIG. 24. Government owned fertilizer plant at Muscle Shoals, Alabama. Wilson Dam and Lake Wilson in the background. (Courtesy, Tennessee Valley Authority.)

Because of the cheapness there of water power, this process has been used to the greatest extent in the Scandinavian countries. Present indications point to the fact that little can be hoped for in the nature of improvements to the arc process.

Cyanamide Process

The cyanamide process was developed in Germany during the period extending from 1897 to 1905. The method is said to have been worked out by two German chemists, Prof. Dr. Adolph Frank and Dr. Nicodem Caro, although the discovery of the cyanamide process is often credited to the chemists Rothe and Freudenberg who were employed by Frank and Caro. Calcium cyanamide was first made on a laboratory scale in 1898. The process of converting cyanamide nitrogen into ammonia was patented by Frank in 1900 and the use of calcium cyanamide as a fertilizer was first attempted in 1901. The first factory for the production of calcium cyanamide was erected at Westeregeln, near Magdeburg, Germany, in 1905, but for several reasons proved to be an economic failure. A second plant was built at Piano d'Orta, Italy, in 1906 and another at Bramberg, Prussia, in 1908, and still another at Trostberg, Bavaria, in 1909. Within the next few years factories were established throughout the world.

The cyanamide process has a distinct advantage over the arc process in that it requires less than one-fourth of the electric energy necessary to fix a unit of nitrogen. The raw materials required—coal and limestone—are relatively cheap and abundant. Furthermore, this process yields a product which is a solid material and which can be sold for fertilizer purposes without further treatment. The cyanamide industry developed rapidly dur-

FIG. 25. Where nitric acid and calcium nitrate are made. Norsk Hydro factory, Rjukan, Norway. (Courtesy, Synthetic Nitrogen Products Corporation.)

ing the World War I period. This was due to the fact that cyanamide nitrogen can be converted into ammonia gas and then into nitric acid. The nitric acid was produced for use in the manufacture of explosives. Although calcium cyanamide can be produced by heating a mixture of limestone with coal in an atmosphere of nitrogen, it has proved more economical to carry out the process in four separate steps. These steps are:

1. The burning of the limestone in a vertical or rotary kiln at about 1100° C. The limestone is decomposed as follows:

$$CaCO_3 + \text{heat} = CaO + CO_2$$

2. The production of calcium carbide by heating coke or coal and lime in an electric furnace. This reaction can be expressed by the equation:

$$CaO + 3C = CaC_2 + CO$$

3. The production of pure nitrogen gas from the air by liquification and distillation.

4. The treatment of the finely powdered carbide with pure nitrogen at 1000° C. to form calcium cyanamide. This reaction can be represented by the equation:

$$CaC_2 + N_2 = CaCN_2 + C$$

In 1909 the first cyanamide plant in North America was built at Niagara Falls, Ontario, Canada. Only one plant has been built in this country—Nitrate Plant No. 2 at Muscle Shoals, Alabama. This plant was built by the United States Government, under the National Defense Act, June 3, 1916, as a wartime measure and with a rated capacity of 100,000 tons per year, though only 2,000 tons of calcium cyanamide had been produced before it was closed after the armistice. For many years the plant was maintained in a stand-by condition. No new plants employing this method of fixation were built during the period of World War II. At present no calcium cyanamide is being produced in this country, but considerable quantities are manufactured at Niagara Falls, Canada, and much of this is imported into the United States.

Direct Synthetic Ammonia Process

Although the direct synthetic ammonia process has been developed since the arc and the cyanamide processes, so far it has proved to be the best method of nitrogen fixation and has largely replaced the arc and cyanamide processes. As already pointed out, much more nitrogen is fixed by the direct synthetic ammonia process than by all other synthetic processes combined.

The first commercial plant to use the direct synthetic ammonia process was built in Oppau, Germany, in 1910. During World War I Germany greatly enlarged this plant and built another at Merseburg. In the period between World War I and World War II the Haber-Bosch plant (Leuna

Works) at Merseburg, Germany, was the largest synthetic nitrogen plant in the world, although a large part of its original capacity had been diverted to the production of synthetic gasoline. The first plant to employ this process in this country was United States Plant No. 1, built at Sheffield, Alabama by the United States Government just prior to the close of World War I. The first successful plant in this country was built at Syracuse, N. Y., in 1921, by the Atmospheric Nitrogen Corporation. In 1929 this company built another plant at Hopewell, Virginia. It was from this plant that the first shipment of synthetic sodium nitrate was made in 1929.

For years the Hopewell plant was the second largest nitrogen fixation plant in the world, and the largest plant anywhere manufacturing sodium nitrate. It is the only sodium nitrate plant in the United States.

Haber-Bosch Modification of Direct Synthetic Ammonia Process

The Haber synthetic ammonia process of fixing atmospheric nitrogen was developed in Germany by Professor Fritz Haber during the years 1905 to 1908. Prof. Haber worked out a method whereby nitrogen and hydrogen gases would unite to form ammonia when brought together under a certain pressure, at a certain temperature, and in the presence of a catalyst. This process was an improvement on older methods of fixing atmospheric nitrogen in that the power consumption was much reduced. In fact, the reaction involved in the union of nitrogen and hydrogen is exothermic after the equipment is heated to proper temperature. The technical difficulties of the Haber process—that of handling an inflammable gas at high temperatures and pressures—were overcome by Dr. Carl Bosch, of the

FIG. 26. Aerial view of the calcium cyanamide plant of the American Cyanamid Company, Niagara Falls, Canada. (Courtesy, American Cyanamid Company.)

Badische Anilin und Soda Fabrik. It is upon the Haber-Bosch modification of the direct synthetic ammonia process that the present nitrogen industry of most countries is based.

A part of the nitrogen used in the Haber-Bosch method is obtained by the Linde process which consists of liquefying air and then allowing it to warm slowly, thus distilling off the nitrogen. However, most of the nitrogen now being used is secured by burning coal in a forced draft of air. The oxygen of the air combines with the carbon of the coal to form carbon dioxide and carbon monoxide while the nitrogen remains unchanged. The resulting gas is commonly called "producer-gas."

The gaseous hydrogen, which is the chief item of expense in ammonia production, is secured by what is known as the alternate steam air "blow." Coke is ignited to a red heat in large cupolas. This is done with the aid of an air-blast. Super-heated steam is then passed through the coke. Under such conditions the carbon of the coke combines with the oxygen of the water to form carbon monoxide. The hydrogen of the water is thus freed as a gas. The mixture of carbon monoxide, hydrogen and impurities is commonly called "water-gas." The water-gas, together with the remaining steam, is then conducted over a metal catalyst. During this operation the carbon monoxide extracts oxygen from the steam and unites with it and another molecule of water, thus forming carbonic acid and at the same time increasing the content of hydrogen. The reactions that take place may be represented as follows:

$$H_2O + C = CO + H_2$$
$$CO + 2H_2O = H_2CO_3 + H_2$$

The producer-gas and the water-gas are then mixed together with a small amount of steam, and the mixture is conducted over metal catalysts at a pressure of about 150 to 200 atmospheres and a temperature of 500° to 700° C. Here the hydrogen and the nitrogen unite to form ammonia, but the yield of ammonia is only about 8 per cent of that which might be expected if all the hydrogen and nitrogen united.

The mixture of gases is so regulated that the proportion of nitrogen to hydrogen will be about 1 to 3, or that required for the formation of ammonia. After the ammonia is formed it is absorbed in water, and any uncombined gases, together with fresh gases, are reconducted through the ovens. The catalyst usually employed is a platinum sponge, although compounds of boron, cerium, iron, manganese, molybdenum, silver, thorium, tungsten, uranium, and vanadium have been used.

Although it takes only a little over half as much power to fix a ton of nitrogen by the Haber-Bosch process as that required by the cyanamide process, the investment in plant requirements costs twice as much.

Many modifications of the Haber-Bosch process have been developed in recent years and are now employed in many countries but practically all of the new plants built during the period of World War II employ the direct synthetic process. World War II brought about a tremendous expansion in the synthetic ammonia production capacity of the United States. Brand (1944) estimated it to be 1,200,000 tons of nitrogen per year, or about three times our normal consumption.

Oxidation of Ammonia to Nitric Acid

Synthetic ammonia is readily converted into nitric acid by the Oswald process which consists of passing ammonia over a catalyst in the presence of oxygen. The reactions involved are probably as follows:

$$4NH_3 + 5O_2 = 6H_2O + 4NO$$
$$2NO + O_2 = 2NO_2$$
$$3NO_2 + H_2O = 2HNO_3 + NO$$

Table 25 shows the world capacity for production of chemical nitrogen, by processes and countries, for a representative pre-World War II year as reported by O'Brien (1937). Figures for 1945 are not available.

Atmospheric Nitrogen and its Relationship to Soil Fertility

For many years students of agriculture and other scientists as well, were alarmed by the tremendous amounts of nitrogen drained from our soils by the needs of our modern civilization. In 1898 Sir William Crookes delivered a now famous address before the British Association for the Advancement of Science, in which he directed attention to the danger involved in the rapid increase in the world's population as compared with the prospective increase in the world's production of wheat. He also called attention to the immense loss of nitrogen from soils, resulting from natural causes, such as leaching, which had been increased greatly by modern methods of cultivation; and to the immense quantities of nitrogen being removed annually from soils in crops. This fear was soon shared by others. In 1905 Shutt showed that some soils in Canada had lost as much as 35 per cent of their total nitrogen in 22 years after being put under cultivation, and similar losses were noted by investigators in other countries.

It was the notable work of Atwater (1884) in this country, and of Hellriegel (1886) and Wilfarth (1887) in Europe, that furnished a basis for the solution of the problem of soil nitrogen depletion. These investigators showed that the organisms which previously had been found to be living on the roots of legumes, and which had been named *Bacillus radicicola* by Beijerinck, had the power to fix the very inert elemental nitrogen of the atmosphere and to make use of it in their metabolism. As about 75 per cent of the atmosphere is nitrogen, there is a reserve supply of over 70,000,000 pounds above every acre. By a process that is still not fully understood, the soil atmospheric nitrogen that is fixed by legume bacteria (Rhizobium) becomes available to the legume host and possibly to any nonleguminous plant whose roots may be in the immediate vicinity of the legume nodules. In addition, the studies of Winogradsky, of Russia, in 1895, and of Beijerinck, of Holland, in 1901, showed that atmospheric nitrogen may be fixed by some nonsymbiotic soil organisms. This fixation has been thought of as being of minor importance but it may yet be proved to be of as great importance to agriculture as the fixation of nitrogen by legume bacteria. The discovery of the nitrogen-fixing power of soil organisms assured the farmer of a means, although perhaps a rather costly one, of increasing the nitrogen content of the soil largely through the growth of leguminous green manuring crops.

Besides biological methods of fixation, chemical and electrical methods of fixation, as already described, have received much attention by scientists during the last third of a century. As European countries were farther removed from the source of Chilean nitrate and were in more danger of being cut off from this source of nitrogen in time of war, some of them surpassed us in the early years in nitrogen fixation investigations.

Characteristics of Synthetic Nitrogenous Fertilizers

In general the synthetic nitrogenous fertilizers which are made from the primary products of fixation, nitric acid, calcium cyanamide, and ammonia, present four outstanding characteristics that are of major importance to the manufacturers of mixed goods and to the farmer. (1) From these fertilizers may be selected those that are acid, neutral, or alkaline. (2) They are high-analysis fertilizers and can be transported long distances at small cost per unit of nitrogen. (3) They are generally free of toxic substances. (4) Fertilizers carrying nitrogen in the nitrate, ammoniacal, or

Table 25

CHEMICAL NITROGEN: WORLD CAPACITY FOR PRODUCTION, BY PROCESSES AND COUNTRIES (EFFECTIVE CAPACITY AS OF JANUARY 1934)
(Figures are in tons of 2,000 lbs. of nitrogen)

Country	Nitrogen capacity by:									
	Synthetic Processes		Cyanamide Processes		By-product Processes		Shanks and Guggenheim Processes		National Total—All Processes	
	Capacity	Per Cent of World Total	Capacity	Per Cent of World Total	Capacity	Per Cent of World Total	Capacity	Per Cent of World Total	Capacity	Per Cent of World Total
Germany	1,136,080	35.2	198,000	36.7	137,500	22.1	..	..	1,471,580	29.0
Chile	..	..	..	..	..	..	690,000*	100.0	690,000	13.6
United States	341,350	10.6	..	..	200,000	32.2	..	..	541,350	10.6
Great Britain	269,500	8.3	..	..	121,000	19.5	..	..	390,500	7.7
France	274,650	8.5	44,000	8.2	50,000	8.1	..	..	368,650	7.2
Japan	218,460	6.8	69,410	12.9	10,850	1.7	..	..	298,720	5.9
Belgium	205,150	6.3	6,600	1.2	27,500	4.4	..	..	239,250	4.7
U.S.S.R.	181,500	5.6	4,400	.8	16,500	2.7	..	..	202,400	4.0
Italy	123,750	3.8	22,000	4.1	6,050	1.0	..	..	151,800	3.0
Netherlands	122,100	3.8	..	..	11,000	1.8	..	..	133,100	2.6
Norway	116,050	3.6	16,500	3.1	220	..	..	..	132,770	2.6
Canada	35,750	1.1	78,000	14.5	11,000	1.8	..	..	124,850	2.5
Poland	70,510	2.2	41,690	7.7	6,600	1.1	..	..	118,800	2.3
Yugoslavia	29,700	.9	22,000	4.1	100	..	..	..	51,800	1.0

Table 25—(*Continued*)

Country	Nitrogen capacity by: Synthetic Processes		Cyanamide Processes		By-product Processes		Shanks and Guggenheim Processes		National Total—All Processes	
	Capacity	Per Cent of World Total	Capacity	Per Cent of World Total	Capacity	Per Cent of World Total	Capacity	Per Cent of World Total	Capacity	Per Cent of World Total
Czechoslovakia	27,280	.8	7,700	1.4	8,800	1.4	..	..	43,780	.9
Manchuria	39,600	1.2	..	..	2,350	.4	..	..	41,950	.8
Switzerland	9,900	.3	16,500	3.1	550	.1	..	..	26,950	.5
Sweden	5,280	.2	6,600	1.2	3,300	.5	..	..	15,180	.3
Spain	10,340	.3	..	..	3,300	.5	..	..	13,640	.3
South Africa	8,250	.3	..	..	880	.1	..	..	9,130	.2
Hungary	6,600	.2	..	..	550	.1	..	..	7,150	.1
Rumania	(660)	..	5,500	1.0	100	..	..	..	5,600	.1
All other countries†	..	..	..	..	3,350	.5	..	..	3,350	.1
Process and world totals	3,231,800	100.0	539,000	100.0	621,500	100.0	690,000	100.0	5,082,300	100.0

* Capacity for the production of nitrogen as sodium nitrate and potassium nitrate.

† India, Austria, Australia, and Denmark.

amide form, can be bought at little difference in price. The production of cheap synthetically produced nitrogen products has decreased the cost of nitrogen in complete fertilizers. A description of the manufacture, characteristics and use of the synthetic nitrogenous fertilizers found on the market in recent years is given below. Some of these materials have been on the market for such a short period of time that only meager experimental evidence as to their crop-producing power is available. On the other hand, others have probably passed the peak of their popularity and are now passing out of use. A few that were on the market at the beginning of World War II have not been available since, and their manufacturers may replace them with other carriers of nitrogen.

Manufacture of Calcium Nitrate

Calcium nitrate, sometimes called *Norwegian saltpeter*, was the first synthetic nitrogenous fertilizer put on the market. It has never proved to be an entirely satisfactory fertilizer and its production has always been limited. In the early days it was sold under the name of *air saltpeter*. It is made by treating nitric acid with lime. The reaction involved is as follows:

$$2HNO_3 + CaCO_3 = Ca(NO_3)_2 + H_2CO_3$$

After about 5 per cent of ammonium nitrate has been added for the purpose of assisting in crystallizing the salt, the clear solution is evaporated to dryness by spraying with compressed air.

At the present time calcium nitrate is being manufactured in Norway. At the beginning of World War II, calcium nitrate furnished about 8 per cent of the nitrogen consumed by the world's agriculture. No calcium nitrate is produced in this country.

Characteristics of Calcium Nitrate

Nearly pure crystallized calcium nitrate is so very hygroscopic that it will melt if held in the hand. In fact, the commercial product that was first put on the market was so hygroscopic that it was often necessary to ship ir in airtight wooden kegs. Likewise, before it could be distributed properly in the soil, it was often necessary to mix the fertilizer with peat dust or some similar material. Lime was sometimes added as a dryer, but the objection to this practice was the reduction of the nitrogen content and the increase in transportation charges per unit of nitrogen which resulted. At other times it has been mixed with potassium sulfate and magnesium

FIG. 27. Spraying of calcium nitrate for the purpose of graining. (Courtesy, Synthetic Nitrogen Products Corporation.)

sulfate for the purpose of improving its physical condition, but with indifferent success.

The calcium nitrate now found on the market is granular in form and free from dust. At one time calcium nitrate was objected to because the dust released when the material is broadcast often blistered the skin and shrunk the shoes of the workmen. With the present product this objection has to some extent been overcome. As it flows rather freely through the ordinary fertilizer distributor care must be taken to regulate the flow if proper distribution is to be obtained.

Commercial calcium nitrate is usually almost pure white. It is alkaline to litmus and very soluble in water. Calcium nitrate, like many lime compounds, causes leather to shrink. Where large quantities are to be used, therefore, it is well for workmen applying it to have their shoes well greased. The calcium oxide equivalent of calcium nitrate amounts to about

28 per cent. Pure calcium nitrate analyzes about 17 per cent and the commercial product about 15.5 per cent of nitrogen.

Calcium nitrate is shipped in 100-pound even-weight bags which are made of burlap with a paper lining cemented to the burlap with asphalt. In this way the material is kept dry and it should reach the farmer in excellent physical condition.

Calcium nitrate is immediately available to plants. The results that have been secured with calcium nitrate by the various agricultural experiment stations indicate that when applied to general field crops the yields secured are comparable to those that have been secured with other nitrates. It is considered an excellent carrier of nitrogen for those crops that utilize large quantities of calcium. Calcium nitrate appears to be best adapted to the alkali soils of the arid West, the calcareous truck crop soils of the East Coast of Florida, and soils in the northeastern states where climatic conditions are more favorable to distribution.

A double nitrate salt of magnesium and calcium has been offered for sale in France. It analyzes 15.5 per cent nitrogen, 7 per cent water-soluble magnesium, and 16 per cent water-soluble calcium.

How Calcium Nitrate Should Be Used

Calcium nitrate may be drilled or broadcast. If it is broadcast, like other soluble salts, it should be applied only on days when the foliage of

FIG. 28. Plowing under corn stalks and 300 pounds of Granular Aero Cyanamid per acre. (Courtesy, American Cyanamid Company.)

plants is dry, otherwise the material may cling to the plants and result in the killing of those parts with which it comes in contact. Because calcium nitrate will burn the skin, it is most satisfactorily broadcast by hand during cool or dry weather. Since all commercial calcium nitrate is to some extent hygroscopic, the bags should not be opened until the product is to be used. Mehring (1930) found that calcium nitrate was perfectly dry and drilled well in an atmosphere of 40 per cent relative humidity, but at 50 per cent humidity it was soggy and drilled poorly, and at 60 per cent humidity it was entirely liquefied.

Because of its hygroscopic character calcium nitrate is not considered suitable for use in factory-mixed fertilizers. Experience with the present commercial product shows that it may be used in home mixtures if these are applied within a short time after mixing.

Characteristics of Calcium Cyanamide

At the present time calcium cyanamide furnishes about 5 per cent of the nitrogen consumed by the world's agriculture. Calcium cyanamide (cī-ăn-ăm′īde) has been put on the market under the trade names of *Aero Cyanamid* (cĭ-ăn′a-mĭd), *lime-nitrogen*, *Nitrolim*, and *Nitrolime*. Pure calcium cyanamide contains 35 per cent of nitrogen and is white in color. The European products contain 10 to 20 per cent of nitrogen, the American pulverized Aero Cyanamid 22 per cent of nitrogen and the American granular Aero Cyanamid 21 per cent of nitrogen. The American commercial products manufactured at Niagara Falls, Canada contain about 60 to 63 per cent of calcium cyanamide, 15 to 20 per cent of free lime, 11 per cent of free carbon, 2 to 4 per cent of water, and some aluminum, iron, and silicon oxides. Its total lime is equivalent to about 70 per cent of calcium hydroxide. This means that the equivalent of 1400 pounds of hydrated lime is found in every ton of calcium cyanamide. Because of the presence of free carbon, commercial calcium cyanamide is black in color. It is distinctly alkaline. Calcium cyanamide now appears on the American market in both powdered and granular form. The powdered calcium cyanamide is shipped in strong double bags, for it is a very fine powder and if carelessly packed would dust out. It is oiled to lessen its dustiness and is then marketed without further treatment. It has excellent drilling properties. The granular form is made by adding a suitable binding agent to the powdered form, and then graining. The granular product is preferred by most farmers and may in time replace the powdered product for direct application to crops. The granular product may be mixed with dry and well-cured superphosphate in any desired proportions without development of excessive heat or reversion of phosphoric acid to the citrate-insoluble form. Such mixtures however, must be applied with the same precautions as would be used with calcium cyanamide itself.

Changes that Take Place in Calcium Cyanamide During Storage

When calcium cyanamide is exposed to the atmosphere it slowly absorbs water and carbon dioxide, thus forming a surface crust and slightly increasing its volume. Richardson (1932) has reported a 10 per cent increase in weight after 27 months of storage. The water enters into chemical combination with the calcium cyanamide and forms calcium hydrogen cyanamide and calcium hydroxide. Because of this change the water absorbed cannot be determined merely by heating a sample of the material to 100° C. The carbon dioxide unites with the calcium hydroxide to form calcium carbonate. Both of the above changes result in an increase in the total weight and volume of the material.

When calcium cyanamide is stored in bulk for a year or more deterioration of the material may result. Under such conditions dicyanodiamide—a compound which is so stable as to be of very little value as a fertilizer—may be formed, especially if little attempt has been made to keep moisture and air away from the calcium cyanamide. It is for this reason that the manufacturers of calcium cyanamide recommend that only calcium cyanamide from newly opened bags be used.

The faint odor of ammonia that is sometimes given off by calcium cyanamide results from the ammonia that has been absorbed by its graphitic carbon. During the process of manufacture small quantities of metallic nitrides are produced. Later, when the calcium cyanamide is treated with small amounts of water for the purpose of decomposing traces of calcium carbide, these metallic nitrides are decomposed with the formation of ammonia. This ammonia is absorbed by the particles of free carbon that are present, and held for an indefinite period. It is this ammonia which when set free is responsible for the faint odor sometimes noticed.

Use of Calcium Cyanamide in Mixed Goods

Calcium cyanamide is considered one of the best driers known and for this reason about one-fourth of the calcium cyanamide imported into the United States from Canada finds its way into fertilizers. Mixtures of fertilizers containing calcium cyanamide are nearly always dry and pulverant, and they drill well. They rarely cake or set in the bags, nor does bag trouble arise with calcium cyanamide mixtures.

The poor mechanical condition of many fertilizer materials is due to the presence of moisture. The poor mechanical condition of superphosphate when present is due to the presence of free acid. When calcium cyanamide is mixed with the common fertilizer mixtures the chemical changes induced bring about the fixation of water as water of crystallization and the production of heat, and these reactions aid in drying the mixture. At the same time the lime in the calcium cyanamide quantitatively neutralizes any free phosphoric acid that is present. In such cases limestone or other carriers of lime may not be needed for conditioning purposes.

Manufacturers have found that 30 to 60 pounds of pulverized calcium cyanamide per ton of average fertilizer mixture gives satisfactory results. This quantity has been found to generate sufficient heat to dry out the mixture and thus induce a good physical condition. Larger quantities may produce too much heat and may also bring about a reversion of some of the available phosphates. Quantities of calcium cyanamide ranging from 75 to 100 pounds are sometimes used by manufacturers, especially when working on a small scale where conditions are such that the heat generated may escape readily. Likewise, 75 to 100 pounds of calcium cyanamide is sometimes used when bulky materials such as dried fish, animal tankage and cottonseed meal are present. These materials help to cover the calcium cyanamide and thus keep it away from the superphosphate. One pound of calcium cyanamide, if thoroughly incorporated with one ton of fertilizer mixture, will raise the temperature of the mixture about 1° F. Thus 60 pounds of calcium cyanamide is sufficient to raise the temperature of a ton of fertilizer mixture 55° to 60° F. A larger quantity of calcium cyanamide would have three marked influences upon a fertilizer mixture: (1) It would supply the calcium which would allow a reversion of the phosphates from the monocalcium phosphate to the dicalcium phosphate and possibly to the tricalcium phosphate. Until recently this reversion was credited with bringing about a marked reduction of the availability of the phosphates, as availability has been measured. (2) It would furnish sufficient heat greatly to accelerate this normally slow reaction. The reaction takes place very rapidly at temperatures above 140° F. (3) If large quantities of calcium cyanamide were used, it might result in the formation of some dicyanodiamide.

When calcium cyanamide comes in contact with ammonium sulfate, ammonia is driven off. In ordinary mixtures containing superphosphate together with calcium cyanamide and ammonium sulfate, there will be no loss of ammonia if sufficient superphosphate is added to make 6 to 10 per cent available phosphoric acid. Such mixtures are acid and remain acid if the quantity of calcium cyanamide does not exceed 150 pounds per ton. Ammonia cannot be lost from an acid mixture. In assembling materials for mixing, the calcium cyanamide and the ammonium sulfate should be kept separate, or if put in the same pile they should be separated with a little superphosphate until the mixing begins. The granular cyanamide is not satisfactory for use as a conditioner. Some calcium cyanamide can be used to furnish nitrogen and as a conditioner in ammoniated mixtures.

FIG. 29. Granular Aero Cyanamid. Flows like birdshot. (Courtesy, *American Hortigraphs and Agronomic Review.*)

Sodium nitrate can be mixed with calcium cyanamide in any proportion, but when mixed with damp superphosphate it will lose some of its nitrogen in the form of nitric oxides. This loss can be prevented by adding to the ordinary mixture 60 pounds of calcium cyanamide per ton. Basic slag phosphates, potassium sulfate, and similar salts can be mixed with calcium cyanamide without undergoing change.

Chemical analyses of most mixed fertilizers containing calcium cyanamide at the time of application show that none of the original calcium cyanamide remains in the mixture. The nitrogen has been converted from calcium cyanamide to urea and the resulting calcium oxide has been neutralized.

Conversion of Calcium Cyanamide within the Soil

When calcium cyanamide is added to a moist soil three distinct changes take place. (1) The material goes into solution in the soil water. (2) It

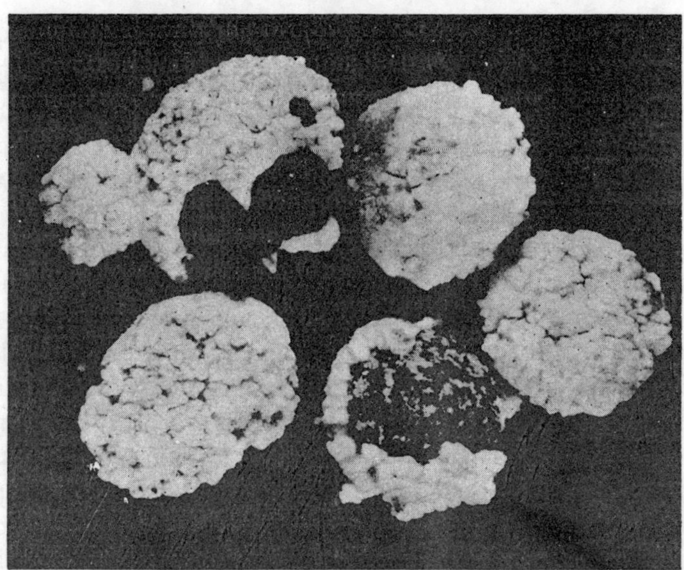

FIG. 30. Photomicrograph (×15) of Cyanamid granules collected 3 months after broadcast application to apple trees. No rain fell on the granules for 2 weeks after application. Note formation of white crystalline crust of carbonate on outside of original black surface of granules. (Courtesy, A. E. Murneek and the Missouri Agricultural Experiment Station.)

decomposes and gives rise to free cyanamide. (3) The free cyanamide reacts with the soil colloids and, in a way that is not thoroughly understood, it is converted, under neutral or acid conditions, into urea, but under alkaline soil conditions some dicyanodiamide ($NH:C(NH_2).NH.CN$) may be formed. That calcium cyanamide may decompose to form dicyanodiamide has been shown by Allison (1929) and others. Urea and dicyanodiamide are the only primary decomposition products that calcium cyanamide can form. The reactions that take place in most humid soils result in the production of urea and are probably as follows:

$$2CaCN_2 + 2H_2O = Ca(OH)_2 + Ca(HCN_2)_2$$
$$2Ca(HCN_2)_2 + 2H_2O = (CaOH)_2CN_2 + 3H_2CN_2$$
$$H_2CN_2 + H_2O = CO(NH_2)_2$$

In the soil the conversion of calcium cyanamide into urea usually takes place in from two to seven days, the action being accomplished more rapidly in fine-grained, moist soils than in coarse, dry soils. The quantity of the calcium cyanamide present in the soil also influences the time required for conversion.

The hydrogen-ion concentration of the soil solution determines whether calcium cyanamide will be transformed to urea or dicyanodiamide. Urea is always produced if the hydrogen-ion concentration of the soil solution has a pH of 7.0 or less; that is, if a neutral or acid condition prevails. If the soil solution is distinctly alkaline, dicyanodiamide as well as urea may be formed. The rate of formation of dicyanodiamide does not, however, become appreciable until the hydrogen-ion concentration reaches a pH of 8.0. Since the hydrogen-ion concentration of nearly all soils of humid regions has a pH of less than 7.0 it is apparent that the soil conditions in these regions are always at an optimum acidity for the production of urea. As a matter of fact, dicyanodiamide assumes a position of importance only under abnormal conditions, such as when the calcium cyanamide is not properly mixed with the soil, or where large quantities are used in soils of low buffering capacity. When excessively large applications are made, and the material is not thoroughly mixed with the soil, the lime content of calcium cyanamide may bring about a localized alkaline area and the production of some dicyanodiamide.

Calcium cyanamide is toxic to plants. The cyanamide injury to vegetable crops results in an etiolation of leaf margins that is somewhat similar to that produced by a deficiency of potash. The extent of the injury depends upon the quantity of the toxic compound absorbed and the nature and age of the plant. Dicyanodiamide is more stable and less easily decomposed in the soil than is free cyanamide.

Guanyl-urea ($NH_2.NH{:}CNH.CONH_2$) or dicyanodiamidine, frequently mentioned in connection with dicyanodiamide, is sometimes produced in the soil from the latter. It is nontoxic to plants even in comparatively large applications, although it appears to retard nitrification slightly. It has no influence on ammonification.

Influence of Calcium Cyanamide on the Macroorganisms and Microörganisms of the Soil

When calcium cyanamide is used in applications of 500 pounds or more, the presence of caustic lime, as well as of free cyanamide, in the soil just after the application may result in the destruction of many microorganisms such as bacteria, actinomyces, fungi, molds, as well as such macroörganisms as nematodes, and other worms. In fact, a partial sterilization of the soil always follows an application of calcium cyanamide. Watson (1917), of the Florida Experiment Station, has secured results which indicate that while nematodes may be partially controlled by the application of calcium cyanamide they cannot be eradicated. European experiments have shown that calcium cyanamide may reduce the ravages of finger-and-toe disease (clubroot), and in this country Haenseler and Moyer (1937) found that calcium cyanamide was very effective in controlling clubroot in cabbage and turnips. It has been suggested that the excellent results tobacco growers sometimes secured with calcium cyanamide are due to the ability of calcium cyanamide to kill various molds, fungi, and bacteria which are believed to produce the condition of soils that is often locally known as "run down," "sick," or "worn out." Brooks (1942) found that 1800 to 2000 pounds of calcium cyanamide per acre of muck soil was effective in destroying the fungus that produces the pink rot of celery and the white mold of snap beans.

Calcium Cyanamide as a Herbicide and Defoliant

A dusting grade of Aero Cyanamid is sold in the United States as a herbicide and defoliant. This form is used principally to kill broad-leafed weeds such as kale, mustard, charlock, pigweed, and chickweed in spring-sown small grains and in asparagus.

When calcium cyanamide is used in large quantities alone or with urea in plant beds or compost it is very effective in killing weed seed, as shown by De France (1943) and others.

Within the last few years the dusting of cotton with calcium cyanamide, just prior to picking, for the purpose of causing the cotton plant to shed its leaves, has been used extensively in certain parts of the South. The

FIG. 31. Applying Aero Cyanamid defoliant to cotton by plane. (Courtesy, American Cyanamid Co.)

presence of foliage at picking time interferes with the picking of seed cotton and may lower the grade of the cotton lint. This is especially true where the new cotton pickers have come into use. In fact, it is anticipated that the practice will become a common one as the new mechanical cotton pickers are adapted to new regions.

The calcium cyanamide should be applied with a mechanical duster at the rate of 15 to 30 pounds per acre. Smaller amounts on weedy cotton may not be sufficient while larger amounts may cause the bolls to "freeze" and not open. The cotton leaves usually fall in from 5 to 10 days. Practically all the cotton bolls should be open and ready to pick within two to three weeks after application.

Use of Calcium Cyanamide

Calcium cyanamide appears to be an entirely satisfactory fertilizer for many crops under many conditions, but is not adapted to as general use as sodium nitrate and similar inorganic nitrogen salts. On the sandy soils of the Coastal Plain and at the South Carolina Sand Hill Experiment Station, calcium cyanamide has not proved entirely satisfactory as the sole source of nitrogen for cotton. When the plants are about half grown they develop a leaf injury which is characteristic and quite distinct from the symptoms of potash, magnesium or manganese deficiencies. In Europe calcium cyanam-

FIG. 32. Tobacco bed of P. D. Altman, Gallivants Ferry, South Carolina. (*Left*) Bed treated with Granular Aero Cyanamid will require no hand weeding. (*Right*) Bed untreated. (Courtesy, American Cyanamid Co.)

ide is extensively used particularly on grass lands, and is most commonly applied directly to the soil as a single fertilizer material.

In its powdered state calcium cyanamide is dusty and is disagreeable to handle unless it has been oiled. Workmen who constantly handle this material occasionally show skin inflammation. This injury does not result from the handling of the granulated product. The greatest objection to the use of calcium cyanamide is probably the fact that it cannot be relied upon to give satisfactory results under all circumstances, although it is highly effective where the conditions are favorable to its use. Many of the objections to calcium cyanamide are offset by the fact that the cost per pound of nitrogen in calcium cyanamide is usually less than in most other nitrogenous fertilizers. If the farmer is willing to inform himself as to its characteristics, and if he uses the fertilizer as recommended by the manufacturers, satisfactory results should be secured in most cases. Moyer and Blair (1930) found that the recovery of nitrogen from calcium cyanamide by oats, rape, barley and corn compared favorably with that recovered from sodium nitrate and ammonium sulfate.

Residual Influence of Calcium Cyanamide

The materials that are left in the soil after the nitrogen of calcium cyanamide has been used by the plant are calcium carbonate, particles of carbon, and small amounts of the oxides of aluminum, iron, and silicon. The calcium carbonate that is left in the soil would amount to slightly more than the weight of the calcium cyanamide applied, and while this with applications commonly employed would very probably be less than the average annual loss of lime from the soil, it nevertheless represents a substantial quantity of lime for plant use.

How Calcium Cyanamide Should Be Applied

The most satisfactory method of applying calcium cyanamide is to plow it under when the soil is broken. Unless the soil is poorly drained satisfactory results can always be expected from such a practice.

When calcium cyanamide is applied uniformly over the surface of the soil, or well-mixed with the upper three or four inches of the soil at least 10 days before planting, the desired chemical reations usually take place. Some investigators have suggested that the waiting period should be about three days for every 100 pounds of calcium cyanamide applied per acre. The best results are secured when calcium cyanamide is applied to loam soils, for such soils offer the most favorable conditions for the essential

Fig. 33. Top-dressing applications of calcium cyanamide should be applied broadcast rather than in streaks. (Courtesy, American Cyanamid Co.)

conversion of the cyanamide to urea, ammonia, and finally to nitrates. When calcium cyanamide is applied to sandy soils, or soils devoid of organic matter, the length of the waiting period before planting should be double that required on fertile soils.

If it is impracticable to delay planting until 10 days after application, satisfactory results can be secured by thoroughly mixing the calcium cyanamide with the soil and planting so as to avoid direct contact of fertilizer and seed. Thorough mixing of the calcium cyanamide with the soil must always be advocated because calcium cyanamide produces free cyanamide, and also because it contains lime in the oxide form. These compounds are very harmful to microörganisms, seeds, and plant roots if the concentration is excessive in any one place. For this reason calcium cyanamide unmixed with other fertilizer materials should never be drilled in the row in direct contact with the seed.

Side-dressings of calcium cyanamide are not generally recommended except for such crops as sugar cane and corn, but when used should always be made with a distributor. The powdered product should not be applied by hand to many broad-leaved crops. The material is very fine and sticks to leaves more than does the granular product or other coarser grained fertilizers. In Germany and France calcium cyanamide is used in large amounts as a nitrogen top-dressing for the cereal crops and for the additional purpose of killing any broad-leaved weeds that may be present. Applied early in the spring little or no injury is caused to the grain crop.

Smith, Heinze and Murneek (1943) found that the best results are secured when calcium cyanamide is drilled in the usual way and followed with a cultivator that mixes the soil and the fertilizer, especially if the weather is dry and if the granular form of the fertilizer is used. Broadcasting has been practiced successfully on established crops such as fruit trees, asparagus, permanent pastures and winter wheat when the applications were made in late fall or early in the spring while there is an abundant supply of moisture in the soil. The smaller the rate of application the less the necessity for exercising care in the use of calcium cyanamide. Allison (1929) recommends that for the average soil not more than 200 pounds of calcium cyanamide be used per acre. He thinks that usually the best results will be obtained by using not more than 100 pounds. Where conditions are especially favorable to its use considerably larger amounts, as much as 500 pounds per acre, may be applied with good effect.

Manufacture of Urea

In this country urea is manufactured extensively by combining pure ammonia with pure carbon dioxide gas under high pressure. Ammonium carbamate is formed first, but it loses a molecule of water and is thus converted into urea. The reactions involved are as follows:

$$2NH_3 + CO_2 = NH_2.COO.NH_4$$
$$NH_2.COO.NH_4 = CO(NH_2)_2 + H_2O$$

The solution can be concentrated in vacuum evaporators and finally dried by spraying into a chamber where it solidifies in pellet form. Urea can also be manufactured from calcium cyanamide by the acid process.

Characteristics of Urea

Urea and urea compounds have been, or are, sold for fertilizer purposes in the United States under trade names such as Floranid, Ureor, Calsi-Ureor, Agramon, and Uramon. The crystal urea now found on the market is a white, crystalline product used almost exclusively for chemical purposes. Chemically speaking, urea is an organic compound. The American Association of Official Agricultural Chemists class urea, as well as calcium cyanamide, as a synthetic nonproteid, organic compound. The fertilizer laws of some states class urea as an "organic," or as a "water-soluble organic" nitrogen source, while the laws of other states class urea as soluble nitrogen along with all inorganic nitrogen sources. Commercial urea is identical with the urea found in animal urine. It contains 46 per cent of nitrogen, an amount greater than that found in any other commercial solid nitrogenous fertilizer material and nearly three times the amount found in sodium nitrate.

Characteristics of Uramon

In 1938 a commercial product consisting of crystal urea which had been conditioned by coating the crystals with inert mineral and organic materials of low fertilizing value was put on the market under the brand name of *Uramon*. It is manufactured by Du Pont at Belle, West Virginia. Since that time this product has become a leader in its field. It is a semi-granular product analyzing 42 per cent of nitrogen. At the present time the cost of nitrogen in Uramon is about equivalent to that in other low cost solid nitrogenous materials, and for this reason alone it has proved popular.

Use of Urea in Mixed Goods

Urea may be used in mixed goods just as satisfactorily as many other sources of nitrogen, provided certain precautions for mixing with other materials are observed. At least 50 pounds of urea may be used in most complete mixtures and in many formulas it is possible to use 100 pounds or over. Urea does not react chemically with potash salts, thoroughly neutralized superphosphate, dolomite, ammonium sulfate, and many other fertilizer materials but it should never be put in mixtures containing appreciable quantities of ammonium nitrate. Nevertheless, urea should only be mixed with ordinary superphosphate or treble superphosphate in which the free sulfuric or phosphoric acids have been neutralized, otherwise a somewhat damp mixture will result. Also, high temperatures of mixtures should be avoided. At temperatures above 130° F. urea may hydrolyze to ammonia and carbon dioxide, but this may or may not be objectionable.

At the present time a great deal of urea is going into complete fertilizers along with ammonia liquors. In most cases this urea does not enter into the chemical reactions involved in ammoniation.

Urea as a Source of Nitrogen for Plants

Urea is completely soluble in the soil solution, but as such is apparently not utilized in large amounts by our common field and horticultural crops. There is some evidence, however, which would indicate that urea may be utilized as such, at least to a limited extent, by some plants. Under most field conditions urea is probably converted to ammonia or to ammonia and nitrates, and in these forms absorbed by plants. The reactions that may take place in the soil before the nitrogen of urea is absorbed and assimilated by plants probably are as follows:

$$CO(NH_2)_2 + 2H_2O = (NH_4)_2CO_3 \quad (1)$$
$$(NH_4)_2CO_3 + 3O_2 = 2HNO_2 + 3H_2O + CO_2$$
$$2HNO_2 + 2O_2 = 2HNO_3$$

The conversion of urea into ammonium compounds within a short time after application to the soil, as illustrated by reaction (1), may be very important, for Allison (1939) found that urea would leach from the soil within four days following application. The loss of nitrogen by leaching is small if the urea has been converted to ammonium carbonate.

Influence of Urea on Soil Acidity

Owing to the rapidity with which certain types of bacteria convert urea into carbamide urea great care must be taken not to apply urea in too large amounts at any one time in close proximity to germinating seeds or roots of plants. Otherwise, injury may result. As urea contains no residual base, soils may eventually become slightly acid following its continued use. The initial influence of urea on the soil reaction is alkaline like that of the proteid organics. Because urea will produce a slight acidity in the soil, the use of small amounts of dolomitic limestone along with the urea has frequently been found to increase the efficiency of the material.

Use of Urea in Tobacco Beds

Urea in the form of Uramon is now being used with calcium cyanamide in tobacco plant beds. The calcium cyanamide will kill weed seeds and the urea will kill plant diseases. To be effective these materials should be applied 60 to 90 days before planting the tobacco seed.

Advantages to Be Derived from the Use of Urea

Theoretically, urea is almost an ideal nitrogenous fertilizer to be applied at planting. It is a cheap source of nitrogen and is completely available to crops. It is converted into ammonium and nitrate forms suitable

FIG. 34. Tobacco at Tifton, Georgia. Plat on left received an adequate supply of a complete fertilizer. The plat on right received no fertilizer.

for absorption at a time when the crop demand for those sources of nitrogen is the greatest. The nitrogen of urea is resistant to leaching and, as has been stated above, it increases soil acidity only slightly.

Calurea

Calurea is the trade name of a product made up of a mixture of calcium nitrate and urea and probably has the formula $Ca(NO_3)_2.4(NH_2)_2CO$. Because urea has the property to replace water of crystallization, as illustrated below, it often produces a poor physical condition when improperly used with superphosphate.

$$Ca(NO_3)_2.4H_2O + 4(NH_2)_2CO \rightleftarrows Ca(NO_3)_2.4(NH_2)_2CO + 4H_2O$$

Calurea contains 34 per cent of nitrogen and soluble calcium equivalent to 14 per cent of calcium oxide. About one-fifth of its nitrogen is in the nitrate form and four-fifths in the amide form. It is a dry and odorless white crystalline product of excellent physical condition.

Use of Ammonium Hydroxide as a Fertilizer

The use of ammonium hydroxide as a fertilizer material has proved successful recently. MacIntire (1943) has shown that when ammonium hydroxide was applied at the rate of 50 pounds of nitrogen per acre at the time of seeding Sudan grass it proved as efficacious as equivalences of the sulfate and nitrate. The practice of adding ammonium hydroxide to irrigation water has been designated as "*nitrogation.*"

Field application of ammonium hydroxide has been carried out successfully from a drum at the time the land is disked to corn and sorghum at the Georgia Experiment Station. Bledsoe and Olson (1944) secured results comparable with those from sodium nitrate. The equipment necessary for application is shown in Fig. 36.

Ammonium Sulfate Nitrate

Ammonium sulfate nitrate is made by mixing hot solutions or moist salts of ammonium sulfate and ammonium nitrate and then drying. The double salt which results is a soft crystalline material containing about 26 per cent of nitrogen of which about one-fourth is in the nitrate form and three-fourths in the ammoniacal form. The German product which was imported into this country following World War I was known by the trade name of *Leunasalpeter*. It was named for the Leuna nitrogen fixation plant in Germany. The French product is sold under the trade name of

FIG. 35. Plant of the Solvay Process Company, Hopewell, Virginia. This plant has been manufacturing synthetic (Arcadian) sodium nitrate since 1929 and is the largest plant in America. (Courtesy, Barrett Division, Allied Chemical and Dye Corporation.)

Lesulfonitrate d'ammoniaque. When ammonium sulfate nitrate is allowed to lie in storage it sets into a cake and must be broken up for further use. Fertilizer tests conducted at many of the agricultural experiment stations in this country have shown that the product compares favorably with many of the leading nitrogenous materials. The use of ammonium sulfate nitrate, like all ammoniacals, increases the acidity of the soil.

Ammonium Chloride

Ammonium chloride, sometimes called muriate of ammonia, is synthesized by combining ammonia with hydrochloric acid. It contains about 26 per cent of nitrogen, and has a very good physical condition. Ammonium chloride may be used in mixtures of superphosphate and potash without caking. It is produced in Germany by a modification of the Solvay process for making sodium carbonate. Briefly, sodium chloride is treated with ammonia and carbon dioxide to form sodium bicarbonate and ammonium chloride, and the resulting ammonium chloride is then separated from the solution. This procedure is similar to the process used in the manufacture of ammonium sulfate from gypsum. Ammonium chloride has a greater acidulating effect on the soil, per pound of nitrogen applied, than has ammonium sulfate.

Ammonium chloride has been shown to be inferior to ammonium sulfate for those crops that are injured by chlorine. Results reported by Russell (1932) at the Rothamsted Experiment Station show that ammonium sulfate is superior to ammonium chloride as a source of nitrogen for potatoes, but less effective for barley. Skinner and Buie (1926) found ammonium chloride to be equal in value to ammonium sulfate in the production of cotton.

Ammonium Bicarbonate

Ammonium bicarbonate has been used as a fertilizer to a limited extent in Europe. It carries about 17 per cent of nitrogen. It is somewhat unstable and may lose to the atmosphere some of its ammonia before it can be adsorbed by the soil. This is especially true when the ammonium bicarbonate is applied as a top-dressing to calcareous or alkaline soils. Because of its unstable character large applications may bring about soil toxicity, for large amounts of free ammonia in the soil have been shown to be highly toxic to some field crops. It is very doubtful that the product will ever be marketed to any great extent for fertilizer purposes.

Synthetic Sodium Nitrate and Ammonium Sulfate

The manufacture and use of synthetic sodium nitrate and ammonium sulfate are discussed in Chapters 2 and 3 respectively. Arcadian synthetic sodium nitrate has been in commercial production since 1928. In this period it has exerted a profound effect on the American nitrogen market.

FIG. 36. Machine for applying fertilizer in solution designed and built at the University of Tennessee. (Courtesy, *The Fertilizer Review.*)

Processes Dealing with Production of Fertilizers Containing Two or More Nutrients

Processes dealing with the production of compounds containing two or more of the plant nutrients, nitrogen, phosphoric acid, and potash, have attracted considerable attention, because the resulting products are very high-analysis fertilizers. It now seems that fertilizers of this type will be used to an increasing extent in the future. Generally speaking, the cost of producing these fertilizers is high but to the farmer the cost per unit of plant food in these materials is low. Gericke (1922) has shown that the pairing of nutrient elements in ions of opposite charge, such as we find in many of these new fertilizer salts, is an important factor in securing the maximum growth of crops.

Manufacture of Monoammonium and Diammonium Orthophosphate

Both the monoammonium and the diammonium orthophosphates have been made for fertilizer use. In this country the common procedure employed in their manufacture is the extraction of phosphoric acid (H_3PO_4) from phosphate rock, the addition of ammonia to the phosphoric acid in correct amounts necessary to secure the desired phosphate and the drying of the slurry in a rotary kiln. The pure monoammonium phosphate carries 61.71 per cent of phosphoric acid and 12.1 per cent of nitrogen. It is slightly acid in reaction. The pure diammonium phosphate carries 53.35 per cent of phosphoric acid, and 21.07 per cent of nitrogen. It is slightly

FIG. 37. Showing acid effect on soil of diammonium phosphate. (*Background*) Diammonium phosphate and lime. (*Foreground*) Diammonium phosphate without lime. (Courtesy, South Carolina Agricultural Experiment Station.)

FIG. 38. Airplane view of an ammonium-phosphate plant, Warners, N. J., in New York Harbor. (Courtesy, American Cyanamid Company.)

alkaline in reaction. The diammonium salt is slightly less stable and more soluble than the monoammonium salt and is more difficult to make, but it carries about twice the quantity of ammonia per unit of phosphoric acid. Diammonium phosphate has the disadvantage that it loses ammonia in humid atmospheres. There is at present no domestic production of diammonium phosphate for fertilizer use.

Another commercial method of preparing crude monoammonium phosphate has been developed in Europe. It consists of replacing with ammonium sulfate about one-third of the sulfuric acid used in acidulating phosphate rock. The reaction involved is probably as follows:

$$Ca_3(PO_4)_2 + (NH_4)_2SO_4 + 2H_2SO_4 = 3CaSO_4 + 2NH_4H_2PO_4$$

Mehring (1930) found that diammonium phosphate did not have as good a distributing property as monoammonium phosphate at 80 per cent relative humidity. At 90 per cent relative humidity the diammonium phosphate gave off ammonia and became a pasty mass. Because of its unstable nature and its poor physical condition diammonium phosphate has not been used as extensively as a fertilizer material as has monoammonium phosphate. Allison (1939) found ammonium phosphate to be quite resistant to leaching. Triammonium phosphate is very unstable and is not an article of commerce.

Ammo-Phos

Ammo-Phos is the trade name formerly used to designate a series of products made up of commercial monoammonium phosphate, either alone, or in combination with ammonium sulfate or other materials, to give various ratios of nitrogen to phosphoric acid. Two grades were sold extensively prior to World War II, Ammo-Phos A and Ammo-Phos B. Ammo-Phos A contains 10.7 to 11.0 per cent of nitrogen and 48 per cent of available phosphoric acid. Ammo-Phos B contained 16 per cent of nitrogen and 20 per cent of available phosphoric acid. Sufficient quantities of lime were sometimes added to overcome all acid residual effects. From 85 to 90 per cent of the available phosphoric acid of Ammo-Phos was soluble in water, and as a result the phosphoric acid of Ammo-Phos had considerably greater mobility in the soil than that of relatively less soluble superphosphate.

Ammo-Phos was a light gray, granular material which very much resembled well dried superphosphate. It was packed for shipment in burlap bags of good quality which held 200 pounds net. The material did not readily absorb water from the air and could be kept indefinitely without reversion.

Use of Ammonium Phosphates

The ammonium phosphates should prove popular in those sections where it is a common practice to apply fertilizer mixtures in which the proportion of phosphoric acid to nitrogen is relatively high. Although the crop response on acid sandy soils deficient in calcium has not always been satisfactory, especially with dicalcium phosphate, on the whole, excellent yields have been obtained from their use on most field crops. Ammonium phosphate has proved to be an excellent top-dressing for small grain.

Willis and Rankin (1930) and Willis and Piland (1931) concluded that the poor results sometimes secured from the application of ammonium phosphates were due to the liberation of excessive quantities of ammonia around the roots of seedlings. In an attempt to overcome injurious effects of this type the manufacturers of ammonium phosphate in this country mixed gypsum or lime with their products which were to be sold for use on certain soils and under certain crops. Calcium sulfate is more soluble than calcium carbonate and has proved to be best for this purpose especially if it is not desirable to change the pH of the soil solution.

As commercial products of ammonium phosphate contain no mineral basic elements they have a tendency to produce soil acidity after long use. This has been shown to be true by Sprague and Evaul (1930). Theoretically, diammonium phosphate will produce more acidity in the soil than an equivalent amount of ammonium sulfate. Because the use of both monoammonium and diammonium phosphate results in an increase of soil acidity the best results on acid soils may not be secured unless the acidity of the soil solution is adjusted to meet the demands of the crop grown.

Ammoniated Superphosphate

Ammoniated superphosphate is manufactured as explained in Chapter 10, and is now available to farmers. It usually contains 3 or 4 per cent of nitrogen and 16 per cent of phosphoric acid. Some producers add potash so that the product is often put on the market as a complete fertilizer.

Leunaphos

Leunaphos formerly was manufactured at the Leuna nitrogen fixation plant in Germany and has been used most extensively in Germany. It is composed of diammonium phosphate and ammonium sulfate, and contains 18.5 to 20 per cent of nitrogen and 49 per cent of available phosphoric acid. It has never appeared on the American market.

Phosphazote

Phosphazote is manufactured in Switzerland and France. It is a mixture of urea and superphosphate and usually contains about 7 per cent of nitrogen and 15 per cent of phosphoric acid, though the proportions of nitrogen and phosphoric acid may vary within wide limits.

Phosphazote is produced by first extracting crude calcium cyanamide with water in a closed vessel and in the presence of carbon dioxide, and then treating the filtered solution of free cyanamide with sulfuric acid The sulfuric acid hydrolyzes the free cyanamide to urea and guanylurea. The resulting urea sulfuric acid is then concentrated under pressure and ground phosphate rock is added. The resulting product very much resembles superphosphate. It has an excellent mechanical condition.

Potazote

Potazote is a French product consisting of a brownish-yellow mixture of potassium and ammonium chlorides, containing about 11 to 13 per cent of ammoniacal nitrogen and 20 to 24 per cent of potash. It is made by treating sylvinite with ammonia and carbon dioxide in a manner similar to the Solvay process of manufacturing sodium bicarbonate.

Potassium Nitrate

Potassium nitrate is synthesized in Germany; is obtained as a by-product in the refining of caliche in Chile; is found as rather pure crystals in bat guano caves; and to a limited extent is leached from manure composts in India. In Germany potassium nitrate is made by treating muriate of potash with nitric acid and adding ammonia which forms ammonium chloride as a co-product.

The continued use of potassium nitrate may bring about a deflocculation of clay particles such as that noted in some soils following heavy and continued applications of sodium nitrate. It tends to produce in acid soils first a neutral and then an alkaline condition. This action is similar to that brought about by the use of calcium or sodium nitrate, but is not as great as that resulting from the use of calcium cyanamide. Potassium nitrate contains about 14 per cent of nitrogen and 44 per cent of potash. It is less hygroscopic than most of the other fertilizer nitrates.

Potassium Ammonium Nitrate

Potassium ammonium nitrate is a double salt containing both nitrogen and potash, and is formed by a double reaction between potassium chloride and ammonium nitrate. It contains about 16 per cent of nitrogen and 28 per cent of potash. About half of the nitrogen is in the nitrate form and half in the ammoniacal form. It is not as hygroscopic, nor does it possess the

FIG. 39. Aerial view of the Belle, West Virginia, Plant of E. I. du Pont de Nemours and Company, Inc. It is here ammonia solutions were first manufactured. (Courtesy, E. I. du Pont de Nemours and Company, Inc.)

explosive properties of ammonium nitrate. It will probably play an increasingly important part in the manufacture of concentrated fertilizers.

Potassium Orthophosphate

Potassium orthophosphate is prepared by heating potassium hydroxide with monocalcium phosphate. It contains 30 to 50 per cent of potash and 32 to 53 per cent of phosphoric acid. As potassium orthophosphate ionizes more easily than monocalcium phosphate its plant nutrients may be slightly more available to crops.

Potassium Metaphosphate

Potassium metaphosphate (KPO_3) is prepared from pure potassium chloride and phosphoric acid. The product contains 40 per cent potash and 60 per cent phosphoric acid. Vegetative tests have shown that both the potash and the phosphoric acid are available to crops.

Sodium Potassium Nitrate

Sodium potassium nitrate is found on the market under the name of *Nitropo*. It is refined from the crude Chilean caliche and is a mixture of sodium and potassium nitrates. Some of the Chilean sodium nitrate deposits have a potassium nitrate content as high as 15 per cent and many contain as much as 5 to 7 per cent. Nitropo contains 20 to 80 per cent of potassium nitrate. It analyzes about 15 per cent of nitrogen and 15 per cent of potash.

Nitrophoska

Nitrophoska is the trade name given to a series of products which consist of a supposedly combined mixture of diammonium phosphate, ammonium nitrate or urea, and muriate or sulfate of potash. It is manufactured in Europe, has been widely advertised, and has attracted attention because it is a complete fertilizer. Nitrophoska differs from ordinary salt mixtures in that it has been grained and a partial decomposition of some of the salts has taken place before the ammonium nitrate is added. All the plant nutrients of nitrophoska are in an immediately available form. The following grades of nitrophoska have been put on the market: 14–11–26, 15–30–15, 12–24–12, 16–16–16, 16½–16½–21½, 15–11–26½, and 10–20–20, the potash of which is derived from potassium chloride; 15½–15½–19, the potash of which is derived from potassium sulfate; and 10–20–15, one-third of the potash of which is derived from potassium chloride and two-thirds from potassium sulfate, and one-half of the nitrogen is derived from urea. As most grades of nitrophoska carry ammonium nitrate, they tend to absorb water when the humidity of the air is high, and to lose water when the humidity of the air is low. This may lead to caking and to the necessity of regrinding the material before it can be used, unless it is applied soon after the bags are opened.

Ammo-Phos-Ko

Ammo-Phos-Ko is a trade name formerly given to a mixture containing Ammo-Phos and potassium sulfate. Like Nitrophoska, it was a complete fertilizer. It was made by mixing monoammonium phosphate and ammonium sulfate with potassium sulfate. The following six analyses have been available to the American farmer: 12–24–12; 9–18–18; 10–30–10; 12–16–12; 12–12–16; and 13–13–8.

Miscellaneous Chemicals Proposed for Fertilizers

Among the miscellaneous chemicals that have been suggested for fertilizer use are ammonium formate, formamide, melamine nitrate, melamine phosphate, melamine sulfate, guanidine sulfate, diguanidine phosphate, triguanidine phosphate, phytin, and nucleic acid.

Value of Double-strength Mixed Fertilizers

The early use of high-analysis complete fertilizers revealed, in many cases their deficiency in the secondary elements, calcium, magnesium, and sulfur, and in some cases in the rarer elements, iron, boron, zinc, and copper. Deficiency of calcium and sulfur may be overcome by adding calcium sulfate to the mixture. Quantities which will furnish as little as 50 pounds an acre have proved to be very effective even on acid soils of low buffer capacity. The addition of dolomitic lime will add calcium and magnesium, and has been found to improve acid mixtures and to make "double-strength" fertilizers equal, in most cases, in crop-producing value to those of "single-strength." It is a credit to the fertilizer industry that recent developments in fertilizer manufacture, such as granulation, ammoniation and the production of physiologically neutral mixtures, have to a large extent overcome the difficulties that first attended the use of the fixed-nitrogen fertilizers in the manufacture of high-analysis mixtures.

Relative Value of Air-derived Nitrogenous Fertilizers

The relative crop-producing values of some of the air-derived nitrogenous fertilizers which have been on the market for some years have been studied for various crops by many of the state experiment stations and the

Table 26

RELATIVE CROP-PRODUCING VALUES OF SOME AIR-DERIVED NITROGENOUS FERTILIZERS FOR COTTON ON NINE SOIL TYPES
(YIELD IN POUNDS OF SEED COTTON PER ACRE)

*Source of Nitrogen**	*Portsmouth Sandy Loam (av. 2 yrs.)*	*Cecil Sandy Loam (av. 2 yrs.)*	*Greenville Sandy Loam (av. 2 yrs.)*	*Davidson Clay (av. 2 yrs.)*	*Marlboro Sandy. Loam (av. 2 yrs.)*	*Marlboro Fine Sandy Loam*	*Appling Sandy Loam*	*Dunbar Fine Sandy Loam*	*Coxville Sandy Loam*	*Average*	*Rating*
Sodium nitrate	1270	1051	1695	1450	1620	1495	1313	834	642	1263	100.0
Ammonium sulfate	1290	993	2040	1418	1573	1288	1075	1052	692	1269	100.4
Ammonium nitrate	1172	957	2015	1357	1428	1403	1163	930	700	1236	97.8
Ammonium chloride	1017	905	1415	1360	1428	1472	1088	1152	786	1180	93.4
Ammonium chloride†	905	964	1245	1185	1407	1242	938	950	708	1061	84.0
Ammonium phosphate‡	1115	935	1850	1525	1486	1311	1238	820	676	1217	96.4
Urea	1025	950	1905	1430	1457	1311	1263	810	640	1197	94.8
No fertilizer	368	465	700	530	805	750	425	445	151	515	40.7

* Superphosphate and potassium sulfate were used as sources of phosphoric acid and potash except where noted.
† Muriate of potash, source of potash.
‡ Ammonium phosphate, source of phosphoric acid.

United States Department of Agriculture. The results of such studies have been found to vary with soil, crop, season, and above all with their leaching and acid-forming characteristics. Willis and Rankin (1930) showed that some calcium salt should be used in concentrated fertilizers if ingredients decomposable into free ammonia are present. In fact, they concluded that if the fertilizer is to be practically free of calcium, no source of free ammonia should be included in the formula. They think it would be preferable to use gypsum in those mixtures that are to be applied to acid-tolerant crops and limestone in those mixtures that are to be applied to lime-loving crops.

A representative example of the many tests that have been made with air-derived nitrogenous fertilizers, as reported by Skinner, Williams, and Mann (1929), is given in Table 26, which represents a summary of experiments conducted with cotton on nine soil types by the United States Department of Agriculture, working in coöperation with the North Carolina Experiment Station. Each nitrogen salt that was employed served as the sole source of nitrogen in a mixture containing 6 per cent of ammonia, 8 per cent of phosphoric acid, and 4 per cent of potash. The fertilizer was applied at the rate of 800 to 1200 pounds per acre. It must be recognized, however, that relative ratings in such experiments may be changed by varying soil management practices and especially by adding or withholding lime.

FIG. 40. Applying liquid manure in Switzerland, a ladle at a time. (Courtesy, Deere and Co.)

6

Sources and Uses of Organic Nitrogenous Fertilizers

Nitrogenous fertilizers—sometimes called ammoniates—may be classed into two groups, the organic and the inorganic or mineral. The organic nitrogenous fertilizers, with the exception of urea and calcium cyanamide, are derived from animal and plant sources. The mineral forms are derived from inorganic sources. The two synthetic compounds, urea and calcium cyanamide, are classed by chemists as organic compounds, but many agronomists think that on account of their low cost per unit of nitrogen and high solubility they should be grouped with the inorganic nitrogenous materials. In some states they are classed as nonproteid organics to distinguish them from the proteid organics of animal and plant origin.

Forms of Nitrogenous Fertilizers

The mineral nitrogenous fertilizers are readily soluble in water, whereas the organic nitrogenous fertilizers, with the exception of urea and calcium cyanamide, are not. The term *form* is usually employed to indicate roughly the rate of solubility. Accordingly, the terms *organic form* and *inorganic* or *mineral form*, as applied to fertilizers, carry a meaning as to their solubility.

Organic Nitrogenous Fertilizers Used to a Less Extent than Formerly

In the early days the fertilizer industry was looked upon as a scavenger industry which gathered together waste materials having some fertilizing value, and compounded them into mixtures suitable for the farmer's use. At first, such materials as guano, dried blood and slaughterhouse tankage, were the main sources of organic nitrogen. As the industry expanded, seed meals, garbage tankage, synthetic organic ammoniates, and processed tankages were added. The inorganic sources of nitrogen were relatively unimportant until about 1900, and the consumption of inorganic sources of nitrogen did not forge ahead of the organic sources until the second decade of the present century.

In recent years an increasing proportion of the organic nitrogenous materials formerly used for fertilizers have been going into feed channels at a better price than they can command on the fertilizer market. Most of the nitrogen entering mixed fertilizers at the present time is derived from inorganic sources, although some organic ammoniates are nearly always included. According to Brand of the National Fertilizer Association, in 1913, 42 per cent of the nitrogen in mixed commercial fertilizers came from organic sources, whereas by 1926 the percentage had dropped to 22, and at present it is less than 15.

Although modern fertilizers are of little or no value as sources of organic matter it is not at all improbable that the fertilizer industry will, at some future date, attempt to supply the farmer with organic manures.

Chemical Methods for Determining Quality of Organic Nitrogenous Fertilizers

Fertilizer chemists subject the organic nitrogenous fertilizers to tests of quality by the use of neutral or alkaline permanganate solutions. These tests have been accepted as satisfactory tests of the activity of water-insoluble nitrogen and have been adopted as official methods by the American Association of Official Agricultural Chemists. The methods are empirical but serve to differentiate between the good and the poor organics. Briefly, the methods consist of digesting the water-insoluble residue of a sample of fertilizer in neutral or alkaline permanganate solution under standard conditions, and then determining the percentage of nitrogen that has gone into solution. The characteristics of the organic fertilizer determine somewhat the choice of the use of the neutral or alkaline solution. In the New England states both of these methods are used, whereas in most of the southern states only the neutral method is employed. The alkaline permanganate method, in most cases, with the exception of the organic vegetable ammoniates, gives water-insoluble nitrogen activities which correspond closely with nitrogen availabilities secured by vegetation pot experiments.

In applying the two laboratory methods in the study of the quality of insoluble nitrogen in fertilizers, an activity of 50 per cent by the alkaline permanganate method and 80 per cent by the neutral method is considered of passing quality. With activities below these figures the quality of the water-insoluble nitrogen is classed as inferior. Because the employment of

Table 27

CARBON-NITROGEN RATIOS AND NITROGEN AVAILABILITY RATINGS OF VARIOUS WASHED ORGANIC MATERIALS

Material	C-N Ratio	Permanganate Activities: Alkaline Method	Permanganate Activities: Neutral Method	Vegetative Test (Sudan Grass): Added Nitrogen Recovered in Tops and Roots (per cent)	Nitrification Test: Added Nitrogen Converted to Nitrate (20 days) (per cent)
Seed Meals:					
Soybean meal	4.70	70.1	92.2	59.0	58
Cottonseed meal	5.40	66.9	82.7	53.6	50
Special soybean meal	7.05	49.6	73.9	43.9	50
Castor pomace	9.36	63.0	87.9	51.7	55
Cocoa meal	14.7	28.8	37.1	..	−1
Ground cocoa cake	19.0	33.2	51.5	−25.2	−14
Plant materials:					
Alfalfa hay	20.8	28.4	68.9	0.8	4
Tobacco stems	28.9	19.8	65.4	−25.3	−14
Peanut hull meal	53.5	25.9	42.6	− 2.6	−1
Wheat straw	197.0	..	..	..	−16
Process tankages:					
Hynite	4.87	73.2	80.9	24.1	24
Processed tankage	5.17	69.8	81.6	14.7	21
Agrinite	5.24	68.6	78.8	13.2	18
Smirow	6.30	64.2	71.8	12.9	13
Animal products:					
Hoof meal	3.31	77.1	93.2	50.1	57
Bone meal	3.46	81.9	39.9	8.8	6
Dried blood	3.51	81.1	87.9	56.3*	51
Dry fish scrap	4.42	72.8	86.0	50.1	51
Animal tankage	5.25	67.1	70.7	29.7	26
Acid fish scrap	5.28	68.0	87.9	22.0	33
Manures:					
Peruvian guano	1.28	41.3	96.7	67.8	67
Bovung	24.4	27.9	47.1	−15.6	−10
Horse manure	32.7	28.8	51.6	..	−19
Chicken manure	36.4	38.8	59.9	..	−19

Table 27—(Continued)

Material	C-N Ratio	Permanganate Activities		Vegetative Test (Sudan Grass)	Nitrification Test
		Alkaline Method	Neutral Method	Added Nitrogen Recovered in Tops and Roots	Added Nitrogen Converted to Nitrate (20 days)
				(per cent)	(per cent)
Sewage products:					
Milorganite	5.98	63.4	75.2	50.5	44
Nitroganic tankage	6.20	65.1	83.7	37.2	41
Sewage sludge	13.7	51.1	65.4	8.4	8
Plastics:					
Beetle molded scrap	1.83	48.0	21.2	− 2.4	1
Beetle scrap dust	2.35	67.5	79.7	37.2	20
Ford molded scrap (unwashed)	..	..	..	..	− 5
Ford molding powder (unwashed)	..	..	..	..	− 3
Miscellaneous:					
Cocoa tankage	13.3	30.4	53.8	− 6.9	− 7
Garbage tankage	13.4	30.6	51.1	−10.0	0
Manito humus	13.7	44.4	40.9	− 5.3	3
Standard material:					
Urea	..	..	..	87.9	..

* Added nitrogen recovered from Sudan grass fertilized with unwashed dried blood was 5.40 per cent.

either method above may result in the condemnation of good materials, it is thought by some chemists that both methods should be used before a given material is classed as inferior.

Rubins and Bear (1942) at the New Jersey Experiment Station have determined the carbon-nitrogen ratio and the nitrogen availability of about 30 proteid organics. Their results are given in Table 27.

Source of Dried Blood

When animals are slaughtered at the abattoirs the blood is led away and collected in vats. It is then dried carefully and ground. About 20 to 25 pounds of dried blood is secured from every 100 pounds of the liquid blood. The liquid blood is dried by raising its temperature by means of steam or hot air. During this process the liquid is agitated vigorously. With the increase in temperature the blood first becomes coagulated and then the water is driven off. Two kinds of dried blood are manufactured—red and black. The difference in color is due to the method of heating employed. The black blood has been dried in "dried heat" driers, while the red blood has been dried in "jacketed steam" driers. Subjecting blood to the high temperatures of the dried heat driers results in a partial loss of nitrogen and in the product becoming somewhat carbonized.

The greater part of the dried blood that is produced in this country is used as a stock feed. Only those portions that are less desirable for use as feed are marketed for fertilizer. In this country the demand for dried blood, both as a feed and as a fertilizer, is so great that we are now importing a large part of our requirement from the Argentine, Uruguay, and Canada.

Characteristics of Dried Blood

Dried blood has a characteristic odor and is dark red to black in color. Like all colloids, it will absorb water from the atmosphere, but dried blood does not absorb excessive amounts of water, and under trade conditions is a commercially stable compound.

The nitrogen content of dried blood varies from about 8 to 14 per cent, depending largely upon its method of manufacture. Besides carrying a high percentage of nitrogen, the better grades of dried blood carry .3 to 1.5 per cent of phosphoric acid and from .5 to .8 per cent of potash. The phosphoric acid and the potash content of dried blood are rarely considered when calculating the amounts of fertilizer ingredients to be incorporated in home mixtures.

Foreign materials such as tankage, bone, and even pulverized roasted leather have been found occasionally in dried blood. The detection of these products by casual examination is very difficult. A chemical analysis should detect the presence of bone, and nitrogen activity tests may show the presence of low-grade materials. If bone is present the phosphoric acid content of the blood will be high. In general most laboratory and field tests which have been made by the American experiment stations have shown that the availability of dried blood is higher than that of any other organic nitrogenous fertilizer with the possible exception of Peruvian guano. Although the nitrogen-availability of blood is high, it is not as high as that of such inorganic nitrogenous compounds as sodium nitrate or ammonium sulfate. Dried blood is generally given a rating of 80; that is, a unit of nitrogen in dried blood is accepted as being 80 per cent as efficient in its crop-producing power under average conditions as a unit of nitrogen in sodium nitrate.

Source of Meat Meal and Tankage

At the larger slaughterhouses of the country many animals are rejected for human consumption because of a disordered condition. These animals, as well as large quantities of waste from the carcasses of other animals, are used as a source of meat meal and tankage. These materials are placed in steel tanks and subjected to steam pressure of 50 to 60 pounds for a few hours. They are then ground and put on the market under the names "meat meal," "tankage," or "bone tankage." The cooking or rendering process puts the bones in a condition in which they can be ground easily, and at the same time it causes the fats to separate from the meat so that they can be removed readily. The process of cooking under live steam pressure and subsequent drying under high heat completely sterilizes the tankage and precludes any possibility of infection from any disease organisms that might have been in the slaughtered animals. Not over 10 per cent of the material that goes into meat meal and tankage is derived from condemned animals. The greater part is derived from the offal of edible carcasses that have been "inspected and passed" by government inspectors. Tankage that is derived largely from meat is designated as meat meal, and the tankage that contains a high percentage of bone is designated as bone tankage. More than 95 per cent of all animal tankage that is produced at the packing plants of this country is consumed as animal feed. Such meat meals as are sold for fertilizer are now being sold largely under the name of tankage.

Characteristics of Meat Meal

Meat meal carries 9 to 11 per cent of nitrogen and .1 to 3.5 per cent of phosphoric acid. It ranks among the high-grade organic materials, its availability being but little below that of dried blood. The amount of phosphoric acid present depends upon the amount of bone that has been allowed to remain with the meat.

Characteristics of Bone Tankage

Bone tankage varies greatly in its chemical composition. It may contain from 3 to 10 per cent of nitrogen and from 7 to 20 per cent of phosphoric acid. The lower grades in particular may contain high percentages of phosphoric acid. Material that contains over 6 per cent of nitrogen is not called bone tankage by the trade, but generally is designated as "blood and bone" or "low-grade tankage." The only bone tankage now being used in the manufacture of mixed fertilizers is ***renderer's tankage*** and an occasional lot of ***packer's tankage*** which is not fit for use as stock feed. The nitrogen in bone tankage has about the same availability as the nitrogen in meat meal, and the phosphoric acid has about the same availability as that found in steamed ground bone.

Process Tankage

At the present time there are found on the market under trade names many process tankages, the exact sources of which are not made public. These products are made from crude, inert, nitrogenous materials of

Table 28

RELATIVE AVAILABILITY OF SOME COMMON ORGANIC FERTILIZERS

Source of Nitrogen	*Average Relative Nitrogen Availability Dried Blood at 80*		*Comparative Nitrogen Activity by the Chemical Methods*	
	Basis: Dry Matter Yield	*Basis: Nitrogen Recovered*	*Alkaline Permanganate Method*	*Neutral Permanganate Method*
Dried blood—red	80.00	80.00	79.34	97.88
Dried blood—dark	79.88	73.26	75.34	73.88
Mowrah meal	27.28	29.05	40.92	64.03
Hoof and horn meal	74.95	69.47	82.85	89.66
Process tankage	61.56	55.57	62.64	86.38
Process tankage "Smirow"	69.66	63.15	78.20	90.11
Garbage tankage	47.44	46.73	39.92	69.54
Omega tankage	57.39	51.78	75.39	86.72
Nitrolene	72.16	68.20	84.35	78.78
Milorganite	66.32	63.15	67.59	83.99
Rapeseed meal	64.06	64.41	60.11	91.59
Whale guano	70.09	68.20	77.11	97.03
Nitrogenous tankage	65.61	59.36	75.89	83.49
Agrinite	63.04	59.36	78.87	89.57
Hynite	63.87	59.36	78.73	87.52
Avail tankage	45.56	46.73	58.78	68.81
Peruvian guano	83.57	87.15	62.59	98.11
Darling's tankage	62.87	59.36	71.48	90.96
Supernite	71.46	65.68	78.66	96.85

animal source by processing under steam pressure, with or without the use of acids, for the purpose of increasing the activity of the nitrogen. Their production was first started in 1910. The availability of processed tankage as compared with dried blood is usually rather high. Vegetation and laboratory tests of some of these tankages, as compared with similar tests with other common organic materials, as determined by Haskins and his co-workers (1929) are given in the accompanying table. The table shows the relative yield of dry matter secured when various organic ammoniates were applied as fertilizers in equivalent amounts of nitrogen, and the nitrogen recovered in the above-ground parts of Japanese millet, and also the activity of the nitrogen in each product as determined by the two permanganate tests.

Source and Characteristics of Ground Leather

Scrap leather from the manufacture of shoes and leather goods is usually processed; that is, ground and roasted or steamed, and sometimes treated with a little sulfuric acid, before it is put on the fertilizer market. It is used chiefly in the manufacture of mixed fertilizers and is in reality a processed tankage of good quality.

The steamed and roasted ground leather may vary in composition from 6 to 11 per cent of nitrogen and from .5 to 1.0 per cent of phosphoric acid. The availability of the steamed leather is slightly superior to that of the roasted leather.

Sources of Hoof and Horn Meal

Hoof and horn meals are also by-products of the slaughterhouses and result from the processing, drying, and grinding of hoofs and horns. The processed meal is a very high-grade nitrogen source. Its nitrogen content may be as high as 15 per cent or more. Haskins (1929), after three years of testing this material at the Massachusetts Experiment Station, reported it to be of good quality.

Source of Hair and Wool Waste

Hair and wool waste, as well as leather scrap, may often be secured by farmers who live near towns where these products occur as waste from manufacturing plants. Usually they can be secured at a low cost per ton and at such costs may be used advantageously by local farmers for building up the soil content of organic matter. The composition of hair and wool waste varies greatly. This is especially true of wool waste for it often contains considerable cotton waste. The nitrogen content of wool waste may range from 2 to 8 per cent and the potash content from .25 to 5.50 per cent. Hair also shows a wide range in composition owing to the presence of foreign materials and to a variable water content. While these materials may be used satisfactorily in building up the fertility of the soil, they should not be used with the idea of securing immediate returns from the investment. When hair and wool waste are used in the manufacture of complete fertilizers, they must be processed or treated with sulfuric acid. The processed material furnishes organic nitrogen of high availability.

Sources of Guano

Guano is made up of the excrement of seafowls, and sometimes of turtles and seals; together with the remains of the bodies of dead birds, such as bones and feathers; and other foreign material, such as sand and gravel. It was one of the first commercial fertilizer materials with which the American farmer became acquainted and to this day in many sections of the South the word *guano* is used synonymously with the word *fertilizer*. The word guano is derived from a Spanish word meaning dung. The sea-

FIG. 41. Guanay on nest. This bird is called by Murphy, "the most valuable bird in the world." (Courtesy, Nitrate Agencies Company.)

fowls that take the lead as guano producers are cormorants, pelicans and gannets. It has been estimated by Murphy (1925) that a square mile of occupied territory on some of the Peruvian Islands supports as many as 5,600,000 birds. A thousand or more tons of fish per day would be required to sustain such a colony.

Deposits of guano are found in many parts of the world but mainly on islands, although a few deposits are located on the coast of mainlands. The deposits of greatest economic importance are located on the islands of the Pacific Ocean off the coast of Peru. It is said that the natives of Peru have long been acquainted with the fertilizing value of guano. Deposits of guano may be newly formed or they may be very old. They vary from a few inches to over 200 feet in thickness. In places they are covered with sand to a considerable depth and on some islands, and in humid parts of others, fermentation has taken place and their physical and chemical compositions have been altered considerably. Because of the dry climates of other localities very little fermentation of the guano has taken place.

Guano has been used in Europe since the twelfth century, but the first exportation from Peru to European ports did not take place until 1810, and a second shipment was not made until 1840. The first importation of Peruvian guano into the United States is said to have been made by a Mr. Skinner of Baltimore, who received two barrels in the year 1824. Importations slowly increased from 1832 until they reached a maximum of 50,000 tons in 1856. As the deposits became depleted, the annual importations declined until during the World War I period no importations were made into this country. At one time guano was Peru's principal export staple, but today the product has dwindled to negligible significance as an item of her international trade.

In recent years the Peruvian government has made efforts to protect the colonies of seafowls and to exploit the fresh deposits of guano as a government monopoly in such a manner that a continuous supply of this valuable material will always be available. The Peruvian islands are now worked in rotation so that it is two to four years before guano is again shipped from the same island. After removing the deposits men and boys are sent on the islands with baskets to pick up all the loose stones and trash. As a result of this procedure the guano that is now being exported is in a reasonably good condition and is a much better product than the early shipments. The bulk of the guano is produced by government monopoly and is sold to the Peruvian farmers at cost, that is, less the profit made on export shipments. Only 15,000 tons of guano were imported into the United States in 1938 and 5,000 tons in 1939 when importations practically ceased.

Characteristics of Guano

Guano may vary in color from a gray to a dark brown. The physical characteristics vary with the age and the nature of the deposit, as well as with the amount of foreign material present. Guano also varies greatly in chemical composition, depending largely upon the origin of the material, the climatic conditions of the locality in which it originated, and the handling it has received. It has a marked odor which results from a combination of ammonia and other volatile organic compounds. As stated before, the food of seafowl is made up very largely of fish, so that guano may contain a high percentage of nitrogen. Some agronomists regard guano as a fish fertilizer.

The principal nitrogenous compound in the excrement of birds is uric acid. Little uric acid is found in guano for since it was voided the uric

FIG. 42. A well-crowded bird island. These are guanays (*cormorants*), the best and most prolific of the guano-producing birds. (Courtesy, Nitrate Agencies Company.)

acid has been changed largely into ammonium carbonate, although some ammonium sulfate and ammonium phosphate may be produced. Much of the nitrogen in guano is immediately available to plants upon incorporation with the soil and the rest soon becomes so. In guanos that have lost most of their nitrogen by fermentation the phosphoric acid content is very high, so much so that they are often referred to as *phosphatic guanos* and may be used in the manufacture of superphosphate. These guanos contain 20 to 25 per cent of phosphoric acid and 4 to 6 per cent of nitrogen. The phosphoric acid is found in the form of calcium, potassium, or ammonium phosphates. On the other hand, guanos that contain a high percentage of nitrogen may be referred to as *nitrogenous guanos*. The Peruvian guanos now found on the market belong entirely to this class. They contain 11 to 16 per cent of nitrogen and 8 to 12 per cent of phosphoric acid, and 2 to 3 per cent of potash.

While the color and odor of guano are characteristic, these cannot be taken as infallible guides to the value of the product. A chemical analysis is the only sure guide to its value. The product is easily adulterated and in former years this was often done. Sawdust, lime, salt, gypsum, and other substances were commonly added. Adulteration with earthy materials may generally be detected by burning a little of the guano and noting the characteristics of the ash. The ash of unadulterated guano is white and shows no signs of red oxide of iron.

Guano has but little effect upon the physical condition of the soil, although many farmers hold the opinion that it has. Fresh guano has a higher fertilizing value than the droppings of domestic fowls. This is due largely to the differences in the nature of the food consumed by seafowl and poultry.

Dissolved and Rectified Guano

Dissolved guano is a name given to guano to which has been added sulfuric acid. When this treatment is made the ammonium carbonate of the guano is quickly converted into ammonium sulfate, and the insoluble phosphates are converted into soluble ones. The phosphatic guanos often contain a high percentage of phosphorus in the reverted dicalcium and tricalcium forms. If guano has any peculiar properties that are superior to the properties of other similar nitrogenous fertilizers, as some users have suggested, these properties must certainly be destroyed by treatment with sulfuric acid. Little dissolved guano is found on the market at the present time.

Rectified guano is a name given to guano to which any or all of the three plant nutrients—nitrogen, phosphoric acid, or potash—has been added for the purpose of making the material a more balanced fertilizer. This guano is also sometimes referred to as *fortified guano*. It is superior to low-grade guano, but often the material has been sold at prices above that which the cost of the added ingredients warrants. Peruvian guano now on the market is of such high quality that it is never rectified.

Sources of Bat Guano

Small deposits of bat guano are found in many parts of the world. It accumulates in caves, church steeples and similar places where bats congregate. Deposits amounting to a few thousand tons have been found in Missouri and the southwestern part of the United States. The Missouri deposits have been estimated to contain approximately 16,000 tons. During the War of 1812 bat guano of the Wyandotte Cave in Indiana was an important source of the saltpeter that was used in the manufacture of gunpowder.

Characteristics of Bat Guano

Bat guano usually occurs as a dry and odorous material. As the bats live upon insects, and as they do not digest insect wings, the excrement presents a glistening appearance owing to the presence of the insect wings. Bat guano is dark brown in color when dry and black when wet. The nitrogen content of bat guano may vary from 2 to 12 per cent and the phosphoric acid from 1 to 14 per cent. The analyses of 10 samples of bat guano analyzed by Vinson (1915) showed a variation of nitrogen ranging from 2 to 14 per cent and of phosphoric acid from 1 to 6 per cent. Considerable quantities of earthy material may be present which may alter the composition. Bat guano originating under dry conditions also varies widely in composition from the guano originating under humid conditions. Bat guano is generally considered inferior to Peruvian guano in its crop-producing power.

Sources of Seal Guano

Seal guano consists of a mixture of the excrement of seals together with the remains of the carcasses of dead animals. Beds of seal guano, 230 feet thick in many places, have been found in the Lobos Isles off the west coast of South America. Other deposits have been found on the Isle of Gortuga. As these islands are still the habitat of seals the material is accumulating continuously. These deposits have been exploited to some extent and the guano has been shipped largely to Europe. Seal guano is considered somewhat inferior to the better grades of the Peruvian bird guano.

Pulverized Manures

Pulverized sheep, goat, cattle and poultry manures are now found on the market in rather large quantities. The sheep, goat, and cattle manures analyze about 1 to 2 per cent of nitrogen, 1 to 2 per cent of phosphoric acid, and 2 to 3 per cent of potash; while poultry manures analyze 5 to 6 per cent of nitrogen, 2 to 3 per cent of phosphoric acid, and 1 to 2 per cent of potash.

Sources of Fish Fertilizers

The use of fish as a fertilizer dates back some centuries. It has been said that when the white man came to America he found the American Indian following the practice of fertilizing his corn by placing a fish in each hill.

FIG. 43. Onions, spinach, and leaf lettuce grown on plats of Michigan Muck soil which received the following treatments per acre: (1) Horse manure, 20 tons; (2) 3–9–18, 800 pounds; (3) Bovung, 20 tons; (4) sheep manure, 20 tons; (5) 3–9 18, 800 pounds; and (6) chicken manure, 11 tons. Sheep manure proves best of the manure treatments, but none equalled the commercial fertilizer. (Courtesy, Paul M. Harmer, Michigan Agricultural Experiment Station and American Potash Institute.)

The manufacture of fish fertilizers was first undertaken in Norway and that nation leads all others in its manufacture at the present time.

Formerly, fish fertilizers were often referred to incorrectly as fish guano. They consist either of the ground, whole and largely nonedible fish, or the remains of edible fish that have been used for other commercial purposes. Often the oil, as is the case with the menhaden, is removed before the refuse is ground and marketed. The principal fish that are used for fertilizers are the cod and herring (menhaden), but such miscellaneous aquarian fauna as whale, shrimp, crab, and starfish, are sometimes used. Menhaden fisheries are found along the South Atlantic and Gulf coasts, but the largest operations are conducted along the North Carolina coast. Occasionally farmers who live near the coast can procure fish scrap from the fishermen, and in some cases whole fish at a cost that will justify buying the material for incorporation in the soil. Such material contains 30 to 80 per cent of water, 2 to 8 per cent of nitrogen and 2 to 6 per cent of phosphoric acid. Although the material decays rapidly, yet because of its physical condition the immediate efficiency is relatively low.

Whale guano consists of the dried ground flesh of whales and contains 8 to 10 per cent of nitrogen and 10 to 12 per cent of phosphoric acid. A two-year test of whale guano by Haskins (1929), of the Massachusetts

FIG. 44. Cows in modern dairy barn heavily bedded down to conserve all liquid manure. (Courtesy, Deere and Co.)

Experiment Station, has shown that in its crop-producing value it compares favorably with fish or animal tankage.

Along the Atlantic coast king crab is used as a fertilizer to a considerable extent. In its fresh or green state it may be applied directly to the land or composted and then applied. In its fresh state it contains 2 to 3 per cent of nitrogen. It decays readily and is considered of high availability. Sometimes king crab is dried and ground to be used in the manufacture of mixed fertilizers. This product contains about 9 to 12 per cent of nitrogen.

Formerly nearly all fish scrap went into mixed fertilizers, but much of it now finds a market as a hog and poultry feed, and only that portion that cannot be placed in the higher-priced market, or that portion that must be acidulated at the factories, goes into fertilizer mixtures.

Characteristics of Fish Fertilizers

All fish fertilizers have a strong fish odor. So pronounced is this odor that some farmers have erroneously thought it to be of value in driving away injurious crop insects. Fish fertilizers have a gray to a dark brown color, but the color, as well as the chemical composition, will vary widely. If the product is composed of fish waste the percentage of bone is likely to be high, and therefore the percentage of phosphoric acid will be high. The fish meal that is made from the Newfoundland cod may contain as high as 18 per cent of phosphoric acid, but it analyzes only 5 per cent of nitrogen. The Norwegian and American fish meals contain about 7 to 10 per cent of nitrogen and 4 to 8 per cent of phosphoric acid. The availability of the nitrogen of the fish fertilizers is usually slightly lower than that of dried blood. The phosphoric acid appears to be somewhat more available than the phosphoric acid in steamed ground bone.

Acidulated Fish

The fish fertilizers, because of their physical characteristics, do not lend themselves readily to treatment with sulfuric acid. Nevertheless, some of the products have been treated fairly satisfactorily. This treatment reduces the odor, converts the insoluble bone phosphate into a more readily available form, and converts some of the nitrogen to ammonium sulfate. The product is known as *acidulated fish*.

There is also a product known as *acid fish* on the market. This is fresh fish scrap that has had a small amount of sulfuric acid sprayed over it to discourage the breeding of maggots and to arrest decomposition. The material is used in the manufacture of base goods.

Sources of Garbage Tankage

Garbage tankage is the rendered, dried, and ground product derived from waste household food materials. It is a very low-grade fertilizer, but is important because the supply increases with the growth of our cities. In those cases where an attempt is made to manufacture this product, the crude material is dried, usually after an effort has been made to remove the fats. Garbage tankage is variable in its composition and finds its chief use in the manufacture of base goods. It generally analyzes 2 to 3 per cent of nitrogen, 1 to 3 per cent of phosphoric acid, and .5 to 1.5 per cent of potash.

Prince and Winsor (1926) have compared the availability of garbage tankage analyzing 3.13 per cent of nitrogen with other commonly used nitrogen carriers where equivalent amounts of nitrogen were applied to barley, rye and sorghum. Their results are given in the following table.

Table 29

COMPARATIVE AVAILABILITY OF NITROGENOUS FERTILIZERS FOR BARLEY, RYE, AND SORGHUM AS DETERMINED BY NITROGEN RECOVERY TESTS USING SODIUM NITRATE AS A BASIS OF COMPARISON

Fertilizer	*Barley* (*per cent*)	*Rye* (*per cent*)	*Sorghum* (*per cent*)	*Average* (*per cent*)
Sodium nitrate	100.00	100.00	100.00	100.00
Urea (C. P.)	105.10	96.90	94.10	98.70
Urea (commercial synthetic)	98.20	93.60	100.70	97.50
Ammonium sulfate	94.20	86.20	84.30	88.20
Standard tankage	52.00	48.70	59.40	53.40
Fish	52.10	42.10	53.50	49.20
Garbage tankage	13.00	9.50	26.40	16.30
Garbage tankage finely ground	10.90	10.20	15.30	13.10

Haskins (1929) has reported that about 84 per cent of the total nitrogen of garbage tankage is in a water-insoluble form and that tests, both laboratory and vegetative, showed the nitrogen to be of an inferior quality.

Activated Sewage Products

In some cities of Europe and America a material known as *activated sewage sludge* is now being made available to farmers. It was first used in mixed fertilizers in 1927. Sewage sludge is a product resulting from the treatment of sewage. It is made from sewage which has been freed from grit and coarse solids and aerated after being inoculated with microörganisms. The resulting flocculated organic matter is filtered, dried in rotary kilns, and then ground and screened. The activated sewage sludge marketed by the sewage commission of the city of Milwaukee is sold in this country under the trade name of *milorganite*. It contains 6.0 per cent of nitrogen and 2.5 per cent of available phosphoric acid. Rehling and Truog (1939) found that milorganite contained many of the rarer elements. The activated sewage sludge produced in Pasadena, California, is sold under the trade name of *nitroganic*. Recently, the city of Chicago has started large scale production of sewage sludge. Sewage sludge is a high-grade fertilizer and eventually may be produced in large quantities. It has a faint nauseating odor to which some workmen object. Its relative availability as shown by Rubins and Bear (1942) is given in Table 27.

In addition to activated sewage sludge a small amount of Imhoff sewage sludge is coming on the American market. It is of lower grade than the activated sludge.

Cottonseed Meal

Cottonseed meal is a by-product of the cottonseed oil industry. After the oil has been removed from cotton seed the cake is ground and sold as a feed or fertilizer. It contains about 6.0 to 9.0 per cent of nitrogen, 2.0 to 3.0 per cent of phosphoric acid, and 1.5 to 2.0 per cent of potash. Cottonseed meal is the principal proteid organic fertilizer material incorporated in home mixtures.

In the South, it is a common practice for farmers to trade their cotton seed for cottonseed meal, the cottonseed meal to be fed or returned to the soil as fertilizer. Such a practice has been advocated by Collings (1926) and others, and when followed, the loss of fertility from Cotton Belt soils may be greatly curtailed.

Cottonseed meal is an excellent feed for cattle, but has only comparatively recently become widely recognized as such. Because of the demand for cottonseed meal for feed, a smaller percentage of the annual production of cottonseed meal is finding its way into the fertilizer market than formerly. It is to be hoped that its use as a fertilizer will be discontinued.

Castor Pomace

Castor pomace is a by-product of the manufacture of castor oil. It contains about 5 to 6 per cent of nitrogen, 2 per cent of phosphoric acid, and 1 per cent of potash. Unlike the meals secured from the seeds of flax

FIG. 45. Crotalaria is a rank-growing legume which, when turned under, adds much high nitrogen organic matter to the soil. (Courtesy, *The Fertilizer Review.*)

and cotton, castor pomace is poisonous to animals and is not fed, being used only for fertilizer purposes. Fortunately, castor pomace ranks high as an organic source of nitrogen.

Linseed Meal

Linseed meal is the resulting product secured after extracting oil from flax seed. It contains about 5.5 per cent of nitrogen, 1.7 per cent of phosphoric acid, and 1.3 per cent of potash. Because it is an excellent feed its market value usually makes its use for fertilizer purposes prohibitive.

Rapeseed Meal

To a limited extent rapeseed meal has been imported into this country from England. It analyzes about 5 to 6 per cent of nitrogen. Haskins (1929) has reported that at the Massachusetts Experiment Station tests of its availability showed the product to compare favorably with cottonseed meal and castor pomace.

Other Vegetable Fertilizers

Besides cottonseed meal, linseed meal, and castor pomace, there are other meals and pomaces which can sometimes be obtained in small quan-

FIG. 46. A properly built farm manure pit. The litter carrier saves time and prevents losses. (Courtesy, Deere and Co.)

tities to be used by farmers as fertilizing materials and by manufacturers of mixed fertilizers in the preparation of base goods. Among these are apple pomace, cocoa-cake meal, copra-cake meal, cranberry pomace, peanut meal, pumpkin pomace, rubber seed meal, tomato pomace, velvet bean meal, soybean meal, and wine lees. As would be expected, the water content of the pomaces is generally high and their composition is very variable. The use by farmers of the pomaces as fertilizer is only of local importance.

Reinforced Non-leguminous Green Manures

Some agronomists think that one of the greatest needs of many of the soils of the United States is an increase in their content of organic matter. For this reason agronomists often recommend that the organic matter of the soil be increased by turning under barnyard manure, crop residues, and green manures. For this purpose the most satisfactory green-manuring crops are those that will produce a rapid growth and make heavy yields. Because of their comparatively narrow carbon-nitrogen ratio the legumes have proved to be better suited for green-manuring than have the non-legumes, but on many soils it is difficult to obtain a good growth of legumes. On the other hand, good yields of rye or other nonleguminous plants can often be produced more satisfactorily than can crops of legumes, but as nonleguminous green-manuring crops have a wide carbon-nitrogen ratio they decay very slowly when they are turned under, and unfavorable results are often noted from their use.

For the above reasons many farmers are now finding that the application of cheap carriers of nitrogen to nonleguminous green manuring crops before plowing down will result in organic matter of a narrow carbon-nitrogen ratio, and a product comparable to the organic matter of legumes or high-grade barnyard manure. Furthermore, many farm wastes, such as cornstalks, leaves, straw, or even grass sods, may be treated in a similar way and turned into the soil with good results.

Conservation and Use of Barnyard Manure

Strictly speaking, barnyard manure is not a commercial fertilizer, but its use on the farms of America is so universal that it seems appropriate to include in this chapter a short discussion of its conservation and use.

In many sections of the country the farm manure produced on the average farm aggregates a substantial total, and because of this, steps should be taken to conserve the plant food represented. Salter and Schollenberger (1939) estimate an annual production of manure in the United States equivalent to 1.02 tons for each acre in crops and pasture. The

FIG. 47. A too common sight. Farmers should abandon this poor practice of manure storage. (Courtesy, Wisconsin Agricultural Experiment Station.)

proper care of manure costs no more labor and little more money than the many poor methods now in common use.

Many authorities concur in the belief that fully one-half of the value of the farm manure produced in the United States never reaches the field. Because proper protection is not given farm manure by the average farmer much of its soluble nutrients are lost through seepage, and much of its organic matter is lost through oxidation and other forms of decay.

Considering as a whole the farm practices of the United States, it seems that the most needed reforms for conserving farm manure in this country are (1) better care in the barn during the production of manure, (2) the adoption of a regular and systematic program of daily handling and spreading on the land, (3) the use of safeguards during temporary storage, and (4) the selection of the most satisfactory times in the rotation to apply manurial treatments.

Manure should be stored under a shed. A great deal of the plant nutrients in manure is soluble in water, and manure exposed in the open may readily lose much of its value even when stored in the open for only a few months or even weeks. If one must feed his stock in an open lot, care should be taken to use as small a lot as feasible. Most feed lots are much larger than is required for the accommodation of the farm animals. By decreasing the size of the feed lot an increase is made in the depth of the manure which collects in the lot, and as a result a decrease in the loss by leaching would be effected.

Sufficient bedding or litter should be used in stalls to absorb the approximate 600 pounds of liquid found in each ton of manure. Wheat and oat straw, shredded cornstalks, and pine needles are capable of taking up twice their weight of water.

Farm manure varies greatly in composition depending upon such factors as (1) kind of animal, (2) age, condition, and individuality of animals, (3) food consumed, (4) litter used, and (5) the handling and storage the manure receives. Farm manure is characterized by its variability, its low analysis, its unbalanced nutrient condition, its moist condition, its rapid fermentative processes, and its residual influence on the soil. A representative ton contains about .5 per cent of nitrogen, .25 per cent of phosphoric acid, and .5 per cent of potash. It is definitely a low-analysis fertilizer.

Because ammonifying organisms in farm manure readily bring about ammonification during the period of storage much of the nitrogen may be lost to the atmosphere as ammonia. This loss can be greatly reduced if 25

FIG. 48. Uniform spreading of manure gives the best returns. (Courtesy, Deere and Co.)

to 30 pounds of ordinary superphosphate per ton of manure is mixed with the manure. The superphosphate unites with the ammonia to form a more stable compound of nitrogen.

A great deal of loss commonly sustained during the storage of manure can be avoided if the manure is hauled daily, or at least at frequent intervals, directly to the fields and spread on the land. If it should rain following the field application of manure the leaching of the manure in the field will take the soluble portion into the soil. If a dry period follows the field application the manure quickly dries to a safe moisture content so that the decay of the product soon ceases.

On many farms a wet condition of the fields, or a peak labor load, may often make the regular hauling and spreading of manure impracticable and inadvisable so that provision must be made for temporary storage. When this is necessary the manure should be kept under cover, but provision should be made to keep the manure wet in order to reduce as far as possible the loss resulting from rapid oxidation. If the manure cannot be kept in the barn it should be kept under a shed, on a watertight floor which is surrounded by four half-high continuous walls. Experimental evidence shows, however, that the spreading of manure well in advance of breaking the land results in the highest yields of corn, so that holding the manure at the barn in storage awaiting plowing time is not the best practice. Top-dressing winter grain with manure helps to hold snow and prevents rapid changes in soil temperature, which in turn reduces winter injury.

Manure should be spread evenly over the soil if the best results are to be secured. This can be done only if the manure is fine and is distributed with a manure spreader. A manure spreader not only spreads the manure evenly, but it processes or conditions the manure as it spreads it.

Light applications of manure give greater returns than heavy applications, that is, greater returns will be secured by applying 20 tons of farm manure to four acres than to two acres. The application of 8 to 10 tons of manure per acre is considered an average application, 15 to 25 tons a heavy application, and less than 8 tons a light application.

The residual effects of barnyard manure may be seen for many years. At the Rothamsted Station, England, the residual effects have been noted for 20 years following a period of 20 years during which manure was applied.

Forty-year results secured by Prince, *et al.* (1941) at the New Jersey Agricultural Experiment Station showed that the organic matter content of the soil was maintained only when manure was applied at the rate of 16 tons per acre annually, and where use was made of lime and green-manure.

Synthetic Manure

In recent years there has been a marked decrease in the supply of farm manure in this country, hence attempts have been made to develop methods of producing manure artificially from straw, hay of poor quality, shredded cornstalks, and other farm wastes. Such attempts have met with varying degrees of success. When materials which are composed mainly of cellulose are mixed with certain fertilizer salts, and then kept in a moist condition, a decay sets in which produces within a few weeks' time a product that resembles farmyard manure in appearance, in its action when applied to the soil and, what is more important, in its influence on crop growth. One ton of air-dry straw will make about 3 tons of moist manure. Mixtures of fertilizer salts for this use have been patented and are now sold on the market under trade names. They are compounded with the idea of making of the mixture of straw and salts an ideal medium in which those microörganisms that break down cellulose may grow luxuriantly. In addition to nitrogen salts and limestone, the mixtures generally carry small amounts of phosphorus, magnesium, sulfur and sodium. The potassium necessary for the microbiological activity is supposed to be furnished by the straw.

Albrecht (1927), of the Missouri Experiment Station, has used one of the patented mixtures successfully, but he has suggested a homemade mixture consisting of 45 pounds of ammonium sulfate, 15 pounds of superphosphate, and 40 pounds of finely ground limestone, and has recommended that this mixture be used at the rate of 150 pounds per ton of straw. When such a mixture was added to wheat straw, and the straw made into flat piles for the purpose of catching rainwater, a product of manure-like appearance and behavior resulted in one trial in about 16 weeks, and in another trial in about 8 weeks. As a top-dressing for wheat the artificial manure proved to be superior to applications of straw or ordinary barnyard manure.

The process of manufacture of synthetic manure was first developed by the Rothamsted Experiment Station, England, where Russell (1926) reported that synthetic manure has been prepared most economically when 3.5 parts of ammonium sulfate were added for each 100 parts of straw, and the mixture allowed to disintegrate in the open. He found that at first the nitrogen is rapidly assimilated by microörganisms and the dry matter lost to the extent of about 20 per cent of the dry weight of the straw. The decomposition then slows down, probably owing to the exhaustion of nitrogen and the soluble carbohydrates which facilitate the decomposition, until about 50 per cent of the total dry matter has been lost. He found that very little nitrogen is lost during this second phase of the decomposition, although all of it had been converted into more complex compounds.

Turk (1936) found that 1 ton of dry straw produced about 2½ tons of wet manure. He produced synthetic manure at a cost of $.70 for the chemicals necessary to make a ton of wet manure. Martin and Wang (1944) have determined the loss during composting of artificial manures made from various sources. Their results are given in Table 30.

Collison and Conn (1929) have concluded that under New York conditions the making of artificial manure hardly seems practical. On the other hand, Brown and Smith (1930), of the Iowa Experiment Station, have secured good results from the use of artificial manure, and advocate its manufacture and use.

Peat, Humus, and Muck

Peat, humus and muck are names used by the fertilizer trade to describe the remains of vegetable materials that have collected under anaerobic conditions in swamps or in depressions along the banks of streams and

Table 30

DECOMPOSITION OF ORGANIC CONSTITUENTS DURING COMPOSTING IN GRAMS PER 100 G. OF ORIGINAL MATERIAL

Constituents Analyzed	Oat Straw + Inorganic Salts			Oat Straw + Clover Hay			Cow Manure			Peat + Timothy Hay			Cornstalks + Inorganic Salts			Leaves + Timothy Hay		
	At Start	At End (110 Days)*	Loss or Gain, %	At Start	At End (110 Days)*	Loss or Gain, %	At Start	At End (110 Days)*	Loss or Gain, %	At Start	At End (110 Days)*	Loss or Gain, %	At Start	At End (70 Days)*	Loss or Gain, %	At Start	At End (250 Days)*	Loss or Gain, %
Total material	100	44	56	100	34	66	100	47	53	100	62	38	100	49	51	100	38	62
Ether-soluble portion	2.4	0.34	− 86	2.4	0.23	−91	1.6	0.23	−86	1.3	0.17	−86	1.4	0.13	− 91	2.8	0.76	−73
Water-soluble organic matter	9.0	2.34	− 74	10.5	4.60	−56	9.0	3.41	−62	5.0	2.66	−46	15.5	2.66	− 83	11.7	2.25	−81
Hemicelluloses	16.5	5.55	− 66	15.5	4.17	−73	15.7	2.82	−82	10.5	4.17	−60	20.0	3.69	− 81	13.0	3.48	−73
Cellulose	33.0	4.19	− 87	30.8	5.10	−83	18.9	1.60	−92	7.8	0.88	−90	27.4	3.94	− 86	15.7	1.82	−88
Lignin	13.4	11.55	− 14	13.6	8.00	−41	20.8	11.00	−47	38.0	22.40	−38	9.5	10.12	+ 7	25.1	11.30	−55
Crude protein	1.7	5.11	+197	3.8	4.05	+ 7	9.1	8.25	−10	14.5	12.15	−16	2.0	4.90	+147	5.5	3.76	−32
Ash	5.5	9.65	+ 75	5.77	5.77	+ 4	17.3	17.38	0	10.7	12.91	+21	6.3	14.05	−124	10.2	10.50	+ 3
Total N	1.36‡	1.03	− 24	1.24	0.85	−26	2.16	1.44	−33	2.38	2.00	−13	1.28*	1.09	− 15	1.13	0.70	−38
Total N†	..	2.34	..	..	2.50	..	..	3.07	..	..	3.22	..	..	2.23	..	..	1.84	..

* On basis of 100 g. of original material.
† Percentage in end-product.
‡ Includes 0.7% N added with inorganic salts.

lakes. These deposits may be almost purely of vegetable origin, or they may contain varying percentages of sand, silt and clay.

Peat usually contains a very high percentage of organic matter. A good grade of peat usually has not over 5 per cent of mineral matter including the ash constituents. Peat is of value chiefly as a soil conditioner for horticultural uses and in poultry houses because of its absorbent properties. The highest quality and most absorbent peat is that derived from Sphagnum moss, of which approximately 75,000 to 85,000 tons annually have been used in the United States in recent years. Peat formed from marsh grasses and other aquatic plants is less absorbent.

Humus is the name applied to peat deposits that have decayed to such an extent that the structure of the fiber is no longer evident. Humus consists chiefly of organic matter, with usually not over 5 to 15 per cent total mineral material. Humus may be of Sphagnum origin, or derived from any type of marshy land plants.

Muck is the name applied to thoroughly decomposed organic deposits containing appreciable amounts of mineral matter, especially sand, silt and clay, and is of value chiefly when hauled from nearby deposits and applied to poor soils.

On a water-free basis these materials usually contain from 1.5 to 3.0 per cent of nitrogen; 0.25 to 0.5 per cent of phosphoric acid; and 0.5 to 1.0 per cent of potash. While the plant food value does not justify the usual processing, transportation and selling expense necessary to market these materials commercially, the absorbent and conditioning properties for the special uses to which peat and humus are adapted have maintained a substantial volume of business in these materials.

The state of Minnesota contains the largest reserves of peat found in the United States, followed in order by Wisconsin, Florida, and Michigan.

Table 31

RELATIVE VALUE OF MUCK AND SODIUM NITRATE FOR CORN

Nitrogen Applied per Acre	*Yield of Corn Fodder—Average of 4 Pots, (grams)*	*Increase Sodium Nitrate 100 (per cent)*
None	50	..
24 lbs. in muck	55	5
48 lbs. in muck	61	10
24 lbs. in sodium nitrate	127	68
48 lbs. in sodium nitrate	165	100

Proul (1921), of the Indiana Experiment Station, has determined the relative crop-producing value of muck and sodium nitrate when applied to corn on an equivalent nitrogen basis. His results are given in Table 31.

Duff Briquette Fertilizers

In recent years there has come on the market a variety of pellets, tablets and briquette fertilizers made of mineral salts alone or in combination with organic matter. In a sense these things are not new. The Chinese for centuries have been fertilizing plants with briquettes made from night soil and marl.

When materials of this kind are placed in the soil they offer a continued source of plant nutrients to plants although there is some danger of chemical injury to the roots. The danger of plasmolysing plant cells is somewhat reduced when they contain duff, manure or clay.

Excellent results have been reported from the use of duff briquettes on shrubs and trees by Wilde and Whittenkamp (1942). They prepared their briquettes from the following materials:

5 pecks of shredded hardwood-hemlock duff, approximate reaction, pH 5.3; total N content, 1.5%; and base exchange capacity, 70 M.E. per 100 grams.

1 peck of pulverized Superior clay, having 35 to 45% of particles smaller than 0.005 mm. in diameter; approximate base exchange capacity of 30 M.E. per 100 grams; and a pH of 5.5.

2 pounds of ammonium sulfate (20% N).

2½ pounds of potassium nitrate (13% N; 44% K_2O).

1½ pounds of ammonium phosphate (11% N; 48% P_2O_5).

These amounts of materials were sufficient for the preparation of 100 $2 \times 2 \times 4$-inch briquettes, or 800 $1 \times 1 \times 2$-inch briquettes.

Ammoniated Peat

When peat and ammonia are heated under pressure the peat can be made to absorb the ammonia to the extent of about 20 per cent. The product is known as ammoniated peat and though it has not been developed commercially it has considerable promise as a fertilizer. About one-third of the ammonia in ammoniated peat is water-soluble. Davis and Scholl (1939) have reported ammoniated peat to give better results than urea or ammonium sulfate when applied to carrots.

FIGS. 49 AND 50. (*Above and below.*) Showing the effect of .42 grams nitrogen in various organics on the growth of Japanese millet. (Courtesy, Massachusetts Agricultural Experiment Station.)

Soot

Soot represents the condensed smoke of coal fires, and is made up largely of carbon which has absorbed varying amounts of ammonia and oxides of sulfur. For fertilizing purposes soot may be considered as an impure form of ammonium sulfate. The nitrogen content of soot may vary widely. Soot that has been derived from flues where soft coal has been used as fuel will test about .5 per cent of nitrogen. Soot can often be secured locally by farmers, but largely because of the small amounts obtainable it is seldom used as a fertilizer except for garden crops.

Manufacture of Base-mix or Base-goods

Many manufacturers of complete fertilizers make up what is called "base-mix," "base-goods" or "wet base-goods." This is usually made by mixing organic nitrogenous fertilizers with phosphate rock and sulfuric acid. The nitrogenous fertilizers commonly used are low-grade materials commonly referred to as rough ammoniates, such as hair, leather, fur, felt, feathers, and wool waste. In their natural state these materials are of little value as carriers of plant nutrients, but when processed or made into base-mix they are converted into readily available forms suitable for crop use. This is one of the ways in which the manufacturer of complete fertilizers performs an economic service. During the last few years, however, the manufacture of wet base-goods in this country has greatly declined.

The manufacturers use base-mix as a base to which is added high-analysis fertilizer materials in sufficient quantities to compound the various grades of mixed fertilizers which they sell on the market.

In the following table are given the various sources of fertilizer nitrogen for a recent year. These figures would indicate that only about 10 per cent of fertilizer nitrogen can be classed as proteid organic.

Table 32

SOURCES OF FERTILIZER NITROGEN IN 1941

	Source	*Tons*	*Total*
Inorganic.	Ammonia—solid	144,448	
	Nitrate—solid	146,100	
	Nitrogen liquors	65,200	
	Synthetic organic	43,082	398,830
Organic.	Vegetable proteins	20,150	
	Animal proteins	4,200	
	Process tankage	9,000	
	Sewage sludge	7,960	41,310
All others.	.	. .	11,168	11,168
Total.	. .	. . .	451,308

Comparative Agricultural Value of the Proteid Organic Nitrogenous Fertilizers

Many investigators have attempted to classify the proteid organic nitrogenous materials on a basis of their relative value for crop production, and also to compare them with the commonly used inorganic nitrogenous materials. Such investigations as have been reported do not furnish much exact information and are of value only in determining roughly the relative worth of the materials tested when they are applied without any attempt being made to correct the acidity resulting from their use. As stated elsewhere, laboratory tests can be used only as indicators.

As is to be expected, the relative crop-producing value of the organic ammoniates varies greatly, depending not only upon the fertilizer, but also upon the crop, soil and climatic conditions under which the crop is grown.

Table 33

EFFECTS OF VARIOUS SOURCES OF NITROGEN IN MIXED FERTILIZERS ON YIELDS OF SEED COTTON—PEE DEE EXPERIMENT STATION, SOUTH CAROLINA, 1919–1925*

Source	*Yield—Pounds of Seed Cotton per Acre*							*Relative Yield*
	1919	*1921*	*1922*	*1923*	*1924*	*1925*	*Average*	
Check	954	920	617	717	595	1472	879	..
Cottonseed meal	980	950	620	700	630	1542	904	87.7
Cyanamid	950	950	830	790	700	1534	959	93.1
Ammonium sulfate	1370	800	620	870	745	1554	993	96.3
Sodium nitrate	1100	730	710	1190	785	1668	1031	100.0
Fish scrap	1280	920	790	850	795	1570	1034	100.3
Dried blood	1310	950	630	990	740	1632	1042	101.1
Ammonium nitrate	..	840	690	800†	690†	1618	..	..

* Yields not obtained for 1920.
† Ammonium nutrate not applied in 1923 and 1924.

Table 34

EFFECT OF VARIOUS SOURCES OF NITROGEN IN MIXED FERTILIZER ON YIELDS OF CORN—PEE DEE EXPERIMENT STATION, SOUTH CAROLINA, 1909–1925

Source	*Yield—Bushels*								*Relative Yield*
	1919	*1920*	*1921*	*1922*	*1923*	*1924*	*1925*	*Average*	
Check	28.8	28.3	28.5	28.4	21.1	27.0	22.1	26.3	..
Cottonseed meal	30.6	29.4	32.9	32.5	24.9	28.5	29.2	29.7	82.0
Cyanamid	28.7	33.6	34.3	32.2	21.9	34.0	31.5	30.9	85.4
Fish scrap	31.1	42.8	32.2	34.9	13.7	30.6	27.4	30.4	84.0
Dried blood	32.0	41.1	37.0	35.9	24.0	36.0	32.2	23.0	93.9
Ammonium sulfate	32.0	42.5	43.2	36.7	30.0	30.4	31.2	35.3	97.5
Sodium nitrate	32.5	47.3	47.3	37.7	26.4	29.0	32.8	36.2	100.0
Ammonium nitrate	..	42.5	43.8	33.2	22.1*	22.2*	35.7	..	..

* No ammonium nitrate applied in 1923 and 1924.

Most organic ammoniates have a slightly acid-forming influence on the soil but some decrease soil acidity. In general, availability tests as commonly conducted have shown that those materials that carry their nitrogen in the nitrate form gives the greatest efficiency for general field crops when grown under average conditions. Basing our estimate upon the many availability studies so far reported, if we assign a rating of 100 for the nitrate fertilizers we can assign a rating of about 90 for the ammonium compounds where lime is not applied, and a rating of about 70 to 80 for the high grade proteid organics. Tables 33 and 34 are given as examples of the results that have been secured from field tests in many parts of the country. These data secured by Skinner and Buie (1926) show the relative influence of some of the more commonly used organics, as well as inorganic nitrogenous fertilizers, on the yields of seed cotton and corn respectively. The check plats have received an annual treatment of 1000 pounds per acre of an 0–8–4, and all other plats 1000 pounds per acre of a 4–8–4. The source of nitrogen has varied with the plat.

In Table 35 are given, according to Paden (1939), 12-year results from the Clemson College tests with various sources of nitrogen which were conducted in concrete rimmed plats. A picture of these plats is shown in Fig. 1.

In many experiments, as in those cited, the relative value of the nitrogenous fertilizers has been based upon the increase in crop yield that is produced by their use. As crops absorbed additional nitrogen when supplied with quantities greater than are necessary for maximum yields, and

FIG. 51. High yield of carrots grown on peat land receiving 400 pounds per acre of 0–9–40. (Courtesy, Wisconsin Agricultural Experiment Station.)

Table 35

AVERAGE YIELDS OF SEED COTTON FROM VARIOUS SINGLE SOURCES OF NITROGEN ON UNLIMED AND LIMED SOIL, 1928–1939
(CONCRETE FRAMES)

Source of Nitrogen	Plot Number	Soil Treatment	pH‡	*Rate of Application of Fertilizer* — 1,000 lbs. — 6–10–4* Average 3 yrs. 1928–30	480 lbs. — 5–10–5* Average 5 yrs. 1931–35	480 lbs. — 5–10–5 Average 4 yrs. 1936–39	Average 12 yrs. 1928–39	*Average Increase from Lime 12 yrs. 1928–39* — Pounds per Acre	Per Cent
No fertilizer	1, 22	Unlimed	6.21	1663	1196	1324	1355	131	9.7
	27, 48	Limed	6.54	1847	1442	1270	1486	..	..
No nitrogen fertilizer	2	Unlimed	6.36	1920	1167	1343	1414	..	..
	28	Limed	6.54	2056	1350	1401	1544	130	9.2
Calcium nitrate	3	Unlimed	6.34	2530	1903	1986	2087	..	..
	29	Limed	6.53	2499	2027	2025	2145	58	2.8
Sodium nitrate	4	Unlimed	6.58	2354	1722	1951	1957	..	..
	30	Limed	6.58	2494	1997	2066	2144	187	9.5
Ammonium sulfate	5	Unlimed	5.97	2320	1880	1931	2007	..	..
	31	Limed	6.49	2287	1857	1920	1995	−12	−.59
Leunasalpeter	6	Unlimed	..	2184	1650	..	..	..	..
	32	Limed	..	2481	2106	..	..	..	..
Ammonium sulfate	6	Unlimed	6.04	..	..	1720	..	..	..
	32	Limed	6.51	..	..	2224	..	504†	29.2
Calurea	7	Unlimed	..	2436	1878	..	..	..	..
	33	Limed	..	2475	2084	..	..	..	..
Cal-nitro	7	Unlimed	6.14	..	..	1919	..	..	..
	33	Limed	6.64	..	..	2008	..	89†	4.6
Urea	8	Unlimed	6.44	2205	1783	1901	1928	..	..
	34	Limed	6.51	2191	1880	2048	2014	86	4.4
Ammo-Phos	9	Unlimed	6.14	2306	1762	1912	1948	..	..
	35	Limed	6.25	2179	1872	2015	1996	48	2.5
Nitrophoska	10	Unlimed	..	2522	1958	..	..	..	..
	36	Limed	..	2148	1913	..	..	..	..
Ammonium nitrate	10	Unlimed	6.34	..	..	2102	..	53†	2.5
Calcium carbonate	36	Limed	6.27	..	..	2049	..	..	..
Cyanamid	11	Unlimed	6 41	2244	1901	1888	1982	..	..
	37	Limed	6.54	2194	1919	2017	2020	38	1.9
Acidulated leather	12	Unlimed	..	1901	..	..	..	..	..
	38	Limed	..	2175	..	..	..	..	..
Ammoniated superphosphate	12	Unlimed	6.38	..	1746	1826	..	..	..
	38	Limed	6.48	..	2048	2206	..	380†	20.8
Milorganite	13	Unlimed	6.06	2131	1589	1637	1743	..	..
	39	Limed	6.70	2632	1994	2059	2175	432	24.8
Cottonseed meal	14	Unlimed	6.29	2430	1832	1815	1976	..	..
	40	Limed	6.48	2768	2075	2024	2231	255	12.9
Dried blood	15	Unlimed	6.29	2407	1854	1917	2013	..	..
	41	Limed	6 64	2654	2042	2105	2216	203	10.1
Fish scrap	16	Unlimed	6.26	2415	1917	2012	2073	..	..
	42	Limed	6.46	2485	2060	2124	2188	115	5.5
Tankage	17	Unlimed	6.28	2195	1898	2061	2026	..	..
	43	Limed	6.51	2297	1993	2000	2071	45	2.2
Average for plots receiving nitrogen	..	Unlimed	6.26	2305	1811	1853	1949	..	..
	..	Limed	6.51	2397	1986	2008	2096	147	7.5

* Nitrogen expressed as ammonium equivalent.
† Average 4 years (1936–39).
‡ pH determinations made June 7, 1939.

as the percentage content of nitrogen usually decreases as the yield of a crop increases, the criterion of relative value in many tests should be the amount of nitrogen recovered by the crop rather than the relative yields of the crops secured as a result of their use. Moyer and Blair (1930), at the New Jersey Experiment Station, made a study on this basis of tankage, Cyanamid, ammonium sulfate and sodium nitrate. Equivalent amounts of nitrogen were applied to all pots. Their results are given in the table below. It will be noted that the recovery of nitrogen in these experiments varied from 25 to 61 per cent. Under field conditions the recovery of nitrogen seldom exceeds 50 per cent even for nitrates, and most of the available data indicates that only about 60 to 75 per cent as much nitrogen is recoverable from the insoluble proteid organics as from soluble sources of nitrogen.

Table 36

RELATIVE AVAILABILITY STUDIES OF SOME NITROGENOUS FERTILIZERS. YIELD IN GRAMS PER POT AT AGE OF 13 WKS. NITROGEN CONTENT OF CROP IN MILLIGRAMS

Crop Grown	*Sodium Nitrate*		*Ammonium Sulfate*		*Cyanamid*		*Tankage*	
	Dry Matter, g.	*Nitrogen, mg.*	*Dry Matter, g.*	*Nitrogen, mg.*	*Dry Matter, g.*	*Nitrogen, mg.*	*Dry Matter, g.*	*Nitrogen, mg.*
Oats*	25.17	407	24.80	390	26.30	367	21.67	248
Barley*	36.70	424	40.05	408	40.20	443	31.07	328
Rape*	28.47	335	29.51	349	29.47	356	21.60	246
Corn†	40.95	324	40.37	308	34.12	320	33.20	219
Barley†	29.69	302	27.50	248	30.18	267	23.17	204
Totals	160.98	1792	162.23	1703	160.27	1753	130.71	1245
Recovery of nitrogen in soil	60.94%		58.96%		61.88%		25.10%	
Relative yields of crops	100		101		100		81	
Relative contents of nitrogen	100		95		98		70	

* Grown in Sassafras loam.
† Grown in Sassafras sand.

Use of Proteid Organic Nitrogenous Fertilizers

Because of the wide range in the chemical composition of many of the proteid organic nitrogenous fertilizers, they should be bought on a basis of their chemical analysis. Many of these materials are not procurable by the farmer, for the mixed-goods industry absorbs the greater part of them.

Many are unbalanced in their fertilizing properties and are unfit to be used economically by the farmer. Where attempts are made to use them they should be supplemented with the needed quantities of phosphoric acid and potash. Because these materials do not become available immediately upon application they should always be mixed thoroughly with the soil and should be applied a few days before the crop is planted. Only a few of the organics can be applied successfully as top dressers.

Willis and Piland (1931) have secured results which showed that even low concentrations of free ammonia in the soil may be detrimental to germination and seedling growth. They therefore advise that organic ammoniates or easily hydrolysable ammonium salts be applied to the soil sometime before planting or used in mixtures containing calcium salts. Apparently, calcium salts may prevent the toxicity of free ammonia, provided the limit of tolerance exhibited by the plant is not exceeded.

The proteid organic nitrogenous fertilizers were long held in popular esteem. Because of the changes which they must undergo after they are incorporated with the soil before they become available to plants, they tend to liberate their nitrogen in an available form gradually throughout the season. In this way crops may be furnished with a supply, even though variable, of available nitrogen during their entire growing period. In addition, the organic fertilizers, including barnyard manure, must be credited with having furnished the American farmer with the secondary and rarer essential elements at a time when their essential nature was not realized. Also, Clevenger and Willis (1935) have shown that the natural organic fertilizers perform an additional service in correcting soil acidity, especially at high pH and temperature levels. The proteid organic nitrogenous fertilizers have proved to be good sources of nitrogen largely because they are not very acidic and because the nitrogen they carry is not readily leached from the soil. Nevertheless, their availability is below that of the inorganics and appears to be influenced by their nitrogen-carbon ratios, especially if their lignin content is left out of consideration.

Because of their influence on the physical condition of fertilizer mixtures the proteid organic nitrogenous fertilizers have long been held in high regard. Mixed fertilizers which contain as much as 10 per cent of an organic carrier, such as cottonseed meal, have an excellent drillability. In the main, the waning popularity of the proteid organic fertilizers is due to the cost of nitrogen carried. A unit of nitrogen in the proteid organic nitrogenous materials costs two to two and a half times as much as that found in the commonly used inorganic nitrogenous materials. For this reason the more expensive carriers of nitrogen are rarely used in standard brands of mixed fertilizers.

Skinner (1931), of the United States Bureau of Chemistry and Soils, has reported that as a result of some 25 experiments made in the South with cotton, on a variety of soil types, and covering a period of four years, the highest yields were attained where 80 to 85 per cent of the nitrogen applied was derived from inorganic or synthetic sources and 15 to 20 per cent from proteid organic sources. Other investigators have reported the highest crop yields where all of the nitrogen was derived from inorganic or synthetic sources.

Plant Utilization of Organic Compounds

For a long while it was generally supposed that it was necessary for the nitrogen of proteid nitrogenous materials to be transformed into nitrates by the microörganisms in the soil before it could be absorbed and utilized by plants. Yet it was early demonstrated by Kelly (1911), and others, that with certain plants such as rice, a complete transformation is not necessary, and since that time many investigators have shown that many plants may absorb and utilize their nitrogen in the ammoniacal form, and especially during their seedling stage of growth. In addition, Hutchinson and Miller (1910), and Schreiner and Skinner (1912), have demonstrated that certain plants can absorb and utilize nucleic acid, hypoxanthine, xanthine, guanine, creatine, arginine, creatinine, and histidine. It is also known that phytin and lecithin, which are organic compounds containing phosphorus, exert a favorable influence on plant growth. Beaumont *et al.* (1931) found that asparagine, cystine, and urea may be absorbed by tobacco. Other investigators, however, have expressed doubt as to whether absorption of organic compounds plays a very important part in the nutrition of field crops grown under normal field conditions. On the other hand, Bear (1929) has suggested that the known good effects of organic matter in the soil may be due in part to its being a source of these or similar organic compounds. It is also altogether probable that the soluble organic compounds aid in keeping iron and phosphates in solution and in this way greatly help the growth of some plants on some soils.

7

Sources and Use of the Mineral Phosphates

Phosphorus is widely distributed in the lithosphere and is found most commonly as calcium phosphate. It is generally reported by chemists in terms of $Ca_3(PO_4)_2$, which is known chemically as tricalcium phosphate. All plant and animal materials and most phosphatic fertilizers carry their phosphorus as salts of orthophosphoric acid.

Limestone often carries tricalcium phosphate to an extent ranging from a trace to as much as 80 per cent. The sedimentary phosphate rock of our western states is composed almost entirely of calcium phosphate. The presence of phosphates in many other rocks testifies to their general distribution. Nevertheless, many soils are very deficient in phosphorus and subsoils are more deficient in available phosphorus than surface soils. All soils contain more phosphates than the parent materials from which they are derived. Schollenberger found that in the surface of an average

FIG. 52. Lysimeters of the Tennessee Agriculture Experiment Station and T.V.A. greenhouse in the background. (Courtesy, Dr. W. H. MacIntire and the Tennessee Valley Authority.)

Ohio soil approximately one-third of the phosphorus was in organic combination, while in the subsoil only about one-fifth of the total phosphorus was in this form. In all probability the major portion of the cultivated soils of the United States east of the Mississippi River would be forced out of crop production within a comparatively few years if phosphate fertilizers were withheld. It has been estimated that eventually the soils of the Middle West will need 4,000,000 tons of phosphatic fertilizer per year.

Origin of Calcium Phosphate Deposits

The native phosphate mineral in most of our soils is a crystallized calcium phosphate containing small amounts of calcium chloride and calcium fluoride. This primary soil mineral is called apatite and is of igneous origin. In its massive form apatite is sometimes referred to as "phosphorite." Two varieties of apatite have been recognized, a fluor-apatite—and this is the form usually found—whose chemical composition is represented by the formula $Ca_3(PO_4)_2.CaF_2$ or $Ca_{10}F_2(PO_4)_6$, and a chlor-apatite having the formula $Ca_3(PO_4)_2.CaCl_2$. Both of these compounds exist in a submicrocrystalline condition. The fluorine contributes to the insolubility of phosphate deposits and seems to be essential to their preservation through the ages. In some soils, particularly those of an acid nature, titanium, iron and aluminum phosphates may be found in varying quantities. These phosphates, when present in the soil, would be classed as secondary soil minerals.

FIG. 53. Dr. Cyril G. Hopkins, formerly Chief, Department of Agronomy, University of Illinois. He was a great advocate of the use of phosphate rock and a proponent of a permanent system of agriculture. (Courtesy, Illinois Agricultural Experiment Station.)

Phosphates, as they are found in the larger deposits, are amorphous and may occur in the compact, concretionary, or earthy forms. These forms are commonly referred to as rock or boulder phosphate, pebble phosphate, and soft phosphate, respectively. The terms "rock phosphate" and "phosphate rock" are synonomous and are usually used as blanket terms to refer to any mined phosphates.

Phosphate deposits may be classified roughly as residual, replacement, or sedimentary. Residual phosphates, such as the Tennessee brown phosphate, are derived from beds of phosphatic limestone. Replacement phosphates, such as the Nauru and Ocean Island phosphates, are limestones that have been phosphatized or changed to calcium phosphate by the action of phosphate-bearing solutions derived from guanos or other organic sources. Sedimentary phosphate deposits are those that have accumulated as sediments and later have become interbedded with other rocks, usually limestones and shales. Such deposits are found extensively in North Africa, and they constitute also the Florida Land Pebble and the phosphates of the western United States. They furnish the principal phosphate rock of commerce.

It is generally conceded by geologists that the larger deposits of calcium phosphate, which are the sedimentary deposits, are of organic origin. Such deposits have been laid down on the floor of the sea. Sea water carries a trace of phosphate, and this phosphate is drawn upon by marine organisms and stored away in their bodies. Upon death the remains of these organisms collect on the floor of the sea, and the phosphatic content is concentrated through the decay of the organic matter and the loss of part of its soluble carbonates. In some cases beds of phosphatic limestone which have been formed in this manner have been raised above the surface of the sea and their phosphate content still further concentrated through the leaching of the more soluble of their carbonates. It is believed that it was in such a manner that the phosphates of the southeastern part of the United States originated. Many bones and teeth of prehistoric fish may still be found in some of these deposits. The amorphous forms of phosphate lack such evidences of organic origin and are often referred to as "mineral phosphate," although this term is sometimes used to designate any type of phosphate rock.

World Deposits of Phosphates

Deposits of phosphate of economic importance have been found on every continent and on many of the islands of the sea. In the United States, deposits of workable size are known to exist in Arkansas, Florida, Idaho, Kentucky, Montana, South Carolina, Tennessee, Utah, and Wyoming. Outside the United States, the most important deposits now being worked are in Algeria, Morocco, Tunisia, and Russia.

Phosphate Deposits of Florida

The phosphate deposits of Florida belong to the Later Tertiary period and are located in the northwest central part of the state. They were discovered in 1881. These deposits were first mined near Arcadia on the Peace River as early as 1887, by Col. T. S. Menchald. The phosphates which they yield vary widely as to their physical and chemical characteristics. Four types of phosphate are recognized: (1) Hard Rock phosphate; (2) Soft phosphate; (3) Land Pebble phosphate; and (4) River Pebble phosphate.

Florida Hard Rock phosphate, or what is sometimes called "boulder phosphate," occurs in irregular pockets as hard phosphorite of many sizes embedded in clay, sand, or Soft phosphate. The deposits range in thickness from a few inches to a hundred feet. In mining, the boulders and pebbles are separated from their matrix by washing. Florida Hard Rock phosphate is by far the best grade of phosphate that is produced in this country. As it contains a low content of iron, aluminum and fluorine it is especially suited for the manufacture of superphosphates. In addition it carries from 78 to 80 per cent of tricalcium phosphate. For these reasons it has always been in great favor among foreign buyers.

FIG. 54. Hopewell phosphate mine, Coronet Phosphate Company, near Plant City, Florida. (Courtesy, Dr. G. R. Mansfield.)

Florida Soft phosphate usually contains considerable quantities of calcium carbonate and clay, and for that reason the tricalcium phosphate content ranges only from 40 to 60 per cent. It is powdery when dry but when wet it is very sticky. It is generally considered to be a replacement phosphate. Individual deposits of Soft phosphate of variable size are found in both the Hard Rock and the Land Pebble areas.

When Hard Rock and Land Pebble phosphates are prepared for the market, the Soft phosphate which is present in their matrix is washed into waste ponds where it settles out with the clay and other impurities. These abandoned waste ponds contain a considerable tonnage which is being recovered to some extent for use as a fertilizer. This material is marketed under the name of "colloidal phosphate," "mineral colloids," and "Phos Cal Oids," but is also known to the industry as " waste pond phosphate." It contains only 18 to 23 per cent of tricalcium phosphate of which 2 to 3 per cent might be classed as available. Because of the fineness of Soft phosphate exorbitant claims have been made as to its availability to crops. Because of this, unfortunately, much Soft phosphate has been sold to farmers in recent years at prices far above its actual value.

Deposits of Florida Land Pebble phosphate lie south of the Hard Rock region, largely in Polk and Hillsborough counties. This phosphate occurs as light gray to black pebbles mixed with clay and sand. The deposits have been formed by wave action on phosphate-bearing formations. Land Pebble phosphate is not as hard as the Hard Rock phosphate, nor is it of as high a grade. It contains from 68 to 75 per cent of tricalcium phosphate but frequently carries 5 to 6 per cent of iron and aluminum oxides. The deposits are mined entirely by the hydraulic method, but the overburden is removed to a considerable extent by dragline excavators. They are the most extensively mined deposits in America and comprise over 95 per-cent of the Florida phosphate rock production.

Florida River Pebble phosphate deposits represent stream deposits of mixed phosphate pebbles, sand, and clay. They were among the first deposits to be mined, but as they yield a low-grade phosphate their exploitation practically ceased in 1908.

The composition of three representative samples of Florida phosphate rock, as given by Matson (1915), is shown in Table 37.

The largest producing phosphate fields in the United States are in Florida and Tennessee. The Florida deposits have the advantage of water transportation and have been able to compete in the European markets. Since 1894 they have easily maintained the lead in this country's produc-

Table 37

ANALYSIS OF REPRESENTATIVE SAMPLES OF FLORIDA PHOSPHATE ROCK

	Hard Rock Phosphate (per cent)	*Land Pebble Phosphate (per cent)*	*River Pebble Phosphate (per cent)*
Insoluble (SiO_2)	2.03	4.78	..
Ferric oxide (Fe_2O_3)	0.10	4.15	0.83
Aluminum oxide (Al_2O_3)	2.30	..	1.37
Calcium oxide (CaO)	51.45	..	45.69
Carbon dioxide (CO_2)	2.45	..	6.37
Phosphoric acid (P_2O_5)	35.52	34.43	28.47
Tricalcium phosphate ($Ca_3(PO_4)_2$)	77.50	74.85	61.77

tion of phosphate rock because of their uniformity of grade, regularity of occurrence and low cost of production. Not only have the Florida phosphates been able to compete with other phosphates in European markets, but Florida phosphates are shipped also to the newer markets of South Africa, New Zealand, and Australia.

Phosphate Deposits of Tennessee

The phosphate deposits of Tennessee belong to the Ordovician and Devonian periods. They were first discovered in 1893. These deposits occur as very hard and compact rock, which are in contrast to the loose structure of the phosphates of Florida. They may be classified as Blue, Brown and White phosphates.

The Tennessee Blue phosphate was first discovered in Hickman County, Tennessee, in 1893 and mining began in 1894, but operations were suspended in 1896 after the discovery of the Brown phosphate. In recent years a few mines have been opened, but only one mine is now in operation. As there are extensive reserves of this phosphate which analyze from 68 to 72 per cent of tricalcium phosphate, other mines probably will be opened in the future. The deposits occur as massive gray, blue, or black rock phosphate, and are found in the counties of Maury, Hickman, and Lewis. The beds, consisting largely of deposits of sedimentary calcium phosphate, range from 1 to 4 feet in thickness and are mined underground.

Tennessee Brown phosphate deposits are found in the central part of Tennessee, in the counties of Maury, Williamson, Davidson, and Sumner. It occurs in beds of varying thickness overlying phosphatic limestone. They represent a residual deposit of phosphatic limestone and are mined

FIG. 55. Phosphate drag line, T.V.A. Phosphate mining operations. (Courtesy, Tennessee Valley Authority.)

from open surface pits. There are three types of the Tennessee Brown phosphate: (1) a dense grayish lump rock analyzing 76 to 82 per cent of tricalcium phosphate; (2) a cellular rock resembling brown sandstone and analyzing 72 to 76 per cent of tricalcium phosphate; (3) a soft blackish-brown muck analyzing 72 per cent and lower of tricalcium phosphate. Although the reserves of high-grade Brown rock phosphate in Tennessee are becoming depleted, the low-grade deposits, because of their location with reference to the fertilizer consuming centers, will probably be able to compete in the domestic market for some time to come.

Deposits of Tennessee White phosphate are found in Perry and Decatur counties. They are of more recent origin than the Brown and Blue phosphates. In appearance the White phosphate resembles very much the Hard Rock phosphate of Florida. Although the deposits of White phosphate carry from 72 to 90 per cent of tricalcium phosphate they occur very irregularly and are not as accessible as the deposits of Brown phosphate. For this reason they have been mined only to a limited extent. Mansfield

(1940) estimates the White phosphate reserves of Tennessee to be only 15,000,000 tons.

Phosphate Deposits of South Carolina

The phosphate deposits of South Carolina are confined largely to an area of about 30 miles in width which extends from Beaufort to a point about 20 miles north of Charleston.

Interest was attracted to these phosphate deposits as early as 1795, but the true nature of the material probably was not realized until about the middle of the nineteenth century. Mining was first begun in 1868 and the first small cargo of rock was shipped during that year.

In general, the deposits occur as nodules and boulders embedded in a matrix of sand and clay. These deposits, like those of Florida, belong to the Tertiary period and contain petrified bones and the teeth of terrestrial and marine animals. The deposits vary from 1 to 3 feet in thickness, but do not average much over a foot, so that they are mined largely by means of grab buckets. The nodules vary in size from mere pebbles to masses weighing a ton or more. They vary in color from brown to dark gray. The darker materials are the hardest and contain the highest percentage of tricalcium phosphate. Some of the deposits are found in the beds of rivers and must be mined by dredging. Such phosphate is referred to as "River phosphate." Similar material that is mined on land is called "Land phosphate." The River phosphate occurs largely in loose fragments of nodular aspect, which have accumulated in irregular banks on the river bottoms. These beds are made up of the original phosphatized marl together with fragments of phosphate which have been derived from the land deposits. After the South Carolina phosphate has been dug or dredged it is dried, ground if necessary, and in this condition it is ready to be marketed.

South Carolina phosphate is very free from iron and aluminum so that from this standpoint it is suitable for use in the production of the superphosphates. On the other hand, it analyzes only 53 to 65 per cent of tricalcium phosphate. This low analysis did not act as a barrier to its sale during the earlier years when there was no competition with phosphates of higher analysis, but with the discovery of the high-analysis phosphates of Florida and Tennessee, South Carolina phosphates entered a period of severe competition which ended in the total suspension of production.

South Carolina phosphate played an important part in supplying the demand for phosphate during the early years. The entire American phosphate production from 1868 to 1888 was secured from these deposits.

FIG. 56. Phosphate plant at Mulberry, Florida, of the International Minerals and Chemical Corporation. (Courtesy, International Minerals and Chemical Corporation.)

During this period South Carolina furnished 90 per cent of the world's supply of phosphate rock. In 1888 the Florida production was only about 3,000 long tons, but by 1894 production had increased to 527,000 long tons—a larger production than that of South Caıolina. In 1899 the production of phosphate rock in Tennessee also surpassed the production in South Carolina. From about this time the production of South Carolina phosphate gradually decreased until it ceased entirely in 1925. There was no production during 1921, 1923, and 1924.

The composition of representative samples of South Carolina phosphate, as given by Rogers (1915), is shown in the accompanying table.

Table 38

ANALYSIS OF REPRESENTATIVE SAMPLES OF SOUTH CAROLINA PHOSPHATE ROCK

	Land Rock (per cent)	*River Rock (per cent)*
Insoluble	13.03	13.30
Aluminum oxide (Al_2O_3)	0.40	..
Ferric oxide (Fe_2O_3)	1.38	..
Calcium oxide (CaO)	39.10	..
Phosphoric acid (P_2O_5)	27.23	25.14
Sulfur trioxide (SO_3)	1.45	..
Carbon dioxide (CO_2)	3.05	3.61
Tricalcium phosphate ($Ca_3(PO_4)_2$)	59.44	54.88

Phosphate Deposits of Kentucky and Arkansas

The phosphate deposits of Kentucky are similar to the Brown rock phosphate deposits of Tennessee. Their phosphate content varies from 60 to 75 per cent of tricalcium phosphate. All Kentucky mines ceased producing in 1926.

The phosphate deposits of Arkansas are located in the northern part of the state and belong to the Ordovician age. They are of sedimentary origin. They are dark in color and are referred to often as "Black phosphate." Although small deposits of high-grade rock have been discovered, the phosphate content of the average rock is low.

Phosphate Deposits of the Western States

The phosphate deposits of the western states belong to the Carboniferous period. They appear to be a precipitate resulting from a reaction between phosphoric acid and calcium carbonate in the marine waters of the Permian Sea. The deposits occur in beds ranging from 2 to 6 feet in thickness and are overlaid by limestone and phosphatic shales. Some of the rock is readily crushed, whereas some is very hard. The phosphates of these deposits vary in color from a light gray to a jet black. The phosphate content of the workable beds varies from 65 to 80 per cent of tricalcium phosphate. The rock is mined by underground methods. Although the

Table 39

ANALYSIS OF REPRESENTATIVE SAMPLES OF PHOSPHATE ROCK FROM WYOMING, UTAH, IDAHO, AND MONTANA

	Wyoming (per cent)	*Utah (per cent)*	*Idaho (per cent)*	*Montana (per cent)*
Insoluble	10.00	9.40	2.62	6.00
Silicon dioxide (SiO_2)	..	..	0.46	..
Aluminum oxide (Al_2O_3)	0.89	0.90	0.97	1.10
Ferric oxide (Fe_2O_3)	0.73	0.33	0.40	0.45
Magnesium oxide (MgO)	0.28	0.26	0.35	0.30
Calcium oxide (CaO)	45.34	46.80	48.91	46.90
Sodium oxide (Na_2O)	1.10	2.08	0.97	..
Potassium oxide (K_2O)	0.48	0.58	0.34	..
Carbon dioxide (CO_2)	6.00	2.14	2.42	2.20
Phosphoric acid (P_2O_5)	27.32	32.05	33.61	32.10
Sulfur trioxide (SO_3)	1.59	2.34	2.16	2.10
Fluorine (F)	0.60	0.66	0.40	..
Tricalcium phosphate ($Ca_3(PO_4)_2$)	59.28	69.54	72.93	70.10

Florida phosphates, as well as the Tennessee Brown phosphate, require washing before they are marketed, the phosphates of the western deposits do not.

The deposits of the Western States were first discovered in Utah in 1889. They are located largely in southeastern Idaho, western Wyoming, northern Utah, and western Montana. Mining operations were started at Conda, Idaho, in 1920 but the western deposits are mined only to a limited extent because they are far removed from the fertilizer consuming centers.

In Table 39 is given the composition of representative samples of phosphate rock from the western states, as reported by Gale and Richards (1910) and Pardee (1917).

Apatite Deposits of Canada and Russia

Canadian deposits of phosphate rock are mined in small amounts largely in Ontario and Quebec. These deposits are made up of the mineral apatite together with miscellaneous impurities. Because this phosphate contains much foreign material, and because it is difficult to mine, the cost of production usually is high.

For about 10 years prior to World War II large quantities of apatite were mined annually in the Kola Peninsula region of European Russia and exported to European countries for manufacture into superphosphates.

Table 40

PHOSPHATE ROCK RESERVES IN THE UNITED STATES

	Long tons
Eastern Field:	
Arkansas	20,000,000
Florida	5,081,839,000
Kentucky	863,000
South Carolina	8,798,000
Tennessee	195,151,000
Total	5,306,651,000
Western Field:	
Idaho	5,736,335,000
Montana	391,323,000
Utah	1,741,480,000
Wyoming	115,754,000
Total	7,984,892,000
Grand total	13,291,543,000

The deposits are estimated to contain two billion tons of high-grade apatite.

Phosphate Reserves of the United States

The phosphate reserves of the United States, as estimated by Mansfield in 1940, are given in Table 40. Mansfield admits, however, that further prospecting, particularly in the West, may uncover other deposits. It has been estimated that our phosphate reserves are approximately one-half of the world's reserve of phosphate.

World Production of Phosphate Rock

The relative importance of the phosphate deposits of the world from the standpoint of present exploitation is shown in Table 41.

Table 41

WORLD PRODUCTION OF PHOSPHATE ROCK, 1936, BY COUNTRIES, IN METRIC TONS

Country	Production
Algeria	531,368
Angaur	89,226
Australia	178
Austria	100
Belgium*	100,000
Canada†	477
Christmas Island (Straits settlements)	161,468
Egypt	531,006
Estonia	11,408
France	40,000
Germany	198
Indo China	11,095
Japan	110,668
Madagascar	4,500
Makatea Island	130,759
Morocco, French	1,335,074
Nauru and Ocean Islands	965,708
Netherland West Indies: Curaçao	81,112
Netherland East Indies	12,071
Poland	12,500
Russia	2,213,786
Spain	10,000
Tunisia	1,475,550
United States	3,518,242
Total	11,346,494

* Estimated.
† Apatite.

Figures given by The International Superphosphate Manufacturers Association for 1936—the latest year for which figures are available—showed that the world production of phosphate rock just prior to World War II was somewhat over 11,000,000 tons. Of this total about 30 per cent was produced under French control, 31 per cent in the United States, 19 per cent in Russia, 9 per cent under British control, and about 2 per cent under Japanese control.

Characteristics of Phosphate Deposits of Foreign Countries

From Table 41 it will be observed that the North African deposits, which include the phosphates of Tunisia, Algeria, and Morocco, produced just prior to World War II a higher percentage of the world's phosphate output than did those of the United States. These, as well as the Egyptian deposits, are largely of the sedimentary type. The highest grade material is produced in Morocco, and has been produced practically as a monopoly of the French government. As the phosphates of northern Africa are soft rocks, practically free of such objectionable impurities as iron, aluminum, and fluorine, they are well adapted for use in the manufacture of superphosphate. Most of the Egyptian phosphate is of too low a grade for export. On the whole, the North African deposits are very favorably located with respect to present world markets. Because of this, together with their high grade, they have been in recent years a severe competitor of the Florida phosphates in European markets.

Phosphate deposits of high-grade material are located on numerous islands of the world, the most important of which are the Nauru and Ocean Islands, the Gilbert, Ellis, and Christmas Islands, all British; Makatea Island, French; Rosa and Angaur Islands, formerly Japanese; and Curaçao and Aruba Islands, Dutch. The Pacific Islands generally furnish a very high grade of phosphate rock. Whole cargoes analyzing as high as 88 per cent of tricalcium phosphate have been received from the Ocean Islands. These islands, however, are distant from the large markets.

In England, France, Russia and other parts of the world may be found deposits of coprolites. Coprolites is a name given to certain phosphatic stones about the size of fir cones. They vary in color from gray to dark brown and contain about 50 per cent of tricalcium phosphate. They are of concretionary origin. At one time they were supposed to be the fossilized excrement of certain extinct reptiles. The deposits are only of local importance.

The Russian phosphate deposits, including those of southwestern Siberia, are of enormous extent but are generally of a very poor grade. These beds are largely of sedimentary origin.

Deposits of phosphate rock of varying size and grade are found also in Spain, France, Belgium, Germany, England, Palestine, China, Japan, Australia, New Zealand, West Indies, Peru, and Chile.

World Reserves of Phosphate

Table 42 gives the world reserves of phosphate rock by countries as estimated in 1926 by Mansfield, of the United States Geological Survey. These figures were intended only to give a general picture of the phosphate reserves at the time the estimate was made. If an estimate of the world reserves of phosphate rock were made today (1947) account would have to be taken of the tremendous deposits recently opened up and exploited in Russia, and of new deposits found to be in existence in Florida. Otherwise, the information regarding world phosphate reserves would remain very much as given in Table 42.

Table 42

WORLD RESERVES OF PHOSPHATE ROCK

Country	*Reserves* (*metric tons*)	*Grade* (*per cent*)
United States	6,431,000,000	60–70+
Tunisia	1,000,000,000	58–68
Algeria	1,452,000,000	58–68
Morocco	1,400,000,000	70–78
Egypt	179,000,000	60+
Nauru & Ocean Islands	140,000,000	80–88
Makatea Islands	10,000,000	80+
Angaur Island	3,000,000	80+
Rosa Islands	3,000,000	75
Palestine	4,000,000	47–80+
Total high grade	10,622,000,000	Generally 60+
Russia	5,568,000,000	Less than 50
Spain (Murcia)	10,000,000	Less than 50
Siberia	667,000,000	Less than 50
Total low grade	6,245,000,000	Less than 50
Grand total	16,867,000,000	

World Consumption of Phosphate Rock

The United States normally consumes a larger amount of phosphate rock than any other country. In Europe, France, Italy, and Russia are the largest consumers. Japan is the only important consumer of phosphate rock in Asia.

Although the American deposits of phosphates have for years been used to maintain the crop yield of both Europe and America, the reserve supplies of phosphate rock in this country can still be measured in billions of tons. Nevertheless, the richer deposits of Tennessee Brown rock and the better grades of Florida Hard Rock are nearly exhausted. The deposits of Montana, Wyoming, Idaho and Utah will probably remain practically untouched until the cost of production in Florida and Tennessee rises to such a point that the western phosphates can compete.

Although the reserve supplies of phosphate in the United States are enormous, they should be conserved to the fullest extent. Proper mining operations, proper safeguards with reference to exportations, and proper use as fertilizer materials, are the goals toward which all should work.

Characteristics of Marketed Phosphate Rock

Phosphate rock, after having been separated so far as profitable from the impurities it contained when mined, is usually crushed or ground before it is marketed. The Florida Soft phosphate and some of the small pebble phosphates are sold as mined, without further treatment. Finely ground phosphate rock is commonly referred to as "floats." The name floats was originally given to the fine dust that was formed when phosphate rock was ground, but at the present time the name is often applied to any unacidulated ground phosphate rock. Floats as found on the market is a heavy, finely ground powder, often light gray in color. The material is only slightly soluble in pure water, but its solubility is increased in water carrying acids, the degree of solubility depending upon the kind and the strength of the acid.

In recent years economic pressure has resulted in the marketing of a better quality of floats. This has been brought about by the more efficient removal of impurities from the phosphate rock by mechanical means.

Chemical Composition of Ground Phosphate Rock

The ground phosphate rock found on the market varies in its composition, depending upon its source, but it generally contains 30 to 40 per cent of phosphoric acid and 3 to 4 per cent of fluorine, and in addition varying amounts of lime. The presence of fluorine in phosphate rock greatly influences the availability of its phosphoric acid to plants. This has been shown by Bartholomew (1935) and others. Jacobs *et al.* (1933) showed that domestic phosphate rock contains no significant quantity of phosphoric acid soluble in a neutral ammonium citrate solution, and that less than 30 per cent of the total phosphoric acid is soluble in a 2 per cent citric acid solution.

As the market price per ton of ground phosphate rock is usually less than half that asked for ordinary superphosphate, and as the ground phosphate rock carries twice as much phosphoric acid as ordinary superphosphate, ground phosphate rock as a carrier of phosphoric acid often appears attractive for use on soils to which it is suited.

Influence of Fineness of Grinding of Phosphate Rock on Its Availability

As might be expected, the degree of fineness to which the phosphate rock is ground influences its availability to plants. The more finely the material is ground the greater is the surface of the material exposed to the soil solvents, and thus greater becomes the quantity of phosphoric acid that goes into solution. Conner and Adams (1926) found that reground Tennessee phosphate rock was 7.7 per cent more effective than commercial finely ground Tennessee phosphate rock, but they concluded that it would not be profitable to grind phosphate rock finer than it is ground in its commercial form; that is, finer than necessary to pass through a 100-mesh sieve. Mooers (1929), of the Tennessee Experiment Station, came to the same conclusion.

Soils Adapted to the Use of Ground Phosphate Rock

Results secured at many experiment stations indicate that ground phosphate rock gives better results with some crops when applied to acid soils than when applied to neutral soils. On the other hand, crops intolerant to acid fail to make a satisfactory growth on soils that are sufficiently acid to exert any material dissolving influence on phosphate rock. When applied to acid soils phosphate rock may undergo a series of chemical changes, forming compounds which are available and which can be utilized by plants, whereas when it is applied to neutral soils, and especially to those of a high lime content, the conditions are unfavorable for these chemical reactions. It appears, however, that soils have to be very acid if the availability of phosphate rock is to be greatly increased. Bartholomew (1937) concluded from vegetative tests and from the amount of phosphorus absorbed by plants, that only on extremely acid soils does the acidity of the soil materially increase the availability of phosphorus in phosphate rock. Phosphate rock is a poor source of phosphorus for those crops that are adapted to alkaline soils such as are commonly grown in the western United States.

Soils that contain high percentages of organic matter, particularly readily decomposable organic matter, are better suited for the application of ground phosphate rock than are soils containing medium or low percentages of organic matter. Thus, acid peat and muck soils are often particularly adapted to the use of phosphate rock, as they contain a high percentage of organic matter and are sometimes very acid. This has been demonstrated experimentally on the acid peat soils of northern Germany. Phosphate rock is supposed to be made available largely through the action of carbonic and nitric acids on the phosphate, probably as follows:

$$Ca_3(PO_4)_2 + 4H_2CO_3 = CaH_4(PO_4)_2 + 2CaH_2(CO_3)_2$$
$$Ca_3(PO_4)_2 + 4HNO_3 = CaH_4(PO_4)_2 + 2Ca(NO_3)_2$$

Excellent results with ground phosphate rock have been secured also by Hopkins, at the Illinois Experiment Station, on the brown and black clay loams of the Corn Belt which are high in organic matter. Experi-

FIG. 57. Mining Florida phosphate.

mental evidence indicates that ground phosphate rock should not be applied to light, open, sandy soils unless large quantities of farmyard or green manure also are incorporated in the soil. Rather disappointing results have been obtained where floats have been applied to open, drouthy soils containing small amounts of organic matter.

Thor (1933) found that the downward movement of phosphorus, following an application of phosphate rock, was greater in open-textured soils. He found an appreciable movement of phosphoric acid into the 8- to 16-inch and the 16- to 24-inch horizons of open-textured soils. He believes that the phosphate rock was carried down by gravitational water and that the movement was largely mechanical.

Plants Capable of Utilizing Ground Phosphate Rock

It has been shown abundantly that plants vary in their physiological ability to secure phosphorus from phosphate rock. Legumes appear to utilize the phosphorus of phosphate rock better than do most other crops, a characteristic that may be correlated with the fact that legumes utilize large quantities of both calcium and sulfur. Plants that have a long growing season are usually capable of securing more phosphorus from phosphate rock than are plants that have a short growing season. It might be expected that the character and extent of the root systems of plants would be correlated with their ability to secure nutrients from phosphate rock, but such a relationship has not been found. Plants that are recognized as good foragers, such as millet and sweet clover, generally utilize phosphate

Table 43

RELATIVE CAPACITIES OF PLANTS TO FEED ON PHOSPHATE ROCK AS COMPARED WITH GROWTH PRODUCED WHEN THE PLANTS WERE ALLOWED TO FEED ON SUPERPHOSPHATE

With low feeding power:		*With high feeding power:*	
Plant	*Relative Growth*	*Plant*	*Relative Growth*
Red clover	33.3	Rape	54.1
Wheat	34.4	Alfalfa	62.3
Oats	41.5	Rye	66.9
Corn	41.9	Buckwheat	72.1
Timothy	45.2	Red top	72.3
Soybeans	47.7	Red sorrel	82.9
		Sweet clover	83.0

rock to a better advantage than poor foragers, such as cabbage, root crops, and cotton. Buckwheat, turnips, red top and sweet clover possess a very strong ability to use phosphate rock, while corn and oats have comparatively weak powers in this direction. Because of their ability to utilize the phosphorus of phosphate rock, it may often be practical to grow and turn under such crops as buckwheat, rape, or sweet clover as green-manuring crops after an application of phosphate rock. In this way the phosphorus of phosphate rock may be made more quickly available for succeeding crops which otherwise could not utilize phosphate rock so readily.

Bauer (1921) has determined the relative capacities of a number of plants to feed upon phosphate rock and superphosphate. His results are given in Table 43, and are expressed as percentage of growth produced, as compared with the growth produced following the use of superphosphate.

Influence of Liming on the Availability of Phosphate Rock

The application of lime to a soil to which phosphate rock has been applied retards the processes by which the phosphate rock is converted into a more available form. This has been noted by Scarseth and Tidmore (1934), Salter and Barnes (1935) and others. Conner and Adams (1926) are of the opinion that this may be accomplished either by the bringing about of a reversion of the soluble forms of phosphate as soon as they can be produced in the soil, or by a more rapid absorption of the carbon dioxide of the soil air by the lime than by the floats, thus reducing the amount of carbonic acid in the soil solution. The latter influence would take place quickly only if the lime used was freshly burned or hydrated. If the lime is applied in the carbonate form, the absorption of the carbon dioxide takes place more slowly. Nevertheless, the general opinion is that heavy lining should be avoided when insoluble phosphates are used.

Some investigators have reasoned that as liming increases nitrification and results in the production of increased quantities of carbonic and nitric acids, it should hasten the production of available phosphoric acid. Russell (1932), however, thinks it is doubtful whether insoluble phosphates are brought into solution as a result of nitrification when calcium carbonate is present. McGeorge (1935) has reported that the reduction of the pH of irrigation water by the addition of small amounts of sulfuric acid greatly increased the absorption of phosphorus by plants from alkaline calcareous soils.

Influence of Acidity and Composition of Cell Sap on the Availability of Phosphate Rock

Some soil workers have suggested that the ability of plants to utilize insoluble phosphates may depend upon the internal acidity of the plant. Yet hydrogen-ion concentrations of plant juices have been determined by Bauer (1923) and others, and no relationship has been found to exist between the acidity of the cell sap of the plant and the availability of the phosphate rock. Buckwheat and sweet clover, which are generally conceded to be strong feeders on phosphate rock, vary widely as to the hydrogen-ion concentrations of their cell sap. The acidity of the cell sap of buckwheat varies from pH 1 to pH 5, while the acidity of the cell sap of sweet clover varies from pH 6 to pH 7.

Truog (1916), of the Wisconsin Experiment Station, has presented some data indicating that plants containing a relatively high calcium oxide content have a relatively high feeding power for the phosphorus in phosphate rock and vice versa.

Neller (1935) found that crops with a sap relatively high in phosphorus, when grown in soils receiving no treatment with a phosphate carrier, responded unfavorably, if at all, to additions of phosphorus.

Considerable work of fundamental importance bearing upon the foraging power of plants remains to be done. Bauer (1923), of the Illinois Experiment Station, believes that when this has been done phosphate rock will be more highly regarded as an economic source of phosphorus for crop production than it is at present in many quarters.

Influence of Accompanying Salts on the Availability of Ground Phosphate Rock

Many investigators have attempted to determine the influence of accompanying chemical salts, particularly those that are used as fertilizers, on the availability of phosphate rock. As stated before, Conner and Adams (1926) have shown that the presence of calcium carbonate decreases the availability of phosphate rock. On the other hand, ammonium salts and chlorides, especially muriate of potash, have been found to increase the availability of phosphate rock. Bauer (1930) found that ammonium nitrate had a very marked influence on the availability of phosphate rock for corn, whereas sodium nitrate had no appreciable influence. Bartholomew (1933) was able to secure an increase in the availability of phosphate rock through concomitant application of magnesium sulfate.

Availability of Phosphate Rock When Mixed with Manure

Although it is generally thought that the mixing of phosphate rock with manure helps to render the phosphate rock somewhat more available to plants, the evidence for such a conclusion is somewhat confusing. Hartwell and Pember (1912), as well as Tottingham and Hoffman (1913), and Brown and Warner (1917), in this country, and Th. Remy in Germany, have reported experiments indicating that the efficiency of phosphate rock is not materially increased by mixing it with stable manure. On the other hand, Hopkins (1910), of the Illinois Experiment Station, quotes the Ohio experiments at Wooster to prove the advantages of mixing phosphate rock with manure. However, Bray (1937), of the Illinois station, believes that the beneficial effects which have been reported are derived from the separate effects of two sources of phosphorus, rather than from the association of the phosphate rock with the manure.

Value of Phosphate Rock in the Reinforcing of Manure

The unbalanced character of stable manure as a complete fertilizer has long been recognized. Stable manure carries too much nitrogen in proportion to its phosphoric acid content. The addition of phosphates to manure aids in balancing its fertilizing properties, as has been shown by the well-known Ohio experiments. In the following table is given an average of 26-years' results secured at the Ohio Experiment Station (Wooster), showing the influence of this practice, as reported by Williams (1924).

Table 44

COMPARATIVE EFFECTS OF PHOSPHATE ROCK AND SUPERPHOSPHATE WHEN APPLIED WITH MANURE

Treatment	*26-year average acre yields:*		
	Corn (bu.)	*Wheat (bu.)*	*Clover (cwt.)*
Superphosphate and manure	67.7	28.5	48.97
Phosphate rock and manure	67.4	27.3	46.75
Gypsum and manure	63.6	25.3	40.71
Manure alone	61.8	23.8	41.51
Unfertilized	36.7	14.1	29.37

In this experiment phosphate rock and superphosphate were mixed with manure and applied on red clover sod, which was plowed under for corn in a rotation with wheat and red clover. Manure was applied at the rate of 8 tons per acre and the phosphate rock, superphosphate, and gypsum applied at the rate of 40 pounds per ton of manure. As superphosphate contains gypsum, in addition to calcium phosphates, plats were included to which gypsum and manure were added.

Availability of Phosphate Rock When Composted with Soil, Manure, and Sulfur

Lipman and McLean (1918), of the New Jersey Experiment Station, have shown that under farm conditions ground phosphate rock can be converted, in part, to a more available form by composting with manure and sulfur. In such a compost the sulfur-oxidizing organisms convert the sulfur to sulfuric acid, which in turn acts upon the phosphate rock and converts a part of it into an available form. In Table 45 are given some results of composting as secured by Lipman and McLean.

Table 45

INFLUENCE OF COMPOSTING PHOSPHATE ROCK WITH SOIL, MANURE, AND SULFUR

Composition of Compost	*Total Phosphoric Acid Rendered Available in 44 Wks., Per Cent*
3400 lbs. soil 400 lbs. floats 200 lbs. sulfur	75.87
2800 lbs. soil 800 lbs. floats 400 lbs. sulfur	39.65
3200 lbs. soil 200 lbs. manure 400 lbs. floats 200 lbs. sulfur	75.87
2600 lbs. soil 200 lbs. manure 800 lbs. floats 400 lbs. sulfur	69.16
3400 lbs. soil 200 lbs. manure 400 lbs. floats	4.09

At the New Jersey Experiment Station also workers have shown that the amount of phosphate rock rendered available can be considerably increased by inoculating the compost with active strains of sulfofying organisms.

Noll, Irvin, and Gardner (1935) have reported 12-year results from the Pennsylvania Jordan plats showing that increased yields of crops resulted from an application of gypsum or flowers of sulfur to the plats treated with phosphate rock.

Application of Ground Phosphate Rock

The Illinois Experiment Station has probably had more practical experience with the use of phosphate rock than any other agricultural experiment station in the world. Because of the cheapness of phosphate rock as compared with other carriers of phosphorus, the Illinois Experiment Station has conducted a very active propaganda in its favor. They recommend that 1000 to 2000 pounds of the phosphate rock be applied once every three or four years, or once during a rotation until the phosphoric acid content of the soil has been raised to about 5,000 pounds per acre. It should be maintained at this level. They also recommend that the phosphate rock be ground to such a fineness that it will pass through a sieve having 100 meshes to the linear inch, or 10,000 meshes to the square inch. They advise that the phosphate rock be broadcast and preferably turned under with a green-manuring crop. If lime is to be applied for the same crop, they recommend that it be broadcast and harrowed into the surface layers of the soil after plowing under the phosphate rock.

In general, ground phosphate rock should not be applied in small annual applications. It may be applied either with a lime spreader or through the fertilizing attachment of a grain drill. Mooers (1929), of the

FIG. 58. Influence of ground phosphate rock on the growth of wheat. (Courtesy Illinois Agricultural Experiment Station.)

Tennessee Experiment Station, recommends that after ground phosphate rock has been applied for two rotations the rate of further applications may be reduced by half. He thinks that when the price of superphosphate is twice as much per ton as the price of phosphate rock more farmers might well consider the use of phosphate rock. This station has reported good results from the use of ground phosphate rock the first year following application. These results are somewhat contrary to the results secured by many of the other southern experiment stations.

It is very probable that a close study of the economics of fertilizer application in most localities of the United States will reveal the fact that under average conditions it is more profitable to apply a superphosphate than phosphate rock. Thorne (1919), of the Ohio Experiment Station, in summarizing the 25-year results of the Wooster Experiment Stations hows that finely ground phosphate rock returns a profit on its cost, but that the effectiveness of superphosphate is so much greater as to make the use of the phosphate rock relatively unprofitable. Nevertheless, as the relative availability of the phosphate rock is influenced by a number of factors, the data at hand do not justify one in concluding that superphosphate is the more economical for use under all conditions.

While applications of phosphate rock may aid materially in building up the phosphorus reserve of the soil, it does not act quickly, and immediate results should not be looked for. Furthermore, the criterion of good fertilization should not be the phosphorus content of the soil but rather the response of the crop. When a comparison of the value of phosphates is made by the farmer, it should always be made on the basis of equal money values.

Consumption of Ground Phosphate Rock

Regardless of its high content of phosphoric acid, ground phosphate rock or floats has not been used extensively as a fertilizer in any section of Europe or America, with the exception of the Corn Belt states of the United States, and of Illinois in particular. In 1939 only 86,131 tons of phosphate rock were used for direct application to the soils of the United States and two-thirds of this came from the Tennessee Brown-rock deposits. Over half of this tonnage was used in the state of Illinois. Although ground phosphate rock was shown, many years ago, to have some fertilizing value when applied to some crops when grown on certain soils, yet in most of the fertilizer using sections of the United States its action has been found to be too slow for the most profitable production of the common farm crops as grown under average conditions.

8

Sources and Uses of Bone Phosphate and Basic Slag

Bones are known to have been employed as fertilizer in England as early as 1653. In fact, their value as a fertilizer material appears to have been recognized in England much earlier than in any other country.

Bone fertilizers have always been held in higher esteem than phosphate rock. This might be explained in part by the fact that fertilizer materials of plant or animal source have, for a long time, been considered to be superior to those of an inorganic source.

Sources of Bone

During the latter part of the nineteenth century deposits of bone were sought, particularly by the English, and collected from all parts of the world for use as fertilizer. Justus von Liebig has stated that even such battlefields as Leipzig, Waterloo, and the Crimea, as well as the catacombs of Sicily served for a time as sources of supply. In the early days in this country large quantities of buffalo bones were collected from the western plains for fertilizer purposes. At present the principal source of supply of bones is the abattoirs, although small amounts of waste bone are secured from the manufacture of bone buttons, knife handles, and similar articles.

Considerable quantities of bones are shipped to Europe from the Argentine and India. As virulent anthrax organisms have sometimes been discovered on old bones there was formerly a certain amount of prejudice in some sections against the use of raw bonemeal. Many countries require a certificate of sterilization before cargoes of bone may be imported.

Chemical Composition of Bone

Bone is made up largely of calcium phosphate scattered through an organic matrix. The formula generally ascribed to the phosphate found in bone is $Ca_3(PO_4)_2$. This phosphate is commonly referred to in the fertilizer trade as B. P. L., or bone phosphate of lime. Hendricks and Hill

(1942), however, have presented evidence that seems to indicate that the phosphate found in bones may be a hydrated tricalcium phosphate type of compound containing sodium, magnesium, and carbonate as essential constituents. Apparently it is an apatite whose fluorine has been replaced by carbonate and whose calcium has been partially replaced by magnesium and sodium. Klemet (1938) showed that while the apatitelike unit structure of tricalcium phosphate hydrate is $(Ca_9)(PO_4)_6(H_2O)_2$ that of human bones is $(Ca_{8.50}Mg_{0.25}Na_{0.19})[(PO_4)_{5.07}(CO_3)_{1.24}](H_2O)_{2.0}$. Nevertheless, many investigators hold that the phosphate compound of bone is a hydroxyapatite. The organic matrix of bone is made up largely of fatty substances, but it also contains small amounts of nitrogenous compounds. The amount of phosphoric acid and nitrogen contained in bone depends upon the kind and the age of the bone. As a general rule young bones contain less phosphoric acid and more nitrogen than older bones. Fluorine may also be present and may be present in large amounts if the animal ingested large quantities of fluorine.

Raw Bonemeal

Green bones are ground and sold under the name of raw bonemeal. The fatty materials found in raw bonemeal tend to delay the decomposition of the material when it is added to the soil. Raw bonemeal of good grade contains about 2 to 4 per cent of nitrogen, and from 22 to 25 per cent of phosphoric acid.

Steamed Bonemeal

Most of the bonemeal that is found upon the fertilizer market is steamed bonemeal. That is, the meal is made from green bones that have been boiled or steamed at high pressure for the purpose of removing the fats, or from which the fat has been extracted with naphtha or other solvents. As fat does not contain nitrogen, phosphoric acid, or potash, and as it has a commercial value, it should be removed, especially since its presence hinders the ready decay of the bonemeal after it is incorporated in the soil. During the boiling or steaming process a large amount of the fatty materials is separated from the bone, and the gelatinous materials are then removed by subjecting the bone to high steam pressure. The gelatinous material is used in the manufacture of gelatin and glue, and for other purposes. Its removal brings about a small loss in the nitrogen content of the resulting bonemeal, but an increase in the percentage of phosphoric acid.

FIG. 59. Sugar cane in Louisiana. (*Left*) Unfertilized. (*Right*) Fertilized. (Courtesy, United States Department of Agriculture.)

The steaming of bone changes its physical composition so that it may be ground more easily and put into a finer physical state for distribution in the soil. Bonemeal is sometimes sold under the name of bone dust. The name may indicate a degree of fineness, but the name is not always associated with the degree of fineness. Steamed bonemeal generally is ground more finely than raw bonemeal. The process of steaming also greatly increases the ammonium citrate solubility of the phosphate. Steamed bonemeal contains 1 to 2 per cent of nitrogen, and 22 to 30 per cent of phosphoric acid.

Steamed bonemeal may be so light in weight per volume that if it must be broadcast during windy weather it should be mixed with fine damp soil or sawdust. It is superior to the mineral phosphates in its immediate crop-producing powers. This superiority may be due to its physical characteristics and to its organic origin. Bonemeal rates well as a source of phosphoric acid. The availability of its phosphoric acid appears to be slightly lower than that of basic slag. The effectiveness of bonemeal is increased by its nitrogen content and also by its content of secondary and rarer essential elements. Best results are secured when it is applied to acid soils containing an ample supply of organic matter. Parker and Tidmore (1926) have reported that liming the soil has a depressing influence on the solubility of phosphorus in steamed bonemeal, and Schreiner, Merz, and Brown (1938) have reported that the availability of the phosphorus in bone is increased by the acid decomposition products of its organic matrix.

Precipitated Bone

Precipitated bone is a by-product produced in the manufacture of glue stock from bones. It contains 40 per cent of phosphoric acid, largely in the form of dicalcium phosphate. It is obtained by neutralizing the hydrochloric acid solution of processed bone with calcium hydroxide.

Bone Ash

Bone ash has been imported to some extent into this country from South America. It is made by burning bones for the purpose of reducing their bulk for shipment. During the burning process most of the nitrogen is driven off. The material analyzes 30 to 40 per cent of phosphoric acid, and contains a small amount of magnesium and fluorine. It is not extensively produced for, or used in, the fertilizer industry.

Bone-black or Bone Char

Bones are sometimes heated in a closed retort in a manner similar to that followed in the coking of coal. The residual charcoal is known as bone-black. Sugar refineries use considerable quantities of high-grade bone-black for the purpose of clarifying sugar. After it has served this purpose it may be used as a fertilizer. It contains about 30 to 35 per cent of phosphoric acid, 1 to 2 per cent of nitrogen, and 10 per cent of carbon. When used in mixed goods its carbon content may act as a conditioner.

Dissolved Bone

Dissolved bone or *bone superphosphate*, or what is in Europe sometimes called *vitriolated bone*, is made by treating bonemeal with sulfuric acid. The action of sulfuric acid on bonemeal is similar to that which takes place in the manufacture of the superphosphates when sulfuric acid is added to phosphate rock. The phosphorus in dissolved bone is generally conceded to have the same crop-producing value as superphosphate. It is today rarely encountered in the fertilizer trade.

Value of Bonemeals

The bonemeals, raw and steamed, appear to act more quickly when applied to well aerated soils. They are generally slower in their action when applied to heavy silts and clays. For many years previous to the present popularity of superphosphate, bonemeal, because of its supposed superior lasting qualities, or residual effect, was held in very high esteem by farmers. It was long a favorite fertilizing material to be used before planting land to grass and for other crops having a long growing season. Because of its lime content the use of bonemeal tends to reduce the acidity of the soil.

Manufacture of Basic Slag

From the standpoint of production and consumption, basic slag ranks next to superphosphate as the world's leading phosphatic fertilizer. Basic slag is secured as a by-product from the steel industry. As the iron ores of Europe carry an objectionable amount of phosphorus, whereas those of America, in general, do not, and as it is necessary that this phosphorus be removed, the quantity of basic slag produced and used in Europe is far greater than that manufactured and used in America. Unless the phosphorus of iron ore is removed during the process of steel manufacture, a high quality steel cannot be made. Steel containing over 2 per cent of phosphorus is very brittle.

Phosphorus may be removed from cast iron by a method that was introduced by Thomas and Gilchrist in 1877 and which was an improvement to the old Bessemer process of steel manufacture. In fact, it revolutionized the manufacture of steel. By this method the molten pig iron containing such impurities as calcium, silicon, sulfur, phosphorus, etc., is introduced into a limed converter and an air blast started. When the mixture reaches a certain temperature the lime melts and unites with the acid impurities, and the resulting compounds, being lighter than iron, rise to the surface of the molten mass, where the gases escape and the solids are poured off as slag. After the slag has cooled it appears as a hard, massive, black cinder. It is then broken into small pieces under stamps, mashed with heavy rollers and finally ground to a very fine powder, in which condition it is placed on the market. Large quantities of this slag, known as Bessemer basic slag, were manufactured and used in Europe before World War I. The best grade of goods sometimes contained as much as 17 to 20 per cent of phosphoric acid.

The method used in the manufacture of Bessemer basic slag meal, as described above, has been superseded by the open-hearth method. In this method the impurities are oxidized by means of the iron ore (iron oxide) itself as it is reduced simultaneously to metal. Iron ore and lime are heated in an open heater and the molten pig iron then added. From time to time additional quantities of lime and ore are introduced, and occasionally to some batches a quantity of fluorspar—about 75 pounds per 10 tons of lime—may be added to reduce the viscosity of the slag. Excessive quantities of either the lime or fluorspar tend to diminish the solubility of the resulting phosphatic slag. The addition of fluorspar brings about the production of an artificial apatite which resembles the natural fluor-apatite.

The open hearth basic slag, as basic slag made according to this process is called, contains a smaller percentage of phosphoric acid than the older Bessemer product, and carries a small amount of calcium fluoride, a compound toxic to plant growth. In addition, the fluorine is objectionable in that its presence in the soil, when the material is applied as fertilizer, may eventually bring about some reversion of the available phosphates to an unavailable form. The open-hearth slag seldom carries more than 15 per cent of phosphoric acid and often as little as 8 per cent. It contains 50 to 60 per cent of calcium and magnesium oxides.

Basic slag is extensively manufactured in Germany, France, and England, and it is the cheapest source of phosphoric acid available to the farmers of those countries. Just before World War II the world's annual consumption of basic slag was about five million tons. Small amounts of basic slag are manufactured and marketed from the Birmingham district of Alabama. The annual consumption of basic slag in the United States is a little less than 40,000 tons. Basic slag weighs more per cubic foot than any other fertilizer material. It is marketed most commonly under the name of *basic slag meal* but it was formerly also called *basic cinder* and *Thomas phosphate*. The name Thomas was given to the product in honor of the chemist who introduced the improvement to the Bessemer process. In the older literature it was often referred to as *odorless phosphate*, and *iron phosphate*.

Chemical and Physical Properties of Basic Slag

The most commonly accepted formula for the phosphatic compound found in basic slag is a double silicate and phosphate of lime ($(CaO)_5.P_2O_5.SiO_2$), but some chemists consider it a tetra calcium phosphate ($Ca_4P_2O_9$), while others believe it to be a basic siliconoxyapatite.

The name basic slag is derived from the field of metallurgy. Any flux that is added to remove impurities is called a slag. Those that are added to remove basic impurities are called acid slags and those that are added to remove acid impurities are called basic slags.

A committee of the American Association of Official Agricultural Chemists has ruled that material sold in the United States as *basic phosphate slag* shall be finely ground and that it shall contain at least 12 per cent of phosphoric acid, of which not less than 80 per cent shall be soluble in a 2 per cent citric acid solution according to the Wagner method of analysis. Basic slag that does not conform to this rule must be designated as low grade. The guarantee of the grade of *basic slag meal or open-hearth basic slag* which is produced in the Alabama district, and which is now found on the American market, is given in the following table.

Table 46

GUARANTEE OF THE GRADE OF BASIC SLAG MOST COMMONLY SOLD ON THE AMERICAN MARKET

	Per Cent
Phosphoric acid (P_2O_5)	8–10
Iron oxide (Fe_2O_3)	15–18
Calcium silicate ($CaSiO_3$)	6– 8
Magnesium oxide (MgO)	5– 7
Calcium oxide (CaO)	45–48
Manganese	2– 4
Fineness: 80 per cent through 100 mesh	

Basic slag is marketed as a heavy dark brown powder. Owing to its lime content it is very alkaline to litmus. It is very likely to solidify if it is stored in a damp place. Its phosphorus content, soluble in a 2 per cent citric acid solution, is considered available for plants. The availability of phosphoric acid in basic slag increases with the fineness of grinding and with the silicon content, and decreases as the fluorine content increases. Basic slag is not used in mixed fertilizers. Its high lime content is of such a nature that chemical introduction of mixtures with ammonium salts may bring about a liberation of ammonia. Lyon and Buckman (1922) state that when basic slag is applied to the soil its phosphorus content does not revert to a more insoluble form, but is rendered quickly available by carbon dioxide. They have suggested the following reaction as being that which takes place in the soil:

$$(CaO)_5.P_2O_5\ SiO_2 + 8CO_2 + 6H_2O = CaH_4(PO_4)_2 + 4CaH_2(CO_3)_2 + SiO_2$$

Artificial Basic Slag

Many modifications of the common method of producing basic slag have been suggested. The addition of ground phosphate rock to iron ore for the purpose of increasing the phosphorus content of the resulting slag has been attempted, as well as the treating of the molten slag with phosphate rock, which would accomplish the same purpose. These artificial slags have been marketed under various trade names, such as "Rhenania-Phosphate" in Germany, "Base-Phosphate" and "Supra-Phosphate" in Belgium, and "Thermo-Phosphate" in Russia. Nevertheless, the development of these processes has made little progress for this means of making the phosphorus of phosphate rock partially available is not as economical as that of treating the rock with sulfuric acid.

Fertilizing Value of Basic Slag

The fertilizing value of any grade of basic slag is influenced by the fineness to which the product has been ground. It is usually ground to such a fineness that 80 per cent or more will pass through a sieve of 10,000 holes to the square inch. Crop response to basic slag is usually greater than that to ground phosphate rock. Because of the lime content of basic slag excellent results are often secured when basic slag is applied in large quantities to acid soils for lime-loving legumes. On the other hand, the application of basic slag or bonemeal to Irish potatoes may make it more difficult to control the potato scab. The silica content of basic slag may be of some value in helping to reduce the crop's demand for phosphorus, for silicon has been shown to be somewhat effective in this respect. Basic slag also carries varying quantities of magnesium and manganese and it may be used for the purpose of correcting soil deficiencies in these elements. For nutrient crop deficiencies of doubtful origin basic slag is one of the best shotgun prescriptions known, and applications of this fertilizer will often return a crop to a healthy condition. When the use of basic slag is justified the South Carolina Experiment Station recommends from 400 to 1500 pounds per acre.

The relative effectiveness of basic slag, bonemeal, and superphosphate, as determined by Williams (1924) at the Ohio Experiment Station, after 20 years of experimentation at Wooster, Ohio, is shown in the accompanying table. The rotation followed in these experiments was corn, oats,

Table 47

EFFECT OF FERTILIZERS CONTAINING DIFFERENT CARRIERS OF PHOSPHORUS, 40-YEAR AVERAGE—ANNUAL VALUE OF INCREASE PER ACRE

Phosphorus Carrier	*Unlimed Land*	*Limed Land**
Superphosphate	$14.32	$14.06
Steamed bonemeal	11.82	11.52
Basic slag	13.86	11.20

* In excess of increase for lime.

wheat, clover, and timothy. The superphosphate used analyzed 14 per cent of phosphoric acid and was applied in three applications totaling 320 pounds per acre per rotation. The bonemeal and basic slag were applied in amounts equivalent to that of the superphosphate. Nitrogen and potash were applied in addition to phosphoric acid. According to these experiments superphosphate was more effective than basic slag or steamed bonemeal but Williams concluded that on acid soils the lime of basic slag may bring up the effectiveness of basic slag to that of superphosphate.

Results obtained at the Kentucky Experiment Station indicate that basic slag is 85 to 100 per cent as effective as superphosphate. The relative effectiveness of different phosphates, including basic slag, as determined at the Indiana Experiment Station by Reed and Hall (1930), and Wiancko (1928), are given in Table 48. All plats were manured and limed. Superphosphate, basic slag, and steamed bonemeal were applied at the rate of 48 pounds of phosphoric acid per acre, and phosphate rock at the rate of 192 pounds of phosphoric acid per acre, with manure for each cover crop. The rotation consisted of corn, soybeans, wheat and clover.

Table 48

RELATIVE EFFECTIVENESS OF DIFFERENT PHOSPHATES

Source of Phosphorus	*Cost of Fertilizer*	*Net Returns per Acre per Rotation*
Superphosphate	$3.36	$13.53
Steamed bonemeal	3.36	9.20
Basic slag	3.36	13.13
Phosphate rock	4.80	7.10
No phosphorus		

As compared with superphosphate, Salter and Barnes (1935) have reported basic slag to have an efficiency of about 85 per cent for cereals when grown on unlimed land but under the same conditions it was 40 per cent superior to superphosphate when applied to clover. On the other hand, Noll, Irwin, and Gardner (1935) have reported that 12-year results at the Pennsylvania Experiment Station show the relative value of phosphate carriers to be, in descending order, basic slag, bonemeal, superphosphate, and phosphate rock.

9

Manufacture and Use of the Superphosphates

The term *phosphate* is correctly applied to fertilizer materials that contain phosphorus, regardless of its availability to plants. *Superphosphate*, on the other hand, is a term generally used in referring to phosphates, the phosphorus of which is in a form readily available to plants. In this country rather recently the name *superphosphate* has been substituted for the name *acid phosphate* which came into common use as early as 1877 and which at that time replaced the name superphosphate.

At the meetings of the First World International Congress on Chemical Fertilizers, held October 3 to 6, 1938, a resolution was passed recommending to all countries that they adopt the denomination "superphosphate" for all superphosphates containing up to 25 per cent of phosphoric acid and the denomination "concentrated superphosphate" for all superphosphates containing more than 25 per cent of phosphoric acid.

In some localities of the Middle West, farmers incorrectly use the term *phosphate* as synonomous with the term *commercial fertilizer*.

Meaning of the Term Phosphoric Acid

Chemically, the term *phosphoric acid* refers to a compound having the formula H_3PO_4. In fertilizer practice however, the term is used to refer to phosphoric anhydride or phosphorus pentoxide, which has the chemical formula of P_2O_5. Unfortunately, the phosphorus of fertilizer materials is always reported by the chemist in terms of phosphorus pentoxide, although free pentoxide of phosphorus is never found as such in fertilizers. It would be preferable and a step in the right direction if state laws would require that the phosphorus of fertilizers be reported on the elemental basis.

Table 49

COMPOSITION OF PHOSPHATIC MATERIALS USED IN THE MANUFACTURE OF SUPERPHOSPHATE

Phosphatic Material	Location	Nitrogen (N) (per cent)	Silica or Insoluble (SiO_2) (per cent)	Oxide of Iron and Aluminum (Fe_2O_3), (Al_2O_3) (per cent)	Calcium Fluoride (CaF_2) (per cent)	Carbonate of Lime ($CaCO_3$) (per cent)	Phosphoric Acid (P_2O_5) (per cent)	Tricalcium Phosphate* $Ca_3(PO_4)_2$ (per cent)
Bone ash	..	4.06	3.01	..	..	5.28	35.38	76.65
Fat-extracted bone	..	4.14	3.60	..	..	5.24	21.68	46.50
Bone or animal charcoal	..	1.12	2.00	0.37	1.00	8.00	31.99	73.10
Apatite:								
High grade	Canada	..	3.67	.70	3.10	4.13	38.60	88.20
Lower grade	Canada	..	8.92	1.03	3.04	8.05	34.42	78.65
Apatite (high grade)	Norway	..	3.62	1.37	2.62	.29	39.44	86.10
Bat guano	Porto Rico	8.35	..	1.59	..	..	7.57	16.52
Guano in crusts	West Indies	.45	..	5.88	..	..	24.36	53.18
Phospho-guano	Mexico	.40	..	1.14	..	..	39.70	86.66
Phosphorite:								
High grade	Somme, France	..	.79	1.39	..	9.17	35.59	77.69
Lower grade	Somme, France	..	8.55	4.07	..	6.95	29.10	63.53
Lower grade	Liége, Belgium	..	16.14	2.39	5.00	7.07	27.20	62.15
Coprolites	Cambridge, Eng.	..	9.00	..	..	2.30	34.00	77.70
Coprolites	Suffolk, Eng.	..	Not det.	9.00	6.00	10.00	23.63	54.00
Phosphate nodules	Russia	..	Not det.	5.00	6.98	12.23	32.82	75.00
Phosphate nodules	South Carolina	..	4–12	1–4	2–8	11–25	25–28	57–64
Hard rock phosphate	Florida	..	4.13	3.00	4.40	3.63	36.39	83.14

* Known to the trade as bone phosphate of lime (b. p. l.).

Table 49

COMPOSITION OF PHOSPHATIC MATERIALS USED IN THE MANUFACTURE OF SUPERPHOSPHATE—(*Continued*)

Phosphatic Material	*Location*	*Nitrogen* (N) (*per cent*)	*Silica or Insoluble* (SiO_2) (*per cent*)	*Oxide of Iron and Aluminum* (Fe_2O_3), (Al_2O_3) (*per cent*)	*Calcium Fluoride* (CaF_2) (*per cent*)	*Carbonate of Lime* ($CaCO_3$) (*per cent*)	*Phosphoric Acid* (P_2O_5) (*per cent*)	*Tricalcium Phosphate** $Ca_3(PO_4)_2$ (*per cent*)
Land pebble phosphate.......	Florida	..	5–10	1–4	0–3	2–5	30–34	69–78
Brown rock phosphate.......	Tennessee	..	2.5–10	3–8	0–5	0–10	30–38	69–87
Blue rock phosphate..........	Tennessee	..	2.5–70	2.5–7	0–3	0–2	27–32	62–73
White rock phosphate........	Tennessee	..	2–7.5	1.5–3.5	..	..	32–38	73–87
Black rock phosphate........	Arkansas	..	15–40	2.5–9.5	..	..	23–28	53–64
Brown rock phosphate.......	Kentucky	..	2–5	2. –5	..	..	30–35	69–80
Oolitic black phosphate......	Utah, Idaho, etc.	..	1.8–10	.07–1.6	.8–1.35	3.8–13.6	27–36.5	62–83
White phosphate.............	Ocean Islands	..	Not det.	.42	1.0	4.91	38.73	84.65

* Known to the trade as bone phosphate of lime (b. p. l.).

Composition of Native Phosphates Used in the Manufacture of Superphosphates

Superphosphates are generally manufactured from phosphate rock, but bone and phosphatic guanos were extensively used in the early days. At present only a few fertilizer plants make superphosphate from bones, and in such instances only on special order. Superphosphates that are made from organic sources are more easily manufactured and usually have a better physical condition, but they contain a smaller percentage of phosphoric acid than those made from mineral sources.

The phosphatic materials used in the manufacture of superphosphate vary rather widely in their composition. Rarely is any material used that does not contain at least the equivalent of 50 per cent of tricalcium phosphate. The impurities commonly found in the native phosphates are organic matter, calcium fluoride, calcium and magnesium carbonates, and iron, silicon and aluminum oxides. The combined content of iron and aluminum oxides in the phosphate rock marketed in the United States ranges between 1.0 and 5.5 per cent, although some Tennessee Brown rock that is marketed for direct application to the soil may contain as much as 8 to 9 per cent. Native phosphates that contain above 4 per cent of iron oxide are considered very objectionable for the manufacture of superphosphate. In fact, it is desired that they should not contain over 2 per cent. The objection is based upon the fact that greater quantities of sulfuric acid must be added to the phosphate rock when iron oxide is present, and to the fact that superphosphate made from phosphate rock high in aluminum and iron has a poor physical condition. The presence of other iron compounds, such as pyrites, and aluminum silicates and the various fluorides, is objectionable also. Jacobs *et al.* (1933) found that most of the phosphate rock on the American market carried 3 to 4 per cent of fluorine. They think the fluorine originated from contact of the phosphate rock with fluorine-bearing waters.

In Table 49 is given, according to Waggerman (1914), the composition of some phosphatic materials used in the manufacture of superphosphate.

Methods Employed in the Conversion of the Insoluble Phosphorus of Phosphate Rock into a Soluble Form

Many methods have been suggested, patented, and employed to convert the insoluble phosphorus of native materials to a soluble condition that can be used by plants. The fluorapatite structure of many phosphates

FIG. 60. Modern sulphuric acid and superphosphate plant of the Davison Chemical Company, Baltimore, Maryland. (Courtesy, *The Fertilizer Review.*)

that accounts for their unavailability may be broken either by removing the fluorine entirely or by breaking the structure chemically without effecting substantial removal of the fluorine. The most important methods of accomplishing this may be classified roughly as follows: First, acid treatments; second, volatilization of the phosphorus; third, decomposition by means of an alkali or silicate; and fourth, processes used in the steel industry. The fourth process has already been discussed in Chapter 8. Many other processes have been worked out by which to obtain available phosphoric acid from phosphate rock, but few offer possibilities greater than the processes now generally used.

Treatment of Native Phosphates with Acid

When any of the native phosphates are treated with an acid, the relatively unavailable calcium phosphate which they carry is converted into the dicalcium and monocalcium phosphates and in the case of the mineral phosphates the insoluble fluorides are separated from chemical union with the phosphorus. The fluorine reacts with silicate impurities to form silicon tetrafluoride and at some plants may be drawn away from the mixture by ventilating ducts. The amounts of dicalcium and monocalcium phosphates produced will depend upon the quantity of acid that is employed. Nitric, hydrochloric, phosphoric, and sulfuric acids have been used for this purpose, but sulfuric acid has proved to be the most satisfactory, because when sulfuric acid reacts with fluor-apatite, calcium sulfate, a dehydrating agent, is produced. This compound greatly improves the physical condition of the superphosphate, as it causes the product to dry out readily and become pulverable. As usually made, the fluorine remains in the mixture.

The first superphosphate made in the United States was manufactured in Baltimore about 1850. Until 1867, when South Carolina phosphate was first used, all superphosphate was made from bone and imported phosphoguanos.

In former years the process of superphosphate manufacture consisted simply of mixing about equal quantities of ground phosphate rock with sulfuric acid and then breaking up the resulting mixture after it had set. Today, at nearly all plants, an analysis is first made to determine the percentage of impurities in the materials to be treated. Allowance is then made for the impurities, and the correct amount of acid necessary to secure the desired product is added. Ordinary, or normal, superphosphate is usually made to carry about 16 to 30 per cent of phosphoric acid.

The cheaper grades of sulfuric acid, that is, those having a specific gravity of from 1.5 to 1.6 or 52° to 55° Baumé, are nearly always used in the manufacture of superphosphate. They contain about 60 to 70 per cent of sulfuric acid.

The required amount of acid and ground phosphate rock is introduced into a mixer, thoroughly mixed, and dropped into a pit or den. Here most of the reactions within the mixture take place within a few hours, but the material is generally allowed to remain in the den for from 24 to 36 hours to cure, although at some plants it is preferred that it be removed from the den the same day it is mixed. A very high temperature is developed within the den, in some cases exceeding 100° C. During the curing process a shrinkage of at least 10 per cent takes place owing to the volatilization of water, carbon dioxide, and fluorine compounds. The calcium sulfate that is produced acts as a drying agent and within a few days the material is ready to be broken up and dried further by heat, air currents, or by dusting with finely ground phosphate rock. It is then the usual custom to store the superphosphate from 2 weeks to several months to allow it to dry out to a point where it will have little tendency to "set up." In this way, it acquires a physical condition that will not clog the fertilizer drills when it is being sown. In some instances limestone, hydrated lime, peat or muck, low-grade tankage or other organic waste product, is added for the purpose of reducing lumping. Too much acid must not be added to the phosphate rock or free phosphoric acid may be produced. In such cases the product becomes moist, difficult to handle, and the free acid is destructive to shipping bags. The reactions that take place when the ground phosphate rock and sulfuric acid are mixed are probably as shown below:

$$3Ca_3(PO_4)_2 + 6H_2SO_4 + 2H_2O = 4H_3PO_4 + Ca_3(PO_4)_2 + 6CaSO_4.2H_2O$$

$$Ca_3(PO_4)_2 + 4H_3PO_4 + H_2O = 3CaH_4(PO_4)_2 + H_2O$$

$$Ca_3(PO_4)_2 + H_2SO_4 + 2H_2O = Ca_2H_2(PO_4)_2 + CaSO_4.2H_2O$$

Sulfuric acid probably acts only on about two-thirds of the phosphate rock, and the free phosphoric acid, produced in the reaction, on the remainder. It appears that because phosphoric acid is a much weaker acid than sulfuric acid it acts more slowly.

As the calcium sulfate is not removed from the final product, the percentage of phosphoric acid in the superphosphate is only about half that found in phosphate rock. Chemically, the tricalcium phosphate is the normal salt, while the monocalcium and dicalcium phosphates are acid salts.

When superphosphate is made with nitric acid or hydrochloric acid no free phosphoric acid is obtained with the quantities of acids usually added, but the resulting calcium nitrate or calcium chloride is hygroscopic and

FIG. 61. View taken at Wilson Dam, Alabama, showing loading of superphosphate from storage into trucks. (Courtesy, Tennessee Valley Authority.)

will impart to the superphosphate poor physical properties. Calcium chloride is known to be toxic to some plants when applied in appreciable quantities. Nitric and hydrochloric acids are more costly than sulfuric acid.

In recent years the spent acids from the refining of petroleum products by the alkylation process have been used satisfactorily in the production of alkylation-acid superphosphate, and this superphosphate has been found by Jacob and Arninger (1944), and others, to be about of the same value as ordinary superphosphate.

Manufacture of Phosphoric Acid by the Wet Method

As already stated, when excess quantities of sulfuric acid are mixed with phosphate rock free phosphoric acid is produced. This procedure is employed in the production of free phosphoric acid, and is known as the "wet method." Ground phosphate rock is treated with an excess of sulfuric acid and the resulting free phosphoric acid separated by decantation and filtration. The phosphoric acid produced in this way usually contains 25 to 28 per cent of phosphorus pentoxide.

Manufacture of Phosphoric Acid by the Furnace Process

The furnace process of producing free phosphoric acid has great possibilities, but has not been sufficiently developed to come into general use nor to supersede the wet method. Briefly, the furnace process consists of mixing sand, coke, and phosphate rock, and submitting them to high temperatures. When the mixture is so heated the phosphorus of the phosphate rock is replaced by silicon to form calcium silicate, elemental phosphorus being volatilized. The calcium silicate is removed as a molten slag, and the phosphorus is oxidized to phosphorus pentoxide, combined with water, and collected as free phosphoric acid. The reactions involved are about as follows:

$$Ca_3(PO_4)_2 + 3SiO_2 = 3CaSiO_3 + P_2O_5$$
$$P_2O_5 + 3H_2O = 2H_3PO_4$$

Phosphoric acid produced in this way is comparatively pure, analyzing 75 to 90 per cent, and can be used for purposes other than the manufacture of fertilizers.

Both electric and fuel furnaces are employed in the manufacture of free phosphoric acid. When the electric furnace is used the heat required is not

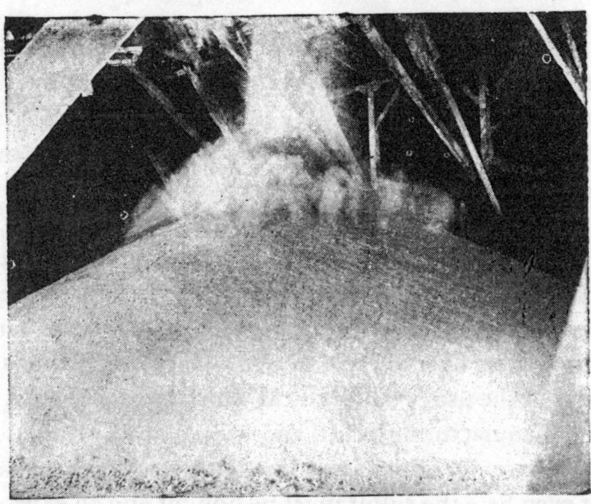

FIG. 62. Discharging three-hour-old superphosphate to storage heap. (Courtesy, Chemical Appliances, Inc.)

obtained directly by the consumption of fuel, so that the gases evolved are relatively concentrated and the phosphoric acid is more easily recovered. When the fuel furnace is used the great quantities of air introduced into the system necessarily produce a large volume of gas which is low in its phosphorus content and therefore makes the subsequent recovery of phosphoric acid more difficult. At present very little phosphoric acid that is produced by the electric furnace process is used in commercial fertilizer manufacture, and less than 1 per cent of all the phosphoric acid used in agriculture in America is produced by the fuel-fired blast furnace.

FIG. 63. Pasture in Vermont. (*Left*) Plat unfertilized. (*Right*) Plat received a complete fertilizer and lime. (Courtesy, National Fertilizer Association.)

Experiments conducted by the Bureau of Chemistry and Soils, of the United States Department of Agriculture, and reported by Royster *et al.* (1937) have shown that the furnace process can be used satisfactorily even with the run-of-mine phosphates. Its employment for this purpose should render a great saving of the phosphoric acid which is usually wasted in the mechanical separation of phosphate rock from its impurities. This waste is considerable, for it has been estimated that in the mining of some phosphate deposits more than a ton of phosphate is unrecovered for every ton that is separated from its impurities. Another saving that the furnace process offers over the acid or wet method of free phosphoric acid manufacture is in the use of sulfuric acid which generally must be bought, hauled to the manufacturing plant, and hauled away again as gypsum in the superphosphate. Of course, at some fertilizer plants the sulfuric acid is made, but even here the sulfur must be bought and an expense incurred in the making of the sulfuric acid. The furnace process also allows the production, without subsequent evaporation, of free phosphoric acid of almost any desired strength. The retail sales of this material have been chiefly in the north central and western states.

In Table 50 is given the location of American concentrated superphosphate, phosphoric acid, and phosphorus manufacturing plants.

Table 50

LOCATION OF AMERICAN CONCENTRATED SUPERPHOSPHATE, PHOSPHORIC ACID, AND PHOSPHORUS PLANTS

Company	*Location*
CONCENTRATED SUPERPHOSPHATE PLANTS	
Anaconda Copper Mining Company	Anaconda, Montana
Armour Fertilizer Works	Columbia, Tennessee
International Agricultural Corporation	Wales, Tennessee
Phosphate Mining Company, The	Nichols, Florida
U.S. Phosphoric Products Corporation	Tampa, Florida
Victor Chemical Works	Nashville, Tennessee
Virginia-Carolina Chemical Corporation	Charleston, South Carolina
Tennessee Valley Authority	Muscle Shoals, Alabama
Consolidated Mining and Smelting Corporation	Trail, British Columbia
PHOSPHORUS OR PHOSPHORIC ACID PLANTS	
American Agricultural Chemical Company, The	South Amboy, New Jersey
American Cyanamid Company	Warners, New Jersey
Monsanto Chemical Company	Anniston, Alabama
Monsanto Chemical Company	Columbia, Tennessee
Oldbury Electro-Chemical Company	Niagara Falls, New York
Victor Chemical Works	Columbia, Tennessee
Virginia-Carolina Chemical Corporation	Charleston, South Carolina

Calcium Metaphosphate

Calcium metaphosphate ($Ca(PO_3)_2$), a salt of metaphosphoric acid and a quenched melt, is one of the products that have recently been developed by the Tennessee Valley Authority. It is a defluorinated phosphate and to make it phosphorus-bearing furnace gases are burned and passed directly into contact with phosphate rock at a temperature of 1200° C. The hot vapors and the phosphate rock unite to form liquid calcium metaphosphate which upon cooling solidifies into a slag. The resulting slag is ground and marketed as such. It contains about 65 per cent of available phosphoric acid and 0.3 per cent fluorine, and appears to be a good source of phosphorus for plants. It is commonly referred to as *metaphos*. Alway and Nesom (1944) concluded from a study made in Missouri, that for alfalfa *metaphos* should be as good a source of phosphorus as superphosphate.

Potassium Metaphosphate

Potassium metaphosphate ($K_4(PO_3)_4.H_2O$) is obtained by reaction of phosphoric acid and potassium chloride at high temperatures. It is only slightly soluble in water but is readily soluble in an ammonium citrate solution, and is a good source of potassium and phosphorus for plants.

Potassium metaphosphate is of interest also because it contains 100 per cent of plant nutrients. It analyses 39.9 per cent of K_20 and 60.1 per cent of P_2O_5.

Serpentine Superphosphate

In New Zealand, during the years of World War II superphosphate was diluted with serpentine. It was claimed that a higher recovery of the phosphorus of this product by plants could be secured than from ordinary superphosphate.

Manufacture of Concentrated Superphosphate

When phosphate rock is treated with free phosphoric acid a superphosphate is produced which may contain as high as 50 per cent of phosphorus pentoxide and generally contains 45 to 48 per cent. It also contains 17 to 22 per cent CaO and 2 to 6 per cent SO_3. Concentrated superphosphate is composed largely of monocalcium phosphate. It may be referred to in the trade as *double superphosphate*, *acid phosphate of lime*, *one-lime phosphate*, *treble superphosphate* and sometimes incorrectly as triple superphosphate. The Anaconda Mining Company sells its superphosphate under the trade name "Treble Superphosphate." The term double superphosphate is derived from the fact that the product is usually secured by double acidulation, sulfuric acid first being used in the production of free phosphoric acid, and then the free phosphoric acid in the production of superphosphate.

Nine concentrated superphosphate plants are now in operation in this country. These plants are able to use the low-grade phosphates—which have little commercial value for the manufacture of ordinary superphosphate—as a source of free phosphoric acid from which concentrated superphosphate can be produced. Because concentrated superphosphate often contains considerable quantities of free phosphoric acid, it is difficult to cure and also difficult to spread satisfactorily. Furthermore, because of its free acidity, it may be injurious to crops for a few days after application to the soil. Manufacturers are now employing it in the compounding of high-analysis mixed goods and it is estimated that the American capacity to produce this product is in the neighborhood of 275,000 tons.

Characteristics of Ordinary Superphosphate

Ordinary, or normal, superphosphate is marketed as a brownish gray powder. It is partially soluble in water, has an acid odor, and is very acid to litmus. It may contain 14 to 25 per cent of available phosphoric acid together with small amounts of citrate insoluble phosphoric acid.

FIG. 64. Where Anaconda treble superphosphate is made. Anaconda Copper Mining Company, Anaconda, Montana. The great stack of this plant is reputed to be the tallest brick structure in the world—585 feet tall. (Courtesy, United Metals Selling Company.)

The granulated product contains 20 to 21 per cent of available phosphoric acid. In the early days, 10 per cent superphosphates were common. Ordinary superphosphate is composed of gypsum or its lower hydrate, and monocalcium phosphate, together with some dicalcium phosphate, tricalcium phosphate, and impurities in the lower grades. Ordinary superphosphate analyzes about 19 to 22 per cent calcium and 10 to 12 per cent sulfur. Both the monocalcium and dicalcium phosphates are considered available to plants. Concentrated superphosphate does not contain gypsum. The following table gives the composition of a representative sample of a 16 per cent superphosphate.

Table 51

COMPOSITION OF A REPRESENTATIVE SAMPLE OF NORMAL SUPERPHOSPHATE

	Per Cent
Gypsum ($CaSO_4.2H_2O$)	50.00
Monocalcium phosphate ($CaH_4(PO_4)_2.H_2O$)	26.60
Dicalcium phosphate ($Ca_2H_2(PO_4)_2.H_2O$)	2.40
Tricalcium phosphate ($Ca_3(PO_4)_2$)	2.20
Silica (SiO_2)	7.00
Iron and aluminum sulfates ($Fe_2(SO_4)_3$)($Al_2(SO_4)_3$)	4.00
Calcium fluoride (CaF_2)	1.50
Water (H_2O)	6.00

Ordinary superphosphate is shipped in bulk or in 200-pound sacks. The sacks are usually treated with a solution of sodium silicate, paraffin, or some other substance that will prevent their ready destruction by the small quantities of free acid that may be present.

A survey made in 1939—a representative prewar year—showed that over 60 per cent of the superphosphate used in the United States as a separate fertilizer material was consumed in the four states—New York, Pennsylvania, Ohio, and Kentucky.

Decomposition of Phosphate Rock by Means of an Alkali or Silicate

A number of processes have been devised whereby the complex phosphate of phosphate rock can be decomposed by fusing the phosphate rock with either an alkali or an alkaline earth and water vapor with or without the removal of fluorine. The resulting products appear to be similar to basic slag, both physically and in their crop-producing powers. One of these products, a calcined defluorinated phosphate rock containing 24 per cent of available phosphoric acid, is produced extensively by sintering phosphate rock with soda and silica. It is sold in Germany under the name of *Rhenania-Phosphate*. Similar products are sold in France and Belgium under the name of *Base-Phosphate* and *Supra-Phosphate*. Waggaman (1914) has been able to secure a slag containing 5 and 6 per cent respectively of citric acid soluble potash and phosphoric acid by fusing a mixture of phosphate rock and feldspar at 1400° C. for about 20 minutes. Whathall and Bridger (1943) fused a mixture of 1.00 pound of rock phosphate (33 per cent P_2O_5, 3.6 per cent fluorine) and 0.46 pound of olivine (45 per cent P_2O_5, 44 per cent SiO_2), which yielded a product containing 22.8 per cent total P_2O_5, 21.4 per cent soluble P_2O_5, and 1.9 per cent fluorine.

The products resulting from sintering phosphate rock with alkali salts are alkaline in reaction and show a wide range of citrate-soluble phosphorus.

Calcined Phosphate

Calcined phosphate is prepared by burning phosphorus and passing the hot reaction products into beds of phosphate rock. The calcium fluor-phosphate molecule is in this way changed to substantially calcium metaphosphate ($Ca(PO_3)_2$) having a phosphoric acid content of about 65 per cent. Searralles (1937) found as a result of both pot and chemical tests at the Pennsylvania State College that calcined phosphate was equal in its crop-producing value to superphosphate and far superior to untreated rock phosphate.

Precipitated Phosphate

Precipitated phosphate is a product which is composed largely of dicalcium phosphate which has been obtained by neutralizing with lime the acid solution of processed bone or phosphate rock. It is fairly available to plants.

Fused Tricalcium Phosphate

As has been pointed out, the presence of fluorine in the apatite molecule is in a large part responsible for the low availability of rock phosphate. Elmore, *et al.* (1942), of the Bureau of Chemistry, Soils and Agricultural Engineering, of the United States Department of Agriculture, has shown that a substantial expulsion of the fluorine ensues from the phosphate if a phosphate containing 5 to 10 per cent silica is fused, followed by quenching of the melt. Almost complete defluorination of phosphate rock can be secured in the temperature range of 1375° to 1425° C. In this process the apatite molecule is converted to an unstable hydroxyl-apatite ($Ca_{10}FOH(PO_4)_6$) and then to a tricalcium phosphate, a compound which is soluble in citric acid and citrate solutions. The resulting fused phosphate contains 30 to 35 per cent available phosphoric acid and many tests have shown it to be an excellent source of phosphorus for plants. Volk (1944) found fused phosphate to be almost as good as superphosphate and calcium metaphosphate for oats and sorghum, and much superior to phosphate rock and waste pond phosphate. His results are given in Table 52.

Reversion of Monocalcium and Dicalcium Phosphate

The water-soluble monocalcium phosphate and the citrate-soluble dicalcium phosphate of superphosphate may partially revert to the tricalcium form when they are allowed to stand for some time. Nevertheless, this tricalcium phosphate and the corresponding phosphate of magnesium are available to plants. Partial reversion also takes place if the product is heated and especially when it is added to some soils. The reversion of citrate-soluble phosphate is not as rapid as is the water-soluble phosphate. This is probably due to the small power of motion of the citrate-soluble phosphate. The superphosphates that are put on the market

Table 52

YIELDS OF OATS GROWN ON CECIL CLAY TREATED WITH VARIOUS COMBINATIONS OF PHOSPHATIC AND NITROGENOUS FERTILIZERS AND LIME

Phosphates Used	*pH of soils after treatment and oat yields in grams of dry weight:*					
	Sodium Nitrate		*Ammonium Sulfate*		*Urea*	
	Yield	*pH*	*Yield*	*pH*	*Yield*	*pH*
None	1.8	5.5	5.9	4.8	1.9	4.9
Superphosphate	27.2	5.7	26.8	5.0	27.4	5.4
Rock phosphate	9.5	5.9	16.8	5.1	15.6	5.6
Waste pond phosphate	7.2	5.9	12.9	5.4	12.0	5.4
Calcium metaphosphate	24.7	5.8	23.1	5.1	24.8	5.5
Fused rock phosphate	21.2	6.2	24.5	5.1	21.3	5.4

today have been prepared with such care that there is little danger of reversion in storage. If stored in a dry place superphosphate may be kept from one season to another with very little danger of depreciation in value.

Recent studies have shown that the phosphoric acid of superphosphate that has been reverted to the tricalcium form is more available to plants than is the phosphoric acid of ground phosphate rock. Also, the freshly precipitated tricalcium phosphate has been shown by Truog (1916) and others to be more available than an older product. MacIntire *et al.* (1937), however, found that a part of the reversion of superphosphate is to a fluor-apatite form and that this product is no more available than ground phosphate rock. Reversion is not likely to take place unless there is more than a trace of flourine in the soil solution.

Work by MacIntire *et al.* (1944) with quenched fusions of rock phosphate show that an 80 per cent removal of the initial fluorine content produced a satisfactory superphosphate composed of readily available alpha form of tricalcium phosphate. Unfortunately, this material is not commercially available, but it should prove to be superior to ordinary superphosphate. Alway and Nesom (1944) have compiled a table showing comparative results to date of calcium metaphosphate and fused rock phosphate. Their results are given in Table 53. MacIntire and Hardin (1940) found that fused rock phosphate could be used to condition superphosphate without detrimental effect.

When superphosphate is added to a soil deficient in lime the reversion appears to be largely to the iron, aluminum, and titanium phosphates, whereas if the soil contains fair quantities of calcium the reversion is quickly to the dicalcium, and slowly to the tricalcium phosphate. According to MacIntire (1942), if fluorides are present, reversion takes place very slowly to calcium fluorphosphate, a compound analagous to apatite of phosphate rock. The reaction is probably as follows: $3Ca_3(PO_4)_2 + CaF_2 \rightarrow Ca_{10}F_2(PO_4)_6$. The reversion of water-soluble phosphate is not as rapid in neutral or alkaline soils as in acid soils. The freshly precipitated calcium, iron, aluminum, and titanium phosphates are soluble in an acid soil solution and under these conditions may be available to certain plants. Under neutral or alkaline conditions in some soils these phosphates appear to become available with difficulty, if at all. This is probably due to the fact that when large quantities of calcic compounds are present the soil water will dissolve those compounds readily and thereby lose most of its capacity to dissolve tricalcium phosphate. That freshly precipitated calcium, iron, and aluminum phosphates may be available under certain

Table 53

COMPARATIVE EFFECTIVENESS OF CALCIUM METAPHOSPHATE AND FUSED ROCK PHOSPHATE IN POT AND FIELD EXPERIMENTS AS COMPILED BY ALWAY AND NESOM

State	Author	Year of Report	Crops	Method of Application of Phosphates	Effectiveness in Comparison with Superphosphate: Calcium Metaphosphate	Effectiveness in Comparison with Superphosphate: "Fused Rock Phosphate"
Pot Experiments						
District of Columbia.	Ross and Jacob	1937	Sudan, millet, wheat, tomatoes, sweet clover	Thoroughly mixed with soil in pots	Approximately equal on acid and neutral soils but inferior on strongly alkaline soil (pH 9.5)	Similar to that of the metaphosphate
Arizona.........	McGeorge	1939	Sudan	Mixed with upper half of soil in pots	Nearly equal	Not used
New Mexico.....	Hinkle	1942	Sweet clover	Mixed with upper 2 inches of soil in pots	On "an alkaline calcareous soil" it increased yield 350% compared with 440% by superphosphate	Not used
Field Experiments						
Montana........	Green	1938	Oats, wheat	Not stated	Much inferior	Much inferior
Tennessee.......	Mooers	1938	Potatoes, corn, wheat, millet, soybeans, cowpeas	Not stated	Equal	Somewhat inferior
Connecticut.....	Brown	1938	Potatoes	In bands at side and slightly below level of seed pieces	Nearly equal	Nearly equal
Massachusetts, New Jersey, Pennsylvania ..	Houghland, *et al.*	1942	Potatoes	In 2 bands 2 inches to side and slightly below level of seed pieces	Much alike	Much alike
Kentucky.......	Roberts, *et al.*	1942	Wheat, corn, lespedeza rotation	In hill for corn, broadcast and worked in before drilling wheat, none on lespedeza	Practically alike	Practically alike

conditions has been demonstrated by the work of Truog (1916), Marais (1922), Parker (1931), and others. Truog (1916), of the Wisconsin Experiment Station, secured results showing that freshly precipitated ferric and aluminum phosphates were, for most of the crops tested, more readily available than the mineral phosphates, although not as available as superphosphate. Truog grew various crops in pots of quartz sand and fertilized them with various phosphates. He then determined the percentage of normal growth of the various crops grown when supplied with each phosphate. The growth of the crops to which the superphosphate had been added was taken as normal and represented by 100. His results are given in Table 54. Nevertheless, as has been pointed out by Fraps and Fudge (1945), while the above pure salts of iron and aluminum are highly available, yet in the soil they usually become hydrated compounds of low solubility and availability to plants.

Table 54

PER CENT NORMAL GROWTH OF VARIOUS CROPS ON PHOSPHATE INDICATED WHEN GROWTH ON SUPERPHOSPHATE IS TAKEN AS NORMAL AND REPRESENTED BY 100

Crop	*Check*	*Super*	*Aluminum*	*Tricalcium*	*Ferric*	*Ferrous*	*Rock*	*Magnesium*	*Manganous*
Oats	6.8	100.0	96.4	70.5	79.9	82.9	9.1	..	..
Buckwheat	3.6	100.0	88.0	70.1	32.5	63.3	70.0	..	..
Rape	0.8	100.0	96.4	76.2	23.4	61.5	46.8	..	..
Corn	8.6	100.0	56.3	26.8	40.3	19.3	10.0	21.3	76.4
Barley	16.7	100.0	104.7	62.2	133.5	79.7	25.8	15.5	125.3
Alfalfa	1.5	100.0	78.6	99.2	93.6	28.1	38.3	7.1	20.8
Clover	1.0	100.0	84.2	64.5	68.9	23.6	6.1	26.7	4.2
Millet	0.7	100.0	86.7	34.8	103.8	31.0	4.1	16.0	73.8
Serradella	0.6	100.0	78.7	90.4	111.7	28.2	3.2	49.9	51.7

It seems therefore—as has been pointed out by Kelley and Midgley (1943)—that phosphate fixation in an acid soil is not due to a simple precipitation of iron and aluminum phosphates, which Truog's work would indicate is somewhat available, but to an "anion exchange" whereby phosphate ions are exchanged for hydroxyl ions of colloidal hydrated iron and aluminum oxides.

Truog (1932) found also that superphosphates are reverted into the not readily available basic iron phosphate when applied to soils that are more than very slightly acid (pH of 6.2). Heck (1935) has shown that when soluble phosphoric acid is applied to laterites it becomes fixed in the soil largely as basic iron phosphates, with perhaps some aluminum phosphates. As laterites are usually low in their calcium content very little of the phosphorus is fixed as the calcium phosphate. Davis (1935), working with Hawaiian soils, found that marked fixation of phosphates occurred in soils in which the replaceable bases had been exchanged for alkali metal ions, and Chandler (1941) showed that the extraction of free iron and aluminum oxides caused a reduction in the phosphorus absorptive capacity in practically every case.

It seems logical to conclude that the reversion of superphosphate in the soil must depend to a great extent upon the mineral composition of the soil, the clay content of the soil, and the reaction of the soil solution and that the reversion to very slowly available forms is dependent upon the presence of fluorides. Perkins and King (1944) have presented data that seem to indicate that the mica groups of minerals in the soil constitute the chief mineral species responsible for phosphate fixation. Their results are given in Table 55.

Table 55

PHOSPHATE FIXATION BY SOME COMMON SOIL MINERALS*

H_3PO_4 Fixed per 100 g. of Mineral		*H_3PO_4 Fixed per 100 g. of Mineral*	
	g.		g.
Kaolinite	0.28	Magnetite (Fe_3O_4)	0.00
Talc	0.01	Hematite No. 1 (Fe_2O_3)	1.14
Pyrophyllite	0.05	Hematite R.O. (Fe_2O_3)	0.07
Muscovite	6.84	Limonite ($2Fe_2O_3.3H_2O$)	0.80
Phlogopite	0.96	Quartz (SiO_2)	0.01
Biotite	5.25	Rutile (TiO_2)	0.29
Margarite	5.78	Ilmenite ($FeO.TiO_2$)	0.01

* All minerals ground to pass 100-mesh sieve (150).

The question has been often raised as to why the superphosphates, if they revert to the tricalcium form when applied to soil containing lime, are superior to the ground phosphate rock. The explanation formerly given was that the soluble phosphates readily go into solution when applied to the soil and that they become thoroughly disseminated before reversion takes place. In this way a dissemination is accomplished that is far greater than could be obtained by thoroughly mixing a finely ground phosphate rock with the soil. However, the character of the freshly precipitated salt and the composition of the reverted tricalcium phosphate—which is quite different from the apatite of phosphate rock—and the increased surface exposed for solution activities probably accounts for its greater availability.

Today it is recognized that phosphate rock carries fluorine and that, as already pointed out, the fluorine is in combination with the tricalcium phosphate. As calcium fluoride is an exceedingly stable combination, this may offer the real explanation for the difference in availability of apatite and tricalcium phosphate. Acidulation of phosphate rock breaks up the calcium fluor-phosphate molecule and this molecule is either not formed again upon reversion in the soil or, as stated above, it is only formed to a limited extent. All soils carry some fluorine, and MacIntire (1942) has reported 0.15 pound of fluorine per acre brought down in rainwater in one locality in Tennessee.

Fluoride-free superphosphate, or superphosphate of low fluorine content, can be made. Such superphosphate would be ideal for ammoniation and far superior for general use to the ordinary superphosphate now on the market.

Rapidity of Fixation or Reversion of Superphosphates in the Soil

Phosphorus does not take part in base-exchange reactions and is not fixed in the soil like potassium, calcium and magnesium. The seat of the phosphate-absorbing power of the soil is in the basoid constituent of the soil clay colloid.

The rapidity of reversion of the water-soluble phosphoric acid of calcium superphosphate depends upon the conditions existing in the soil. A study of the results of investigational work would indicate that the reversion takes place rather rapidly, in many cases immediately, when superphosphate is thoroughly mixed with the soil, but when superphosphate is not well mixed with the soil, reversion may not take place for some months. Scarseth found that a Vaden clay of the Alabama Black Belt reduced the availability of a 250-pound per acre application of superphosphate 12 per cent in 30 days and 85 per cent in one year. It appears that the longer a phosphate remains in the soil the larger will be the percentage of phosphate fixed in a less available form. On the other hand, the more the compounds of the soil become saturated with phosphorus the smaller will be the percentage of the phosphorus fertilizer that may be held in an unavailable form. Experiments on Broadbalk field of the Rothamsted Experiment Station, England, as reported by Hopkins (1910), have shown that fixation of the soluble phosphates takes place within the surface 9 inches of soil. After 50 annual applications of 392 pounds of superphosphate per acre the subsoil was found to be very little richer in phosphoric acid. Likewise, Bear and Salter (1916) found that soluble phosphates applied to field plats at the West Virginia Experiment Station were fixed in the surface 6⅔ inches of the soil. Metzger (1940) found that a large proportion of the easily soluble phosphorus, which accumulated in the soil following the top dressing of various phosphates on alfalfa for 27 years, remained in the surface 4 inches of the soil. At the Wisconsin Experiment Station 300 pounds of superphosphate was applied as a surface application to an acre of permanent pasture. A year later it was found that it had not moved downward beyond 1 inch below the surface and that a greater part of it was contained in the surface ½ inch of soil. Crawley (1902) mixed soluble phosphates with layers of soil of different depths. He then leached the soil layers with water, immediately in one case, and after 15 hours in another. He found that of the soluble phosphoric acid added the amount that remained in the soil after leaching the 1-inch layer was 53.35 per cent when leached immediately, and 99.18 per cent when leached 15 hours after application. The amount that remained in the 6-inch layer was 97.22 per cent when leached immediately and 99.66 per cent when leached 15 hours after application. Harrison and Das (1921) have shown that the fixation of superphosphate takes place more rapidly in a calcareous soil than in a noncalcareous soil. Flor (1943) believes that the chlorotic dieback of flax often seen in the Red River valley of North Dakota is due to a phosphate deficiency resulting from a soil pH of 7.6 and above. Heck (1934) found the greatest phosphate penetration occurred in soils which showed a neutral to slightly acid reaction and which contained little active iron or aluminum. Because of the lower fixation power of sandy types of soil a greater loss of phosphoric acid by leaching may occur from sands than from loams and clays. Hibbard (1935), summarizing the results of much published data, concluded that the fixation of phosphates in the soil may be increased by raising the pH or by adding lime or soil colloids. It could be decreased by lowering the pH or by adding Na^+, NH_4^+, soluble silica, or organic colloids and by employing specific methods of placements and by using pellets instead of fine particles of phosphate fertilizer.

In view of the great attention that has been given the problem of phosphorus fixation by soil scientists, it is of interest to note the conclusions of Kurtz *et al.* in 1946: "This research gives evidence that so-called 'phosphate fixation' is not a serious problem in Illinois soils, and that phosphate fixation, therefore, has no justification as an argument against the use of soluble phosphatic fertilizers."

Organophosphates

Stewart (1936) has suggested that advantage be taken of the solubility of some organic phosphate compounds and that phosphorus be applied to the soil in these forms. He concluded that these compounds are not made insoluble in the presence of calcium, iron and aluminum. Stewart has successfully used calcium glycophosphate, calcium glycerophosphate, calcium sorbitol-phosphate and calcium glucophosphate on tobacco. All these phosphates were able to penetrate a 7-inch column of soil. Hilbert *et al.* (1938), however, found that the organic phosphates could be fixed but that the fixation varied widely with the soil.

Allison *et al.* (1941) found that the organic phosphates showed availabilities similar to those of the common inorganic phosphates but that

Table 56

AVAILABILITIES OF SOME ORGANIC PHOSPHATES AS FOUND BY THE NEUBAUER METHOD IN SAND AND SOILS

Phosphates	P_2O_5 *Added* (*mg.*)	*Percentage of added phosphorus taken up by plant in:*				
		Quartz Sand	*Norfolk Sandy Loam*	*Chester Loam*	*Cecil Clay Loam*	*Las Vegas Loam*
Sodium glycerophosphate...	10	78	84	33	23	47
	25	71	74	33	21	58
Calcium hexose diphosphate	10	58	55	7	..	..
	25	48	60	14	..	..
Sodium nucleate.........	10	84	55	17	..	..
	25	79	71	16	..	..
Nucleic acid..............	10	79	45	4	..	..
	25	66	47	10	..	..
Dipotassium phenylphosphate.	10	52	51	5	..	..
	25	54	48	15	..	..
Potassium diphenylphosphate.	10	73	79	30	..	..
	25	57	53	25	..	..
Dipotassium diphenylpyrophosphate.	10	67	63	12	..	..
	25	65	56	14	..	..
Calcium ethylphosphate....	10	74	75	36	..	..
	25	65	76	35	..	..
Calcium diethylphosphate...	10	50	58	42	23	31
	25	40	52	38	20	26
Triethylphosphate.........	25	3	4	3	..	..

soil type influences the availabilities of the phosphates. Some of their results are given in Table 56.

In 1944 Spencer and Willhite reported Neubauer tests with 4 inorganic and 11 organic phosphates, which showed the chemical availability of the phosphorus in 9 of the 11 organic carriers to be just as high as that in any of the inorganic carriers. Their results are given in Table 57.

Table 57

UPTAKE OF PHOSPHORUS IN NEUBAUER TESTS WITH CHRISTIANA SOIL TREATED WITH INORGANIC AND ORGANIC PHOSPHATES

Phosphate	*Phosphorus (Found)*	*Uptake of P by Cultures Planted after Various Periods of Phosphate-Soil Contact*			
Name	*Per Cent*	*0-day*	*6-day*	*18-day*	*30-day*
		mg.	mg.	mg.	mg.
(No treatment)	..	10.5	11.1	11.2	11.0
INORGANIC PHOSPHATES:					
Treble superphosphate	20.65	12.2	13.1	13.2	12.6
Calcium metaphosphate, TVA #443-A	26.93	11.9	12.9	13.2	12.9
Fused rock phosphate, TVA	12.32	11.5	13.1	13.2	12.3
Sintered phosphate	16.17	11.7	13.3	13.2	12.3
ORGANIC PHOSPHATES:					
Calcium glycol phosphate	16.96	12.2	13.0	12.8	12.3
Calcium glucose-3-phosphate	10.14	11.9	12.7	12.4	12.7
Monoethyl calcium phosphate	18.43	12.1	13.1	12.9	12.3
Diethyl calcium phosphate	17.91	12.1	13.2	13.2	12.7
Triethyl phosphate	17.18	11.5	11.8	11.4	10.9
Monoethyl calcium acid phosphate	21.18	12.8	13.1	13.0	12.9
Mixture (50–50) of monoethyl and diethyl calcium phosphates	18.10	12.5	13.5	13.0	13.0
Mixture (50–50) of diethyl calcium phosphate and monoethyl calcium acid phosphate	19.50	12.4	13.6	12.9	13.1
Calcium salts of phosphates of sugars in blackstrap molasses	6.24	11.2	12.3	12.0	11.7
Crude "calcium ethyl phosphate"	17.70	12.3	13.1	13.0	12.8
Mixture (50–50) of monomethyl and dimethyl calcium phosphates	18.40	12.9	13.0	12.7	12.5

Fertilizers that May or May Not Be Mixed with Superphosphates

For many years the idea has been prevalent that fertilizing materials that carry lime should not be mixed with the superphosphates, as a reversion might be brought about which would make the phosphoric acid unavailable to plants. For this reason it was thought inadvisable to mix such substances as wood ashes, lime, or basic slag with the superphosphates, except under special conditions. It has been shown, however, by Moores (1928) and MacIntire and Shuey (1932), of the Tennessee Experiment Station, that ground dolomitic limestone which is not very reactive may be added as a filler to a complete fertilizer without greatly reverting the soluble phosphates, and similar results have been secured by Conner (1930). In view of this and other work by many investigators, most of the state experiment stations are now advising the use of ground dolomitic limestone, but not calcitic limestone, as a filler for fertilizers. It appears also, that the use of reasonable amounts of lime with superphosphate on acid soils will aid in keeping the applied phosphorus in an available calcium form and thus prevent the formation of the relatively unavailable iron and aluminum phosphates. Albrecht and Smith (1940) think that the higher the degree of calcium saturation in the soil the longer will applied superphosphate remain in the soil in an available form, and the greater will be the efficiency of the phosphate application. Reverted tricalcium phosphate is now accepted as being largely available for most plants.

Sodium nitrate and the superphosphates may be mixed in any desired ratios. In the early days of the fertilizer industry it was advised that sodium nitrate and the superphosphates should not be mixed, for it was said that free nitric acid and finally the oxides of nitrogen, would be produced, which might become a menace to both life and property. The idea probably arose from the fact that formerly commercial mixing plants often reported a loss of nitrogen where the two compounds were mixed. This was due to the fact that their materials were not dry and the soldium nitrate was decomposed by sulfuric acid probably as follows:

$$NaNO_3 + H_2SO_4 = NaHSO_4 + HNO_3$$
$$2NaNO_3 + H_2SO_4 = Na_2SO_4 + 2HNO_3$$

The high temperature accompanying the reaction readily decomposed a part of the nitric acid, producing nitric oxide. If dry sodium nitrate is mixed with superphosphate, likewise dry, no nitric oxide is given off and no loss of nitrogen will be sustained. It is true, however, that if such a mixture becomes moist, free chlorine and fluorine are produced from impurities to such an extent that the mixture cannot be shipped or stored in bags, for there are few substances that will destroy bags more quickly than free chlorine or fluorine.

FIG. 65. Influence of superphosphate on the growth of alfalfa at the Moran experiment field, Kansas. (*Above*) Lime only. (*Below*) Lime and superphosphate. (Courtesy, Kansas Agricultural Experiment Station and *The Fertilizer Review.*)

Ammonium sulfate and potassium sulfate may be mixed with the superphosphates without producing reversion of the phosphates to the citrate-insoluble form, but the mixture will "set up" within a short time and unless applied immediately will have to be reground.

Analytical Determination of Water-soluble Phosphoric Acid

While the monocalcium and dicalcium phosphates have already been spoken of as soluble, they are not both soluble in water. Only the monocalcium phosphate ($CaH_4(PO_4)_2$) will go into solution in pure water. In fertilizer analytical work the water-soluble portion of a phosphate is determined by leaching a known quantity of the material and then drying and determining the loss in weight. This loss in weight largely represents the amount of monocalcium phosphate present in the fertilizer, and is known in the trade as *water-soluble* phosphoric acid. The monoammonium and diammonium phosphates and some organic phosphates, such as glycerophosphates and hexosephosphates, also are water-soluble.

Analytical Determination of Citrate-soluble Phosphoric Acid

The American Association of Official Agricultural Chemists, at its first meeting in 1884, adopted a method for the determination of the availability of water-insoluble phosphates, other than basic slag, which with few minor changes is still employed.

According to this method, as now used, if a 1-gram sample of superphosphate which has previously been leached with water is treated with a standard amount of a neutral solution of ammonium citrate of a specific gravity of 1.09 for 1 hour at a temperature of 65° C., the dicalcium form ($Ca_2H_2(PO_4)_2$) will be carried into solution. As stated before, this citrate-soluble phosphate has been commonly referred to in the trade as *reverted* or *precipitated*, phosphoric acid.

It has been found that a similar treatment of water-leached material will dissolve a considerable amount of the phosphoric acid found in tankage, bonemeal, basic slag, and aluminum and iron phosphates as well as calcium and potassium metaphosphates, the defluorinated phosphates, and such products as Rhenania-Phosphate. Basic slag, however, is universally evaluated by means of the Wagner citric acid method.

Tricalcium phosphate that has originated through the precipitation of monocalcium or dicalcium phosphate is also a reverted phosphate, although it is not the compound that is known in the trade as reverted phosphate. It is soluble in ammonium citrate solution usually to the extent of about 50 per cent, depending upon the conditions under which the material was precipitated and subsequently dried. This precipitated tricalcium phosphate is not of the same crystalline structure as that of the phosphate of phosphate rock. The raw rock phosphate is only slightly soluble in ammonium citrate solution.

For seventy-five years the study and experimentation of the neutral ammonium citrate method has proved it to be an excellent method to be used in fertilizer research and in the administration of fertilizer control laws.

Superior Qualities of Dicalcium Phosphate

Since dicalcium phosphate will not revert in the soil as rapidly as monocalcium phosphate, and since it is available to plants, it has been suggested that all calcium superphosphates should carry their phosphorus in the citrate-soluble form. But the production on a commercial scale of a superphosphate of this nature is rather difficult. Experiments for the solution of this problem are now being carried on. Dicalcium phosphate cannot be obtained by treating phosphate rock with an acid but is made by precipitating phosphoric acid with lime. It has excellent physical properties and is a better source of phosphorus for plants than calcium metaphosphate, calcined or fused phosphates.

A citrate-soluble phosphate containing its phosphorus in the tricalcium and tetracalcium forms may be produced by heating ordinary superphosphate to a temperature of 1400° C., but plant tests conducted with this material have not been reported.

A phosphate carrier which has been described by Nordengren and Lehrecke (1940) may meet the needs of the agriculturist better than any superphosphate now on the market. It consists of coating granules of ordinary superphosphate with any powdered product containing citric acid soluble phosphoric acid such as basic slag or other products produced in smelting or sintering processes. A reaction results between the monocalcium phosphate and the free phosphoric acid of the granules and the citric acid-soluble compound of the coating, forming a crust with a higher percentage of citrate-soluble phosphoric acid than in the interior of the granule. In this way, the free phosphoric acid is reduced without essentially diminishing the water-soluble phosphorus, and when applied to the soil, the reversion of the water-soluble phosphoric acid is also reduced.

Analytical Determination of Available Phosphoric Acid

The water-soluble phosphoric acid, together with the citrate-soluble phosphoric acid, largely represent what is known industrially as *available* phosphoric acid. All phosphoric acid of superphosphate that is not soluble in water or in a neutral ammonium citrate solution of a specific gravity of 1.09 is referred to as *citrate-insoluble* phosphoric acid. A rather recent change in the official method for determining the available phosphoric acid in fertilizers now allows a large part of the reverted tricalcium phosphate to be classed as available phosphoric acid.

Available phosphoric acid in basic slag is that phosphoric acid which is soluble in a 2 per cent citric acid solution. In other words, available phosphoric acid in basic slag is citric acid-soluble. This method of determination of availability is known as the Wagner Method.

Analytical Determination of Organic Phosphorus

It is now recognized that in those soils which are high in organic matter a significant proportion of the total phosphorus occurs in organic combination. Plants are known to utilize this phosphorus, although not as readily as they utilize the phosphorus in calcium or magnesium compounds. Dickman and DeTurk (1938), of the Illinois Experiment Station, have worked out a method for the determination of organic phosphorus based on the liberation of phosphorus by decomposition of organic matter with hydrogen peroxide and subsequent extraction with 0.2 *N* sulfuric acid.

Residual Influences of Superphosphate

The residual influences of superphosphate are noticeable for some years following an application of the fertilizer. Bushnell (1941) reported that an annual application of 400 pounds per acre of 16 per cent superphosphate for 16 years resulted in the accumulation of available phosphorus sufficient to maintain the yield of tomatoes and cucumbers for four years, and the yields of cabbage and sweet corn for eight years. Nagaoka (1904) has reported the residual effects of superphosphates upon rice grown on acid soil of a paddy field. He found that the increases in rice yields due to the use of superphosphates were greatest the first year, decreasing the second and third years, but increasing the fourth year. Baker (1925), working with a number of Iowa soil types, both in the

Table 58

RELATIVE INFLUENCE OF THE CONTINUED USE OF VARIOUS PHOSPHATES ON THE LIME REQUIREMENT OF THE SOIL

	Total Crop Grown During Last 5 Yrs. (tons)	*Lime Requirement, Jones Method, Lbs. CaO per Acre*
Check	17.6	2700
Floats	29.4	2520
Superphosphate	31.5	2340
Extra superphosphate	42.2	2250
Ground bone	33.4	2160
Thomas slag	32.7	2070

greenhouse and in the field, calculated the rate of exhaustion of the phosphoric acid following applications of 44 per cent treble superphosphate to be 12.3 per cent the first year, 8.9 per cent the second year, and 7.1 per cent the third year. This is a total of only 28.3 per cent for the first three years. Russell (1932) concludes that plants do not usually recover more than 20 per cent of the phosphates supplied.

Skinner and Beattie (1917), working at the Arlington, Virginia, farm of the United States Department of Agriculture, reported that the lime requirement of the surface 6 inches of the soil was increased following 5 annual applications of superphosphate at the rate of 200 pounds per acre, but Burgess (1922), of the Rhode Island Experiment Station, concluded that there is no foundation for the statement that soils will become more acid following the continuous use of superphosphate. His results, which show the accumulated influence for a period of 27 years, are given in Table 58. They show that the application of all phosphates will produce a less acid condition in the soil. The phosphates were applied at the rate of 200 pounds of superphosphate, or its equivalent, per acre annually.

Russell (1932) has stated that at the Rothamsted Experiment Station, there has been no indication that the superphosphates ever make a soil acid.

The experimental data available in regard to the influence of superphosphate upon the acidity of the soil are not conclusive, but as would be expected it appears that under most conditions it does not have a material influence. Although monocalcium and dicalcium phosphates are acid phosphates they readily revert and do not materially influence the pH value of the soil. Pierre (1934) has concluded that for most soils of humid regions, the pH values of which lie between 5.0 and 6.0, superphosphate, phosphate rock, and monocalcium phosphate can be considered to have no appreciable effect on soil reaction.

The application of superphosphate to soils that contain soluble aluminum compounds will bring about, at least temporarily, a precipitation of these compounds. This has been shown by Pierre and Stuart (1933). As soluble aluminum compounds are toxic to plant growth, particularly to seedlings, the application of superphosphate to such soils may aid in correcting this toxicity, at least until the plants have passed their seedling stage of growth.

Do Superphosphates Make Soils Hard?

There has been a general belief in some sections that the regular use of superphosphate will cause a soil to become hard. This belief is entirely unfounded and observed instances probably result from a decrease in organic matter due to intensive farming.

Value of Superphosphate as Compared with Ground Phosphate Rock

Experimental evidence shows that in most cases the use of ground phosphate rock under average conditions is not economical. Mooers (1929), summarizing a 10-year comparison of superphosphate and ground phosphate rock at the Tennessee Experiment Station, when applied to a DeKalb soil, concluded that while ground phosphate rock proved to be profitable its margin of profit was less than that for superphosphate.

Thorne (1919), summarizing 25-year results of experiments designed to determine the relative crop-producing value of various phosphates at

Table 59

SUMMARY OF ALL EXPERIMENTS IN WHICH SUPERPHOSPHATE AND GROUND PHOSPHATE ROCK HAVE BEEN COMPARED AT THE INDIANA EXPERIMENT STATION—AVERAGE OF 82 TESTS WITH VARIOUS CROPS

Crop	Carrier of Phosphoric Acid	Average per Acre per Year					
		Yield	Increase	Phosphate (lbs.)	Value of Increase*	Cost of Phosphate	Profit or Loss
Corn,	Superphosphate	43.04 bu.	5.49 bu.	190.5	$ 2.95	$1.52	$1.43
36 tests	Phosphate rock	42.20 bu.	4.65 bu.	532.0	2.52	1.86	0.66
	None	37.55 bu.	..	..	..	..	..
Wheat,	Superphosphate	15.28 bu.	4.31 bu.	190.5	4.63	1.52	3.11
33 tests	Phosphate rock	12.88 bu.	1.91 bu.	532.0	2.12	1.86	0.26
	None	10.97 bu.	..	..	..	..	..
Legumes,	Superphosphate	2998 lbs.	320 lbs.	190.5	1.60	1.52	0.08
9 tests	Phosphate rock	2773 lbs.	95 lbs.	532.0	0.47	1.86	1.39
	None	2678 lbs.	..	..	..	..	..
Potatoes,	Superphosphate	112.0 bu.	29.4 bu.	500.0	14.70	4.00	10.70
9 tests	Phosphate rock	94.2 bu.	11.6 bu.	1000.0	5.80	3.50	2.30
	None	82.6 bu.	..	..	..	..	..
Average of the 82 tests†	Superphosphate	..	..	205.0	4.06	1.64	2.42
	Phosphate rock	..	..	545.0	2.29	1.91	0.38

* Value of increase includes value of corn stover and wheat straw.
† Average length of time tests have been running, 3.47 yrs.

the Wooster, Ohio, Experiment Station, concluded that superphosphate proved the most effective carrier of phosphorus in proportion to its cost. In these experiments basic slag and steamed bonemeal were next to superphosphate in effectiveness.

Very extensive investigations of this subject have been carried on at the Indiana Experiment Station. In 1916 Wiancko and Conner, of this station, summarized the results of all the experiments of the Indiana Experiment Station where phosphate rock and superphosphate had been compared. Their summary is given in Table 59.

As has been pointed out in Chapter 7, the application of ground phosphate rock is usually made in large quantities, a method that appears necessary if satisfactory results are to be secured. This is also true, but to a lesser extent, when the bone products are applied. On the other hand, the superphosphates are generally applied in relatively small quantities, the purpose being the fertilization of the immediate crop. Generally, phosphate rock and the bone products are applied for the purpose of fertilizing all the crops in the rotation and therefore may be used most satisfactorily in sections where extensive farming is practiced. Usually the best results are secured from the slowly available phosphates when they are applied to crops having a long growing season, or which are biennial or perennial rather than annual. It seems desirable that highly available forms of phosphorus should always be applied to crops having a short growing season. In Table 60 is given, according to Karraker, *et al.* (1941), the relative value of various phosphates for a number of field crops, as determined in the greenhouse and in the field at the Kentucky Experiment Station.

Table 60

RELATIVE VALUE OF VARIOUS PHOSPHATES FOR CROPS IN KENTUCKY

Phosphate	*In Greenhouse*	*In Field*	
		Not Limed	*Limed*
Ordinary superphosphate	100	100	116
Triple superphosphate / Monocalcium phosphate	94	97	119
Dicalcium phosphate	101	100	116
Calcium metaphosphate	103	110	126
Fused rock phosphate	100	114	126
Tricalcium phosphate	77	95	92
Rock phosphate	11	71	36
Colloidal phosphate	18	..	..

Relative Value of Potassium and Calcium Phosphate

There is some experimental evidence to indicate that under certain conditions plants can assimilate phosphorus of potassium phosphate more readily than that of the soluble calcium phosphate. Cooper and Wilson (1930) have shown that the phosphorus content of pasture grasses is positively correlated with potassium and negatively correlated with calcium.

Application of Superphosphates

It has been demonstrated by Haskell (1923) that superphosphates applied in any reasonable application are comparatively safe to germinating field crop seed, with very little regard to the method of application. He concludes that a localized application of fertilizer is more beneficial and economical than broadcast distribution. Similar results have been reported by Hicks (1900) and many other investigators.

Because superphosphates in solution do not travel far from the place of application, particularly in clay soils, it is preferable not to apply them as a top dressing, but to place them so that they will be within reach of the feeding roots of plants to be fertilized. Nevertheless, superphosphates have been known to produce marked beneficial effects when top-dressed to pastures and old stands of alfalfa, but the slow downward movement of phosphorus in permanent pastures often prevents a rapid response of vegetation following an application of superphosphate.

Most superphosphates are applied during the spring months, but there appears to be no good reason why they should not be applied at any time during the year when the ground is not frozen.

The reversion of superphosphate in the soil may be limited, in part, if the superphosphate is not thoroughly mixed with the soil. Also, the application of granulated superphosphate will reduce the loss resulting from fixation. Because fixation of phosphates is the greatest in clays, phosphatic fertilizers will remain available longer in sands.

The subject of phosphorus availability needs further study but published results of experimental work would indicate that the availability of phosphorus from the principal phosphorus compounds of the soil is in the following order: (1) calcium or magnesium compounds of phosphorus, (2) organic compounds containing phosphorus, (3) compounds of iron or aluminum and phosphorus, (4) apatite.

FIG. 66. Influence of superphosphate on the growth of alfalfa. (*Right and left*) Superphosphate. (*Center*) No superphosphate. (Courtesy, Kansas State Agricultural College.)

Importance of Phosphorus in Various Systems of Farming

Phosphorus is one of the first elements to become deficient when virgin land is put under a grain or livestock system of farming. In the midwestern states of the United States, where these two systems of farming are largely employed, phosphorus is the first limiting element of plant growth. This is due in part to the fact that phosphorus is concentrated in the grain and the livestock products which are sold from the farm. Whitson and Stoddart (1909) found that soils which had been cropped to grain for 50 years had lost about one-third of their total phosphorus.

When virgin soil is put under cultivation a rapid decay and nitrification of the organic matter is induced, and the resulting carbonic and nitric acids aid materially in making available the insoluble phosphates of the soil. As the supply of organic matter decreases, so do the quantities of carbonic and nitric acids, and in turn the quantity of available phosphoric acid. This natural process has played an important part in increasing the demand for the superphosphates, as well as for the potashes, in the older populated areas.

Influence of Phosphoric Acid on Root Development, Maturation, and Composition of Plants

Many investigators have called attention to the influence of the phosphates in producing well-developed root systems, and especially to their influence on the development of the roots of seedlings. This influence is very apparent with such crops as swedes and turnips. Hoagland (1944) has called attention to the fact that plants with a determinate growth, like wheat, can absorb enough phosphate within a few weeks to insure a reasonably adequate supply stored in the plant for the remainder of the growth cycle, but that plants with an indeterminate growth, like tomatoes, present a different problem.

Noll (1932), of the Pennsylvania Experiment Station, has reported the influence of phosphates on the hastening of maturity of small grain and corn. He found an average difference for 4 years of 3½ days in the ripening of wheat between plats receiving 48 pounds of phosphoric acid in superphosphate, and plats receiving the same amount of phosphoric acid in phosphate rock. Increasing the amounts of both superphosphate and phosphate rock up to 144 pounds of phosphoric acid per acre, caused small but consistent increases in earliness. The effect on earliness of cotton of varying the percentage of phosphoric acid in a complete fertilizer, as determined by Blackwell and Buie (1924) in South Carolina, is given in the table below.

Table 61

EFFECT ON EARLINESS OF COTTON OF VARYING PERCENTAGES OF PHOSPHORIC ACID IN A COMPLETE FERTILIZER. FIGURES ARE PERCENTAGES OF TOTAL YIELDS OBTAINED AT FIRST PICKING

Place	*Soil Type*	*Yrs.*	*Rate Applied*	*0%*	*2%*	*4%*	*6%*	*8%*	*10%*	*12%*
Allendale.	Norfolk sandy loam	3	1000	53.6	66.8	60.2	67.4	70.1	69.3	*
Bishopville. . . .	Ruston sandy loam	3	600	59.1	59.7	62.8	66.3	64.5	67.2	*
Trenton.	Greenville clay loam	3	800	45.2	48.6	46.4	51.2	50.5	48.5	*
Greer.	Davidson clay loam	3	600	36.7	33.2	35.2	41.9	41.1	39.7	45.[illegible]
Gaffney.	Cecil sandy loam	2	600	36.9	50.0	48.2	57.8	54.2	56.7	60.[illegible]
Lancaster.	Durham fine sandy loam	1	600	21.2	*	27.6	28.7	37.6	44.6	35.[illegible]
State Park.	Norfolk sand	1	1000	74.8	76.9	76.9	88.4	88.7	90.6	*
Average.	. .	. .	. .	47.1	53.5	50.9	57.0	57.7	57.7	

* Not represented.

The influence of superphosphate on hastening the maturity and yield of tomatoes as determined by Hepler and Kraybill (1925) is given in Table 62. Similar influences have been noted by Williams (1931) at the North Carolina Experiment Station, and by Skinner (1931), of the United States Department of Agriculture. Noll (1923), however, was unable to secure increases in earliness from rates of application above the needs of the crop as indicated by yields.

Table 62

EFFECT OF SUPERPHOSPHATE ON MATURITY OF TOMATOES

Harvesting dates	*Superphosphate per Acre*			
	None, lbs.	*500 lbs.*	*1,000 lbs.*	*1,500 lbs.*
August 15	108	345	648	456
August 23	143	641	900	724
August 27	225	962	1,274	1,096
September 3	507	1,906	3,122	2,413
September 6	1,067	4,231	6,779	5,472
September 12	2,123	7,653	11,797	9,974
September 18	4,048	13,330	18,069	17,508
October 1	17,565	23,153	25,646	27,422

The apparent influence of phosphoric acid on earliness, as indicated by the responses shown in crop yields, seems to be correlated with the solubility of the phosphate fertilizer. It has been shown by Thorne (1919) that superphosphates have a greatㄩ influence in promoting earliness than steamed bonemeal or basic slag, although the crop yields were approximately the same where each of the three materials were used. Nevertheless, the available evidence in regard to the influence of phosphoric acid on plant growth is not conclusive. It is still doubtful whether the effect of phosphoric acid upon root development is more pronounced than is the effect of the other plant nutrients when they are the factors limiting plant growth. While phosphoric acid may appear to have a more pronounced influence on hastening earliness, nitrogen and potash in particular, and even calcium and other elements also may induce earliness if applied to soils where these nutrients are deficient.

FIG. 67. Plats 9, 10, and 11 of Tier 2 of the phosphate fertilizer experiment at Pennsylvania State College. Plat 9 received 96 pounds; plat 10, none; and plat 11, 48 pounds of phosphoric acid. (Courtesy, Pennsylvania Agricultural Experiment Station.)

Increasing the rates of application of phosphatic fertilizers above the needs of the crop, as is indicated by the yield of the crop, has not been shown to increase earliness further.

In Africa and other parts of the world the soils of large grazing areas have been reported to be so deficient in phosphoric acid that it is reflected in the natural herbage, with very serious consequences to the grazing stock. Brand (1943) has reported that phosphorus is so deficient in the Lima valley of Peru as to give a purplish color to the leaves of corn crops, which in turn imparted a purplish hue to the evening atmosphere.

As might be expected, the phosphoric acid content of crops grown in phosphorus-deficient soils may be increased if superphosphate is added. This is especially evident with such crops as turnips. Hartwell (1913) reports that the percentage of phosphoric acid in turnips at the Rhode

Table 63

PHOSPHORIC ACID CONTENT OF SAMPLES OF PASTURE GRASSES SECURED FROM SCOTLAND AND ENGLAND

Nature of Pasture	*Per Cent*
Cultivated pasture	0.74
Good natural pasture, all grazed	0.67
Poor natural pasture, eaten	0.60
Poor natural pasture, not eaten	0.37

Island Experiment Station is a better criterion of the supply of available phosphoric acid in the soil than is the extraction of phosphorus from the soil by ordinary solvents. Corrie (1930) has secured results of research which were obtained with pasture grasses at the Rowett Research Institute which appear to bear out this generality. These results are given in Table 63.

Loss of Phosphates from the Soil

Many investigators have shown that comparatively small amounts of phosphoric acid are lost from the soil by leaching. This is due in part to the fact that the mineral and organic phosphates do not readily ionize, and in part to the readiness with which soluble phosphates revert to an insoluble form.

Blair and Prince (1924) have reported that in the New Jersey cylinder experiments very little phosphoric acid has been lost from the soil by leaching where average applications of phosphate have been made. The experiments do show, however, that the larger the application of phosphate the greater the loss of leached phosphorus.

Bryan (1933), Hester (1937), Anderson (1927) and others have pointed out that in those sections where the application of complete fertilizers has become a common practice the total phosphoric acid content of the soil tends to increase because generally more phosphoric acid is added to the soil than is removed in crops and lost by leaching. Under the present practices of fertilizer application, more phosphoric acid is being returned to the soils of the South Atlantic states which receive fertilizer than is being removed in crops and drainage. It has been said that the American people are the greatest producers, consumers, and wasters of phosphates in the world.

10

Use of Ammonia Solutions as Sources of Nitrogen in Mixed Fertilizers

Probably the most outstanding development that has taken place in recent years in the manufacture of fertilizers has been the utilization of ammonia solutions, sometimes called liquors, as a source of fertilizer nitrogen. The last 15 years has witnessed the introduction into the fertilizer market of a number of aqueous solutions of ammonia and of mixtures of ammonia with ammonium nitrate and/or urea and other soluble nitrogenous salts. At the present time these solutions are being supplied to the American manufacturers of mixed goods by E. I. DuPont de Nemours & Company, the Barrett Division, Allied Chemical and Dye Corporation, the Hydrocarbon Products Company, the Matheson Alkali Company, T.V.A., and Spencer Chemical Company, and perhaps others. The nitrogen in these solutions has been derived either from the air by the employment of the direct synthetic ammonia process, or from by-product sources. They represent, therefore, the cheapest source of nitrogen available.

Although the adoption, during the last 15 years, of the ammonia solutions as sources of fertilizer nitrogen has been of almost startling proportions, it is now admitted that these sources of nitrogen will probably continue to be of increasing importance in the trade for some years to come. It must be admitted, however, that there are more limitations to the use of solutions than is the case with many solid nitrogenous fertilizer materials.

Ammoniation of Superphosphate

The importance of ammonia solutions as a source of fertilizer nitrogen was recognized as soon as it was established that ammonia would combine with certain of the superphosphates to form stable compounds that could readily be utilized by plants.

When aqua or anhydrous ammonia is mixed with ordinary superphosphate the ammonia will combine chemically with the superphosphate to form a mixture of monoammonium phosphate, ammonium sulfate, dicalcium phosphate, tricalcium phosphate, and gypsum. In other words, the readily volatile ammonia becomes fixed in a form that will allow its transfer to the soil without loss. While the anhydrous ammonia can be, and has been, used for this purpose, and is still available on the market, it has been largely superseded in recent years by more satisfactory solutions.

The ammoniation of superphosphate was first extensively employed in 1928. At the present time approximately 35 per cent of the nitrogen consumed in mixed fertilizers, or more than 25 per cent of all fertilizer nitrogen, is derived from ammonia solutions. Some of the most commonly employed sources of fertilizer nitrogen used in the year 1941 are given in the following table.

Table 64

APPROXIMATE FERTILIZER NITROGEN CONSUMED FROM FIVE LEADING SOURCES IN 1941

Sources	*Total (tons)*	*In Mixed Goods (tons)*	*For Direct Application (tons)*
Sodium nitrate	858,000	69,000	789,000
Ammonium sulfate	531,000	354,000	177,000
Liquors	193,000	186,000	7,000
Cyanamid	116,000	33,000	83,000
Uramon, urea, calurea, etc.	37,000	31,000	6,000

Advantages Accruing from the Use of Ammonia Solutions

There are at least six reasons why the employment of ammonia solutions as sources of fertilizer nitrogen is of such outstanding economic importance. (1) Ammonia solutions are the cheapest and the most concentrated form of nitrogen that is available to the fertilizer industry. This is illustrated in Table 65. (2) The use of ammonia solutions eliminates the need for large quantities of acids, such as sulfuric acid, that would be necessary if the ammonia were converted to ammonium sulfate. (3) The ammoniation greatly improves the mechanical characteristics and the curing and storing qualities of mixed fertilizers containing superphosphate by neutralizing the free acidity of the superphosphate. This practically eliminates the need for other conditioners, and reduces to a minimum the rotting of bags. (4) The resultant potential acidity of fertilizers for the nitrogen supplied in solutions is only one-third that of ammonium sulfate which the solutions largely displace. This reduces the amount of alkaline materials necessary to make neutral or basic fertilizers, and permits formulation of higher grades or the addition of more dolomite to supply calcium and magnesium. (5) The time required to cure superphosphate can be reduced by using an excess of sulfuric acid in the making of superphosphate, since the excess acid is subsequently neutralized by the ammonia. (6) It facilitates the preparation of high-analysis mixtures.

Table 65

WHOLESALE PRICES OF A UNIT (20 LBS.) OF NITROGEN IN FERTILIZER MATERIALS IN STATED YEARS

Year	*Sodium Nitrate*	*Ammonium Sulfate*	*Anhydrous Ammonia*	*Cyanamide*
1885	$2.75	$3.03	..	..
1890	2.33	3.18	..	..
1895	2.36	2.75	..	..
1900	2.37	2.79	..	..
1905	2.97	3.01	..	..
1910	2.76	2.64	..	$3.43
1915	3.04	3.09	..	2.54
1920	4.44	4.08	..	3.40
1925	3.28	2.65	..	2.20
1930	2.49	1.79	$1.40	1.65
1931	2.36	1.34	1.36	1.38
1932	1.86	1.02	1.34	1.08
1933	1.53	1.12	1.15	1.13
1934	1.54	1.18	1.09	1.20
1935	1.47	1.13	1.09	1.20
1936	1.55	1.17	1.09	1.21
1937	1.64	1.32	1.09	1.26
1938	1.68	1.36	1.09	1.29

Composition of Ammonia Solutions Now Available

Six types of ammonia solutions are available on the market. These are, anhydrous ammonia, aqua ammonia (commonly known as "B-Liquor"), ammonia solution in which ammonium nitrate has been dissolved, ammonia solution in which sodium nitrate has been dissolved, ammonia solution in which urea has been dissolved, and ammonia solution to which urea and a water-insoluble source of nitrogen has been added.

The use of anhydrous ammonia has not proved popular, probably because of the necessary precautions that must be taken in handling liquids of high vapor pressure. The most popular solutions are the ammonia solutions containing ammonium nitrate and urea. The former solutions are sold under the trade name of Nitrogen Solutions, and the latter under the trade name Urea-Ammonia Liquor. At present, over 80 per cent of the fertilizer plants manufacturing 5,000 tons or more of fertilizer annually employ ammonia liquors.

In Table 66 is given the composition of nine of the solutions now found on the American market.

The solutions which carry ammonia and nitrate nitrogen are popular with one section of the trade, while the solutions which carry urea and ammonia are popular with another. The urea-ammonia solution known as UAL-37 is unique in that its manufacturers attempted to meet a demand of the trade for a water-insoluble source of nitrogen, and succeeded. When this solution is used for the ammoniation of superphosphate some water-insoluble nitrogen compounds of the urea-formaldehyde class are produced. Although this solution will prove to be popular among certain sections of the trade, the necessity or value of this form of nitrogen may be questioned by agronomists as, indeed, may be that of other forms of insoluble nitrogen.

FIG. 68. Tank cars loaded with anhydrous ammonia leaving synthetic ammonia plant, Hopewell, Virginia. (Courtesy, The Barrett Division, Allied Chemical and Dye Corporation.)

Table 66

PERÇENTAGE COMPOSITION OF SOME AMMONIA SOLUTIONS NOW ON THE AMERICAN MARKET

	B-Liquor	*Urea-ammonia Liquor*					*Nitrogen Solutions*		
		A	*B*	*C*	*D*	37	IIA	III	IV
Urea nitrogen	..	15.1	20.2	12.3	16.4	9.1			
Insoluble nitrogen	.	..	..	..	..	7.4			
Nitrate nitrogen	.	..	..	..	..	..	11.4	9.7	11.7
Ammonia nitrogen	24.7	30.4	25 3	24.7	20.6	20.6	29.2	31.1	25.3
Total nitrogen	24.7	45.5	45.5	37.0	37.0	37 1	40.6	40.8	37.0

The urea in the ammonia liquors A, B, C, and D does not enter into the reactions but is precipitated as very fine crystals intimately mixed with the ammoniated goods.

How Ammonia Solutions Are Shipped

Ammonia solutions for use in the ammoniation of superphosphate are shipped from the synthetic plants in steel tank cars that hold about 90,000 pounds of solution. When the loaded cars reach the fertilizer plant the solution may be used directly from the car or it may be unloaded into stationary tanks for temporary storage. It is generally used directly from the car. This is illustrated in Fig. 69.

Chemical Changes Involved in the Ammoniation of Superphosphate

In the ammoniation of superphosphate the common practice is to introduce the ammonia solution in one operation, although two-stage ammoniation is feasible. The most economical procedure in manufacturing plants producing complete fertilizers appears to be to introduce a certain amount of the ammonia solution whenever superphosphate-containing materials pass through the mixer.

The chemical reactions involved vary somewhat, depending upon whether the superphosphate is ordinary or concentrated superphosphate. The reactions of ammonia and ordinary superphosphate are probably as follows:

$$H_3PO_4 + NH_3 = NH_4H_2PO_4 \quad (1)$$

$$2CaH_4(PO_4)_2.H_2O + NH_3 = 2NH_4H_2PO_4 + Ca_2H_2(PO_4)_2 + H_2O \quad (2)$$

$$NH_4H_2PO_4 + NH_3 = (NH_4)_2HPO_4 \quad (3)$$

$$2Ca_2H_2(PO_4)_2 + 2CaSO_4 + 4NH_3 = 2Ca_3(PO_4)_2 + 2(NH_4)_2SO_4 \quad (4)$$

As shown in the above equations, when ordinary superphosphate is treated with ammonia a portion of the monocalcium phosphate and the small amount of free phosporic acid which is usually present are converted into monoammonium phosphate. Another portion of the monocalcium phosphate is converted into dicalcium and tricalcium phosphates. The amount of tricalcium phosphate produced increases rapidly with the amount of ammonia added above about 2.5 per cent. This has been shown by the work of Kennen (1930) and others. Four per cent ammonia, or 40 pounds of the ammonia per 1000 pounds of superphosphate, is generally considered the maximum practical absorption for ordinary superphosphate. If precautions are taken to introduce the correct quantities of ammonia into the superphosphate, reaction (4), as shown above, will not take place, and the formation of the less soluble tricalcium phosphate will be prevented. The introduction of correct quantities of ammonia to give the above results is called *minimum ammoniation*. The introduction of a quantity of ammonia which is less than that necessary for minimum ammoniation is called *under ammoniation*. Under ammoniation may be objectionable in that if the ammoniated superphosphate is later mixed with ammonium sulphate caking in bags may arise, as is explained on p. 56.

Ordinary superphosphate absorbs only up to about 6 per cent of ammonia, although a slurry of superphosphate may absorb 8 or 9 per cent. Of course, as shown in the above equations, in completely ammoniated superphosphate a very large part of the phosphoric acid has been converted to the less soluble tricalcium phosphate, and practically all of the ammonia is in the form of ammonium sulfate. The quantity of ammonia that can be added without serious formation of citrate-insoluble phosphate amounts to about 3 per cent with ordinary superphosphate and 9 per cent with double superphosphate. The composition of some ammoniated superphosphates, as determined by Parker (1931), are given in the following table.

In the ammoniation of superphosphates it has been found that high temperatures and the presence of moisture encourage reversion of phos-

Table 67

COMPOSITION OF AMMONIATED SUPERPHOSPHATE

Compound	Percentage Composition with Per Cent Ammonia Added as Indicated						
	0.0	*1.0*	*2.0*	*3.0*	*4.0*	*5.0*	*6.0*
$CaH_4(PO_4)_2$	25.0	14.3	3.5	0.0	0.0	0.0	0.0
$Ca_2H_2(PO_4)_2$	4.5	9.5	12 0	11 2	5.0	0.0	0.0
$Ca_3(PO_4)_2$	0.0	0.0	0.0	8.2	17.5	25.7	30.6
$NH_4H_2PO_4$	0.0	8.9	14.3	15.5	12.5	9.1	6.0
$(NH_4)_2SO_4$	0.0	0.0	0.0	2.7	8.5	14.3	20.0
$CaSO_4.2H_2O$	62.0	62.0	62.0	58.3	51.0	42.0	32.0
Inerts	3.0	3.0	3.0	3.0	3.0	3.0	3.0
Totals	94.5	97.7	94.8	98.9	97.5	94.1	91.6

phorus, and in many plants the heat developed in the ammoniation process has been a source of trouble and a limiting factor in the use of ammonia solutions. A rotary cooler has been developed which has proved successful in the dissipation of heat, and its use has resulted in speeding up fertilizer manufacture and the curing process in some plants. Other factors which may hasten reversion, and which must be considered if successful ammoniation is to be secured, are the degree of ammoniation, the presence of limestone, alkaline materials, gypsum, fluorides, urea, free moisture, and above all, storage at temperatures above 130° F.

Triple superphosphate and fused phosphates also may be ammoniated satisfactorily. The chemical reactions involved are about the same as those shown in the reaction of ammonia and ordinary superphosphate, but, in the absence of calcium sulfate the dicalcium phosphate may be reverted to the tricalcium phosphate, probably about as follows:

$$3Ca_2H_2(PO_4)_2 + 2NH_3 = 2Ca_3(PO_4)_2 + 2(NH_4)_2HPO_4$$

If calcium fluoride is absent the later reversion to calcium fluorapatite is prevented.

Availability of Reverted Tricalcium Phosphate

As has been explained above, the tricalcium phosphate formed when superphosphate is ammoniated is a precipitated product. Rather recent chemical and x-ray investigations have shown that this precipitated product is distinctly different from the complex chlor- and fluor-calcium phosphates which are found in phosphate rock. In recent years it has been

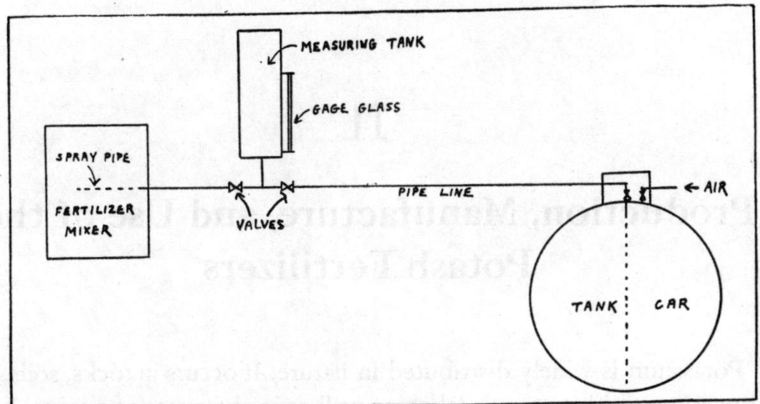

FIG. 69. Illustrating how tank car of ammonia solution may be connected with fertilizer mixer at fertilizer plant.

abundantly shown that the phosphoric acid of precipitated tricalcium phosphate is available to crops but not to the extent of that in the monocalcium or dicalcium forms. The phosphate availability of reverted tricalcium phosphate appears to be about equal to that in bonemeal and basic slag.

Because at least a portion of the reverted tricalcium phosphate has been shown to be available to plants, the American Association of Official Agricultural Chemists in 1932 adopted certain modifications to the official method for the determination of citrate-insoluble phosphoric acid. These modifications now allow the fertilizer manufacturers to use larger quantities of ammonia in their fertilizer mixtures before sustaining a serious loss of available phosphoric acid. The number of manufacturers taking advantage of this cheap form of nitrogen is increasing. Williamson (1935) found the relative increases in cotton yields due to four phosphate carriers to be as follows: superphosphate 100; ammoniated superphosphate (2% N) 100; ammoniated superphosphate (4% N) 90; and precipitated tricalcium phosphate 85.

Availability of the Water-insoluble Nitrogen Derived from Urea-ammonia Liquor-37

Vegetative and nitrification experiments conducted by McCool (1941), at the Boyce Thompson Institute of Plant Research, has shown that the insoluble nitrogen derived from UAL-37 is a good source of nitrogen for growing crops. The tests showed that the availability of the nitrogen does not change in storage, and that its availability compared favorably with many insoluble nitrogen materials used in the manufacture of fertilizers.

11

Production, Manufacture, and Use of the Potash Fertilizers

Potassium is widely distributed in nature. It occurs in rocks, soils, and in the saline residues of salt lakes, as well as in the water of oceans, lakes, and rivers, and in the tissues of plants and animals. It has been estimated by Clarke (1924) that the crust of the earth will average about 2.58 per cent of potash.

Early Potash Industry of the United States

The production of potash from wood ashes was begun in the United States as early as 1608 at Jamestown, Virginia. In fact, one of the incentives of Great Britain for establishing colonies along the Atlantic Seaboard was the acquisition of a source of potash supply from wood ashes to meet the growing industrial needs of England.

The potash first made was a crude product largely compounded of a mixture of potassium carbonates, chlorides, and sulfates, together with lime, magnesia, phosphates, silica, and carbon. Later, a partially refined product called pearl ash found its way on the market.

At one time during the Revolutionary War, Massachusetts advertised a bounty of £100 per ton for potash manufactured in the American colonies. Also at this time potash was accepted in lieu of taxes.

The first patent approved by the United States government was for a process for making potash and pearl ash. This patent was signed by President George Washington on April 10, 1790, and was issued to Samuel Hopkins on July 31 of that year.

The first custom house records show that in the 14 months from August, 1789, to October, 1790, there were exported from the United States 7,050 tons of potash with a value of $661,634.00, and 1,548 tons of pearl ash with a value of $177,459.00. By 1825, the value of the exports of potash and pearl ash had reached $2,000,000, a peak from which our

FIG. 70. Pendleton Farmers' Society Hall in Pendleton, South Carolina—oldest farmers' society hall in the United States. (Courtesy, Gilbeart H. Collings, President, Pendleton Farmers' Society.)

The minutes of the Pendleton Farmers' Society record that it was here, at a banquet held the evening of October 20, 1843, that Thomas Green Clemson offered the following toast:

"In the absence of marl, permit me to propose to the citizens of Pendleton, a more familiar acquaintance with the effects of the application of those soils which contain potash, soda, and magnesia."

The date of this toast is significant in that it was during 1843 that the Rothamsted Experiment Station was established. This was two years prior to the time that Liebig determined potash to be an essential plant nutrient and many years prior to the time that nitrogen and magnesium were definitely determined to be essential for plant growth.

exports rapidly declined owing to an exhaustion of timber in our Atlantic States and a change in the market demands of Europe.

As pointed out by Browne (1935), it should not be overlooked that from an agricultural point of view the manufacture of potash from wood ashes constituted one of the greatest economic crimes in the history of the United States. The shipment of wood ashes for two centuries from the United States to Europe did much to retard the agricultural development of the eastern United States, and also did much to aid the development of European industries at a time when potash was needed and other sources were unknown.

Development of the European Potash Industry

Following the discovery in 1839 of tremendous deposits of soluble potash salts in Germany, and the opening of these deposits for exploitation in 1860, the European potash industry developed rapidly. From that time until the beginning of World War II it dominated the potash market of the world because it controlled the only known large deposit of soluble potash salts, and because these deposits were accessible to the consuming markets of Europe and to the cheap water freight rates to the potash-consuming markets of the United States and the rest of the world. The European potash industry had a virtual monopoly on all the potash reaching the American market until the beginning of World War I in 1914, and it was able to dictate prices on the American market for a decade or more following the cessation of hostilities.

Development of the Present Potash Industry in the United States

During World War I considerable progress was made toward developing a new American potash industry. The stimulus of high prices for potash—50 per cent muriate sold for as much as $500.00 a ton in December, 1915—encouraged private capital to seek new sources and $50,000,000 was invested in the production of American potash. With the close of the war and the resumption of German shipments of potash salts, most of the plants producing potash in the United States were closed. The one outstanding exception was the American Potash and Chemical Corporation located at Trona, on Searles Lake, near Los Angeles, California. It has been estimated that $50,000,000 was spent in making wartime potash, but the annual output never reached more than 54,000 tons of potash.

Between 1921 and 1930 the United States again became dependent on foreign countries for its supplies of potash. Also, during this period popular interest in a permanent source of potash declined. Nevertheless, the United States Congress wisely appropriated $500,000 for exploration and investigation of potash deposits within the United States by the Bureau of Mines and the Geological Survey.

At the close of the war in 1918 American potash came from many sources and was produced by 128 plants. With the resumption of imports in 1919, 74 out of 77 war emergency production units were forced out of business and by 1939 production from kelp, wood ashes, the Nebraska lakes, alunite, cement dust, beet sugar waste, blast furnace dust and other miscellaneous sources had practically ceased and there remained only 6 companies producing potash in the United States. In Table 68 is given the production of potash salts in the United States since 1920. The tremendous increase in the production of American potash during the critical years of World War II is a production record that any industry could exhibit with pride.

Table 68

PRODUCTION OF POTASH SALTS IN THE UNITED STATES SINCE 1920

Year	*Number of Plants*	*Potassium Salts (tons)*	*Equivalent as Potash (tons)*
1921	20	25,485	10,171
1922	12	25,176	11,714
1923	12	39,029	20,215
1924	12	43,734	22,903
1925	9	51,565	25,448
1926	7	46,324	23,366
1927	9	76,819	43,510
1928	9	104,129	59,910
1929	5	107,820	61,590
1930	5	105,810	61,270
1931	6	133,920	63,880
1932	5	143,120	61,990
1933	4	333,110	143,378
1934	8	275,732	144,342
1935	10	357,974	192,793
1936	7	431,470	247,340
1937	7	486,090	284,497
1938	9	534,945	316,951
1939	6	546,757	312,201
1940	7	658,249	379,679
1941	7	986,458	524,875
1942	7	1,267,455	679,206
1943	7	1,428,840	739,141
1944	6	1,578,498	834,568
1945	7	1,588,305	874 243

The present, past, and possible future sources of potash in the United States are many and varied, the principal ones being as follows: industrial wastes, kelp, Searles Lake, Nebraska Lakes, Salduro Marsh, Great Salt Lake, greensand, Georgia shales, alunite, leucite, subterranean deposits of soluble salts in Texas and other states.

In Table 69 is given the name, composition, and potash content of the most important potash minerals of the United States as compiled by Dolbear (1946) and Johnson (1933).

Industrial wastes offer a very promising field as a source of potash. It was estimated by Turrentine in 1926 that 170,000 tons of potash is vola-

Table 69

POTASH MINERALS OF THE UNITED STATES*

Name	Composition‡	Equivalent K_2O† Content (per cent)
Chlorides:		
Sylvite	KCl	63.1
Carnallite	$KCl.MgCl_2.6H_2O$	17.0
Chloride-sulfates:		
Kainite	$MgSO_4.KCl.3H_2O$	18.9
Sulfates:		
Alunite	$K_2(Al(OH)_2(_6)SO_4)_4$	11.4
Polyhalite	$K_2SO_4.MgSO_4.2CaSO_4.2H_2O$	15.5
Langbeinite	$K_2SO_4.2MgSO_4$	22.6
Leonite	$K_2SO_4.MgSO_4.4H_2O$	25.5
Syngenite	$K_2SO_4.CaSO_4.H_2O$	28.8
Krugite	$K_2SO_4.MgSO_4.4CaSO_4.2H_2O$	10.7
Apthitalite (Glaserite)	$(K,Na)_2SO_4$	42.5
Picromerite (Schoenite)	$K_2SO_4.MgSO_4.6H_2O$	23.3
Kalinite	$K_2SO_4.Al_2(SO_4)_3.24H_2O$	9.9
Nitrates: Niter	KNO_3	46.5
Silicates:		
Leucite	$KAl(SiO_3)_2$	21.4
Feldspars:		
Orthoclase	$KAlSi_3O_8$	..
Microcline	$KAlSi_3O_8$	16.8
Anorthoclase	$(Na,K)AlSi_3O_8$	2.4–12.0
Micas:		
Muscovite	$H_2KAl_3(SiO_4)_3$	11.8
Biotite	$(H,K)_2(Mg.Fe)_2Al_2(SiO_4)_3$	6.2–10.1
Phlogopite	$(H,K,Mg,F)_3Mg_3Al(SiO_4)_3$	7.8–10.3
Lepidolite	$K,Li(Al(OH,F_2)Al(SiO_3)_3$	10.7–12.3
Zinnwaldite	$H_2K_4Li_4Fe_3Al_8F_8Si_{14}O_{42}$	10.6
Roscoelite	$H_8K(Mg,Fe)(Al,V)_4(SiO_3)_{12}$	7.6–10.3
Glauconite	$KFeSi_2O_6.nH_2O$	2.3– 8.5
Carnotite	$K_2O.2U_2O_3.V_2O_5.3H_2O$	10.3–11.2
Nephelite	$K_2Na_6Al_8Si_9O_{34}$	.8– 7.1

* Johnson, Bertrand L.: "Potash," Economic Paper 16, U.S. Bur. Mines.
† Figures in above publication as Potassium converted to K_2O, by W. F. Boericke.
‡ Certain of the formulae above may be subject to minor adjustments reported since publication.

tilized and escapes in the furnace gases of the cement and blast furnace industries alone; that if the potash lost in waste from sugar and alcohol industries is added to the above amount, the total potash loss from these industries would be in excess of 200,000 tons, an amount almost equivalent to our importation. Tobacco stems, wool waste, flue dust, and similar wastes from many industries are available and it is believed that eventually they will be completely utilized by the fertilizer industry.

Beginning in 1925 with the discovery of sylvite in New Mexico, a new mining industry was definitely established in America and independence from foreign sources of potash became a reality. Nevertheless, at the beginning of World War II, in September, 1939, domestic producers had just begun to compete strenuously with importers in the American market and to venture into the export trade in a small way. Today 80 per cent of the potash production of the United States comes from the Carlsbad region in New Mexico. A great deal of the remainder comes from the natural brines of California and Utah. The various American sources of potash, as well as some of the principal world sources of potash, will be discussed in the following pages.

Wood Ashes

It has been estimated that there are about 6,000,000 tons of sawdust produced in this country annually. This is sufficient sawdust, if it were properly burned, to manufacture 6,000 tons of potassium carbonate. Of course, the hauling necessary to collect such scattered material would make the cost of the material prohibitive.

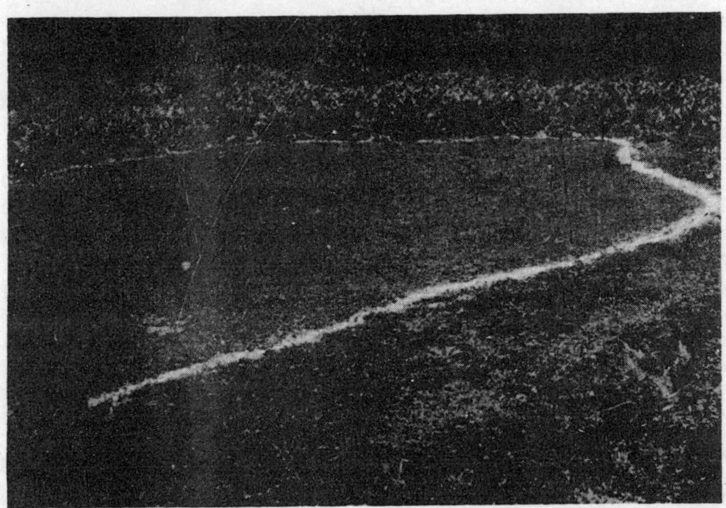

FIG. 71. Area of white clover outlined by lime in contrast with the periphery of moss, ferns, and weeds. This vegetative change was brought about from the application of wood ashes produced by a maple syrup evaporator. (Courtesy, A. R. Midgley and the Vermont Agricultural Experiment Station.)

Wood ashes may contain from 5 to 25 per cent of potash. Ashes resulting from the burning of twigs and small branches are richer in potash than those secured from trunk wood. Softwood ashes contain less potash than hardwood ashes.

It is of interest to note here that the term *potash* is derived from the term "pot ashes." Pot ashes were formerly obtained by leaching wood ashes and evaporating the solution to dryness. The salts secured consisted of an impure mixture of potassium carbonate with other salts.

Shine ash is an ash made from burning waste from flax grown in the United Kingdom. It contains some potash and phosphoric acid.

Although unleached wood ashes contain about 2 per cent of phosphoric acid, 5 to 25 per cent of potash, and 30 per cent of lime, leached wood ashes contain only about 1.5 per cent of phosphoric acid, 1 per cent of potash, and 28 to 30 per cent of lime. The potash found in wood ashes is in the form of potassium carbonate. As potassium carbonate is very alkaline, wood ashes should not be applied in such a manner that they come in contact with germinating seeds or plant roots.

The percentages of phosphoric acid and potash in the ashes of some common vegetable wastes, as determined by Jenkins (1917), of the Connecticut Experiment Station, are given in the following table.

Table 70

PERCENTAGES OF PHOSPHORIC ACID AND OF POTASH FOUND IN THE ASHES OF MISCELLANEOUS VEGETABLE WASTE

Vegetable Waste	*Phosphoric Acid*	*Potash*
Apple parings...........	3.08	11.74
Banana skins....................	3.25	41.76
Cigar ashes.....................	2.57	16.81
Corn cobs.......................	3.14	17.25
Grapefruit skins.........	5.58	30.64
Lemon skins................ ..	6.30	31.00
Orange skins....................	2.90	27.04
Peanut shells...................	1.23	6.45
Potato peelings.................	5.18	27.54

Vegetable or plant ashes derived from kelp, tobacco, seablight and glasswort is available on the European market.

Tobacco Stems

In the manufacture of cigarette, cigar, and chewing tobaccos there results considerable waste, consisting largely of stems and the ribs of the tobacco leaves. This waste material is sometimes ground and sold as fertilizer. The product contains 4 to 10 per cent of potash and 2 to 3 per cent of nitrogen but very little phosphoric acid. The nitrogen is present both in the organic and nitrate forms.

Potash Derived from Wool Waste

Before raw wool can be used by the spinner it must be treated for the removal of a material called "suint." Suint is the dried sweat of sheep and to a large extent consists of potash salts. As crude wool contains about 1.5 to 5.0 per cent of potash by weight, this source of supply of potash is of some economic importance, considering the amounts of wool produced and imported into this country annually.

Source of Flue Dust

Flue Dust was introduced on the market as a source of potash during World War I. It is a by-product of iron smelting and Portland cement manufacture. In the smelting of iron, iron ore, together with coal, coke, and limestone—all of which carry varying, although small, amounts of potash—are put into a blast furnace. When ignited, the tremendous heat produced volatilizes part of the potash compounds. The gases containing the potash compounds are led from the furnace to ovens where they are burned. The potash compounds are deposited in the dust which settles in the flues through which the gas passes. From time to time this dust is collected and sold for fertilizer purposes. The major portion of the potash of the furnace charge, however, is eliminated in the slag. According to

FIG. 72. Influence of potash on the yield of cotton at the South Carolina Sandhill Experiment Station. (*Left*) 200 pounds of potash per acre. (*Right*) No potash. (Courtesy, South Carolina Agricultural Experiment Station.)

Turrentine (1926), the addition of sodium chloride to the blast furnace charge has been found to increase the yield and the water solubility of the potash. Flue dust from this source has been put on the European market in a number of grades ranging from 3 to 13 per cent of potash. The potash is present largely in the chloride, sulfate, and carbonate forms. No potash has been recovered from this source in recent years.

The minerals from which Portland cement is made usually carry varying amounts of potash. In the manufacture of cement the potash is volatilized from these minerals and escapes with the fumes and dust from the cement mill. The process of recovering the dust at the cement plants was first perfected at Riverside, California, in an attempt to stop the nuisance caused by the escape of fumes and dust, and their consequent injury to citrus groves. This source of domestic supply of potash is very promising. Ross *et al.* (1917) analyzed 113 samples of the "raw mix" used in the mills of the United States and Canada, and found that the potash content of raw mix varied from 0.20 to 1.16 per cent. The same investigators found that the percentage of potash volatilized from the raw mix in different manufacturing plants varied from 24.5 to 95.9 per cent, and they estimated that in 1917 for every barrel of cement produced in this country 1.93 pounds of potash escaped. It has also been estimated that at most cement plants at least 90 per cent of this potash is recoverable. Although at the present time some flue dust from cement kilns, an impure sulfate of potash, is being profitably recovered by the North American Cement Company at its plant at Hagerstown, Maryland, many thousands of tons are still being lost from the kilns of both the United States and Canada.

The potash in flue dust from cement plants is recovered as a sulfate and may be produced in high concentrations. This source of potash has the great advantage of being not far removed from the fertilizer markets. It has been estimated that about 71 per cent of the cement plants of this country are located in states where the use of commercial fertilizer is already established.

Potash Derived from Sugar and Alcohol Industries

Potash salts are obtained as a by-product of the distillation of cane sugar molasses at industrial alcohol plants. At one American plant, the United States Industrial Chemicals, Inc., of Baltimore, the waste which contains the potash is burned and a vegetable ash is secured which analyzes about 33 per cent of potash. It has been called *vegetable potash*.

In Europe, a waste containing potash is secured in the refining of ugar-beet molasses. It is sometimes sold as such, or refined to yield an sh containing about 35 per cent of potash.

If all domestic waste sources of potash were utilized it might be ossible to realize 140,000 tons of potash in the United States annually.

eaweed as a Source of Potash

Seaweed has been referred to as the "poor man's manure." It contains high content of potash, but this varies with the variety. It is poor in hosphoric acid and as rich in nitrogen as barnyard manure. In its fresh tate seaweed also carries a very high percentage of water, usually 70 to 0 per cent. Some farmers who live near the coast in New Jersey, Conecticut, and Rhode Island use the material as a green manure, but it an be used profitably in the fresh state only in sections near the coast. n some areas in the tropics it has found a wide use as a fertilizer for ananas.

During World War I kelp and other seaweeds of our Pacific coast vere dried and either ground or burned, and shipped to the eastern narkets as a potash fertilizer. Kelp occurs in great beds all the way from Magdalena Bay, in Lower California, to the Shumagin Islands off the Alaskan peninsula. These plants grow very rapidly and may often attain length of 100 feet. The beds are accessible, as was shown in 1912 by urveys made by the Bureau of Soils of the United States Department of Agriculture. The annual crop is of sufficient size to supply, if necessary, large part of our annual consumption of potash. Not only does seaweed yield potash, but it may also yield iodine, ammonia, carbon, acetone, nd tar. These by-products are valuable and aid in reducing the cost of producing the potash. Kelp as a possible source of iodine alone is probably of as much economic importance as it is as a source of potash, for t the present time we are dependent upon foreign sources for iodine. During World War I kelp was harvested by marine harvesters, dried in lirect-heat rotary kilns, and burned to an ash. The ash contained about 0 per cent of potash.

Potash Derived from Sea Water

Sea water contains an inexhaustable amount of potash. A cubic mile of sea water may contain over 2,000,000 tons of potash. Its recovery has been investigated and progress in the profitable extraction of this potash can be expected. Certainly sea water offers an unlimited supply of potash.

Potash Deposits of Searles Lake

Potash salts were discovered in the brine of Searles Lake in 1911, and during that year the American Trona Corporation, now the American Potash and Chemical Corporation, erected the first plant for the extraction of potash. Commercial production was started in 1915. This company was originally a British corporation. Its successor in 1926, the American Potash and Chemical Corporation, is now the second largest single potash producer in America. Searles Lake is located in San Bernardino County, California. It occupies the floor of a valley that is 1618 feet above sea level and is largely surrounded by hills. It is a body of mud and crystalline salt of about 12 square miles in area and averages about 70 feet in depth. The brine has a concentration of about 34 per cent and contains about 2.5 per cent of potash or a total of 20 million tons of potash in the deposit. The brine also contains about 65 per cent of water, 1.5 per cent of anhydrous borax, 37.2 per cent of sodium sulfate, and 16.5 per cent of sodium chloride. From this brine, potassium chloride from 95 to 98 per cent purity is recovered. The muriate of potash is produced at such a cost thar it has been able to compete with foreign potashes in the American market. It analyses 62 per cent potash. Not only is high-

FIG. 73. Aerial view of the plant of the American Potash and Chemical Corporation, Trona-on-Searles Lake, California. (Courtesy, American Potash and Chemical Corporation.)

FIG. 74. The porous, crystalline salt surface of Searles Lake in California is hard enough to support easily the weight of an automobile or tractor. (Courtesy, American Potash and Chemical Corporation.)

analysis muriate produced but some potassium sulfate is now manufactured by base exchange with sodium sulfate. In addition, phase-rule studies have led to the production of such by-products as borax, soda ash, salt cake, sodium-lithium phosphate, boric acid, sodium sulfate, potassium bromide, crude lithium salt, and bromine. In 1918 the product manufactured by this company contained about 15 per cent of borax. Since that time methods have been devised whereby much of the borax of the crude brine may be removed, so that the muriate of potash from this source now put on the market has a borax content of only about 0.5 per cent.

From 1921 to 1932 natural brines were the major source of potash in the United States but they now produce only about one-third of the domestic production, although additional potash-bearing areas at Searles Lake have recently been opened to further exploitation.

Potash Derived from Brines of Nebraska Lakes

In the western part of Nebraska there are a number of small shallow lakes that contain varying concentrations of sodium and potassium salts. The potash is present mainly as potassium carbonate. The source of the potassium has not been explained. During World War I many plants were erected for the purpose of extracting the potash salts from these brines by evaporation and during this period the brines furnished most of the potash of American origin. The crude salt was ground and shipped to market without further treatment. At the close of the war these plants were forced to shut down as they were unable to compete with foreign potash. The quantity of potash available from this source has been estimated by Gale and Hicks (1920) to be 20 million tons of potash.

Potash Derived from the Great Salt Lake Desert

The desert west and southwest of the Great Salt Lake, Utah, contains 2 exposed salt bodies and these are underlaid with brines. These salts and brines consist largely of magnesium, potassium, and sodium chlorides. The larger salt body near Salduro has for some time been a source of common salt. During World War I potash was recovered from the associated brines, and in 1920 the Utah Salduro Company of Salduro, Utah, was the largest individual producer of potash in the United States. Production in this territory was discontinued for some years beginning with 1921. A small amount of high-grade muriate has been produced

FIG. 75. Loading low-grade potash salts from evaporating pond, Salduro Marsh, Utah. (Courtesy, American Potash Institute.)

recently from this source by the Bounerville Limited through solar evaporation of the brines and the separation of sodium and potassium chlorides by flotation.

The area is covered with a deposit of salt to a maximum depth of four feet. The clay beneath the salt is nearly saturated with sodium chloride and 1.5 to 3.0 per cent of potassium chloride. Briefly, the method of recovery is to dig ditches about 10 feet deep, 3 feet wide and several miles long through the salt flats and extending laterally from a central point at which a pump is located. As the ditches fill with brine by seepage, they are emptied by pumping the brine into shallow evaporating ponds held by dikes of salt where evaporation takes place during the summer months. As evaporation takes place, the brine is conducted from pond to pond until both the sodium chloride and potassium chloride are precipitated, yielding a product containing 30 per cent potassium chloride. The salt mixture is then pulped, subjected to the flotation process for the separation of the two salts and dried in a rotary kiln.

It has been estimated that the recoverable potash in this deposit may amount to 1,000,000 tons.

Nolan (1928) has analyzed 199 samples of the brine and found them to average about 0.3 per cent of potash.

Potash Derived from Porphyry Copper Mine Tailings

Extensive mining of the porphyry copper deposit in Utah and other mining states has produced immense quantities of finely ground tailings analyzing 5 to 10 per cent of potash. Laboratory tests have shown that this potash can be extracted and put on the market as 90 per cent muriate of potash.

Potash Minerals of the Nature of Silicates

Immense deposits of insoluble potash minerals occur in many states. Among the more important of these deposits are: (1) the greensands of New Jersey, Delaware, and Maryland, as well as other Atlantic and Gulf States; (2) the potash shales of Georgia, Ohio, and Minnesota; (3) the deposits of alunite in Utah; and (4) the deposits of leucite in Wyoming.

Many investigators, as well as the Mineral Experiment Station of the United States Bureau of Mines, are now experimenting with methods for the extraction of the potash from the potash-bearing silicates. Many patents have been taken out with this object in view. Some of these concern themselves with processes which depend upon the volatilization of the potash by heating the silicates to high temperatures, while others deal with methods which depend upon leaching the silicates with acids and alkalies.

Potash from Greensand

Greensands occur in our eastern states, notably in the states of New Jersey, Delaware, Maryland, Virginia, and North Carolina. Greensand is found in New Jersey alone in quantities sufficient to supply the potash needs of the United States for hundreds of years. Much of this material can be extracted from open pits or mined with the steam shovel. The deposits range in thickness from a few feet to 30 feet, and carry about 6 per cent of potash. Ashely (1918) has estimated that a 20-foot bed that covers a square mile should yield 1,500,000 tons of potash. The nearness of this huge tonnage to the principal consuming areas makes greensand one of the largest potential sources of potash in the United States. The mineral is supposed to be a product of the alteration of biotite in the presence of sea water. It is classified as glauconite ($KFeSi_2O_6.nH_2O$) and is specifically a hydrated silicate of iron and potassium. Deposits are known to exist on the sea bottom at depths ranging from 300 feet to 2 miles.

The profitable production of soluble potash from greensand depends upon the manufacture of by-products. With this end in view processes have already been evolved. The Moxham process appears to be the most

FIG. 76. Loading potash ore into mine cars with scraper loader. (Courtesy, U.S. Potash Co.)

promising. As reported by Turrentine *et al.* (1925), this process requires that the greensand be mixed with hot, dilute sulfuric acid. Reaction of the materials starts at 85° to 90° C. and this temperature is secured merely by the dilution of the acid with water and the heat of the reaction of the acid on the greensand. The reaction is complete within 6 hours. The resulting materials are alum, aluminum sulfate, alumina, iron oxide, ochers, potassium sulfate, glaucosil, and fuming, or other sulfuric acids. The potash from greensand is extracted as the sulfate. This is fortunate in that potassium sulfate receives a preferential price in the American market.

It would appear, however, that commercial exploitation of these deposits may require the utilization of large tonnages of waste material in the manufacture of Portland cement, sand-lime brick, activated silica, and other by-products.

Formerly greensand was commonly used in the neighborhood of its occurrence as a low-grade potash and phosphoric acid fertilizer. With the advent of commercial fertilizers its use in this way gradually declined. Greensand contains about 1 per cent of phosphoric acid. It is sometimes referred to locally as greensand marl. It contains, as stated elsewhere, about 6 per cent of potash, but in some cases it may also carry a small amount of phosphoric acid.

When greensand is applied to the soil it becomes available very slowly. Although its action on plant growth is not immediately apparent, a slight influence in this direction appears a few years after application and may be noted for some years following. Skeen (1925) has secured results which indicate that after the addition of from 5 to 15 tons of greensand per acre the average soil which is deficient in potash should show an increase in crop growth. Rudolfs (1922) has shown that a small amount of potassium is liberated from greensand when the greensand is composted with sulfur.

Potash from Georgia Shales

The Georgia shales, sometimes called Cartersville slate, occur in deposits extending for about 9 miles in a northeasterly direction from Cartersville, Georgia. The potash shales extend to an unknown depth, and occur in beds a thousand or more feet wide. The composition of these shales as reported by Maynard (1918), are given in Table 71.

These shales are hydrous silicates and are much more easily decomposed than the anhydrous silicates. This source of potash has been somewhat neglected, but the deposits promise successful exploitation. A

FIG. 77. Cotton rust or symptoms of potash deficiency. (*Left*) Two cotto leaves showing symptoms of potash deficiency. (*Right*) Normal cotton lea (Courtesy, South Carolina Agricultural Experiment Station.)

great amount of research relative to methods for the extraction of th potash remains to be done. The deposits have been estimated to cor tain 250,000,000 tons of potash.

Value of the Potassium Feldspars as Sources of Potash

Many investigators have studied the value of such crude feldspars a leucite, alunite, phenolite and nephelite as potash fertilizers. These mi erals have a value as fertilizers if they are finely ground, but their avai ability is very low. It should be kept in mind that when fertilizers a

Table 71

CHEMICAL COMPOSITION OF GEORGIA SHALES

	Per Cent
Potash (K_2O)	8.79
Silica (SiO_2)	57.10
Alumina (Al_2O_3)	19.04
Ferric iron (Fe_2O_3)	5.37
Ferrous iron (FeO)	1.84
Magnesium oxide (MgO)	1.92
Calcium oxide (CaO)	0.01
Titanium oxide (TiO)	0.91
Soda (Na_2O)	0.53
Water plus	3.52
Water minus	0.36

rated on the basis of chemical analysis, insoluble forms of potash, such as the feldspars, are not recognized as of value.

Hartwell (1919) has reported a 3-year test with a material prepared by treating ground orthoclase-bearing rock with calcium chloride and subjecting it to heat. The resulting product contains 4.3 per cent of water-soluble potash, 5.7 per cent of total potash, and 16.0 per cent of calcium oxide. He found that under soil nutrient conditions, which were believed to be suboptimum for potassium only, the rock potash fertilizers were slightly more efficient than high-grade potassium chloride or sulfate when supplying the same amount of water-soluble potassium. His results are given in Table 72.

Table 72

YIELDS OF POTATOES AND ALFALFA PER ACRE OBTAINED FROM THE USE OF ROCK POTASH FERTILIZER

Fertilizer Applied	*Potatoes (bu.)*		*Alfalfa (tons)*
	1915	*1917*	*1918*
Check	86	233	2.40
Potassium chloride or sulfate	122	262	2.57
Rock potash fertilizer	130	285	2.76

Bauer (1921) has determined the relative capacities of plants to feed on feldspar when applied at the rate of 2 tons to the acre, as compared with the growth produced when plants were allowed to feed on potassium sulfate. The plants tested were buckwheat, corn, sweet clover, and oats. His results are given in Table 73.

Table 73

RELATIVE CAPACITIES OF PLANTS TO FEED ON FELDSPAR, AS COMPARED WITH GROWTH PRODUCED WHEN THE PLANTS WERE ALLOWED TO FEED ON POTASSIUM SULFATE

Plant	*Relative Growth*
Buckwheat	23.3
Corn	45.2
Sweet clover	106.5
Oats	107.6

Potash from Alunite

The mineral alunite ($K_2(Al(OH)_2(6)SO_4)_4$), which chemically is a hydrated basic sulfate of potassium and aluminum, is found extensively in at least six of our western states. It is insoluble in water and is usually contaminated with silica and other impurities. The largest deposits of alunite are supposed to have been formed by the action of hot water carrying sulfuric acid on rocks containing potassium-aluminum silicates. The mineral occurs in many foreign countries. Alunite has been utilized in Italy for the manufacture of alum since the thirteenth century.

The deposits of alunite near Marysvale, Utah, have attracted the most attention in the United States, and give promise of commercial development. During World War I, alunite was one of the principal sources of our potash production. The processes of recovery include ignition, leaching, filtration, and evaporation. The obtainable by-products include alum, alumina, potassium sulfate, and sulfuric acid. The profitable extraction of potash, however, depends largely upon solving operating difficulties and upon the production of profitable by-products.

In the following table is given the analysis, according to Butler and Gale (1912), of a selected specimen of the supposedly best material of a deposit near Marysvale, Utah.

Table 74

CHEMICAL COMPOSITION OF ALUNITE

	Per Cent
Potash (K_2O)	10.46
Alumina (Al_2O_3)	37.18
Sulfur trioxide (SO_3)	38.24
Phosphorus pentoxide (P_2O_5)	0.58
Soda (Na_2O)	0.35
Silica (SiO_2)	0.22
Water	12.99

Potash from Leucite

Leucite ($KAl(SiO_3)_2$) is a potash-bearing silicate mineral which occurs in Wyoming as a constituent of the volcanic rocks which go to make up the Leucite Hills of Sweetwater County. Mansfield (1930) has estimated that there are approximately 200,000,000 tons each of potash and alumina in this deposit. The potash, if recoverable on any basis fairly competitive with foreign potash, could supply our national needs for potash for generations to come, but its profitable extraction is dependent upon the manufacture of by-products since by-products would have to be produced and marketed to share with potash the cost of production. The compositions of minerals of the leucite-bearing rocks, as given by Schultz and Cross (1912), are given in Table 75.

Table 75

CHEMICAL COMPOSITIONS OF MINERALS OF THE LEUCITE-BEARING ROCKS

	Leucite	*Sanidine*	*Phlogopite*	*Diopside*
Potash (K_2O)	21.5	16.9	10.70	0.42
Silica (SiO_2)	55.0	64.7	42.56	50.86
Titanium oxide (TiO_2)	..	..	2.09	3.03
Alumina (Al_2O_3)	23.5	18.4	12.18	..
Chromic oxide (Cr_2O_3)	..	..	0.73	..
Ferric oxide (Fe_2O_3)	..	..	2.73	1.19
Ferrous oxide (FeO)	..	..	0.90	1.82
Magnesia (MgO)	..	..	22.40	17.42
Baryta (BaO)	.	..	1.00	..
Lime (CaO)	..	..	0.20	23.23
Soda (Na_2O)	..	..	0.44	0.76
Water (H_2O)	..	..	2.35	0.31
Fluorine (Fl)	..	..	2.46	..
Other substances	..	..	0.06	0.03
	100.0	100.0	100.80	99.16

American Deposits of Soluble Potash Salts

The Permian salt basin in the United States extends from southwestern Texas and southeastern New Mexico in a northeasterly direction through the Panhandle section of Texas and Oklahoma into Colorado and Nebraska. For a long time the general geological similarity of this basin and the potash-bearing salt basins of Germany and France had led geologists to hope that commercial beds of sedimentary potash salts might ultimately be found in this area.

Material evidence of a subterranean deposit of potash in the Permian salt basin in Dickens County, Texas, was reported in 1912 by J. A. Udden of the University of Texas. In the Permian period, about twc hundred million years ago, that portion of the Great Plains now comprising the western portions of Texas, Oklahoma, and Kansas, southeastern Colorado, and eastern New Mexico was covered by a shallow, intermittently landlocked sea. The remains of this sea occur today as a salt deposit about 373 to 2737 feet below the present surface. This salt deposit varies in thickness from 200 to 4,400 feet, the average thickness in the New Mexico portion of the basin being about 2,000 feet. Potash salts, primarily polyhalite, have been found in that portion of the deposit known as the Staked Plains area, an area of some 40,000 square miles. These deposits range in thickness from 1.5 to 9.0 feet and contain 9 to 13 per cent of potash. Drilling for potash in the basin was begun in 1915 by the United States Geological Survey but the well did not show any significant occurrence of potash and the strata or beds were too thin to be commercially feasible. Search for potash, however, was continued and oil-well cuttings from many parts of the country were examined by the United States Geological Survey. As a result of these examinations polyhalite was found in well cuttings from Midland and Dawson Counties, Texas, in 1921, carnallite from Utah in 1924, sylvite from a dry well in Eddy County, New Mexico, in 1925. The last discovery gave impetus to further work in the Permian basin by the United States Government, which resulted in 23 core tests being drilled during the ensuing five years and in an increased activity by certain private concerns. The minerals carnallite, langbeinite, polyhalite, and sylvinite were found in many core tests.

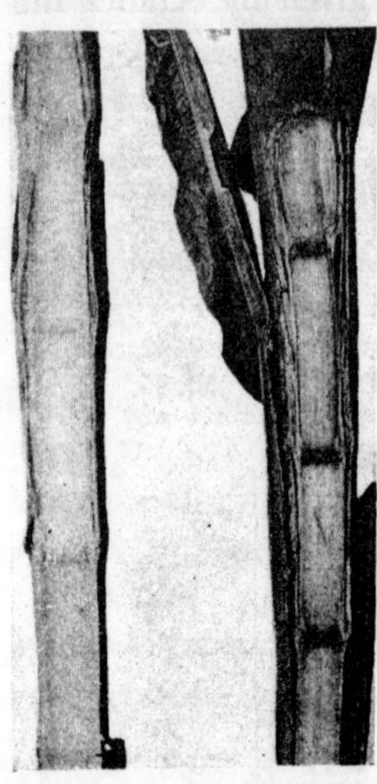

FIG. 78. (*Left*) Joint tissues of corn remain healthy in plants adequately supplied with potassium. (*Right*) The joint tissue discoloraations are caused by accumulations of iron compounds. This is the typical symptom indicating potassium starvation. The symptom is confirmed by chemical tests of these tissues. (Courtesy, Indiana Agricultural Experiment Station.)

According to Wroth (1930), the first mineral to be discovered in government test wells, in quantities sufficient to allow economical exploitation was polyhalite. Pure polyhalite contains 15.6 per cent of potash. Up to the present time 6 potash minerals, analyzing from 5 to 30 per cent of potash, that are also found in German and Alsatian potash deposits have been found in Texas and New Mexico. These are polyhalite ($2CaSO_4.MgSO_4.K_2SO_4.2H_2O$), sylvinite ($KCl.NaCl$), carnallite ($KCl.MgCl_2.6H_2O$), kainite ($MgSO_4.KCl.3H_2O$), langbeinite ($2MgSO_4.K_2SO_4$), and

FIG. 79. Refinery of the U.S. Potash Company. (Courtesy, U.S. Potash Company.)

leonite ($K_2SO_4.MgSO_4.H_2O$). Dolbear (1946) estimates a reserve of 60,000,000 tons potash alone in the polyhalite of the Permian basin. However, the polyhalite is not as easily refined as sylvinite. Turrentine (1943) estimated that the area underlain by sylvinite and other water-soluble potash beds covers an area of 3,000 square miles, principally in New Mexico, but extending slightly across the southern border of New Mexico into Texas. Large bodies of sylvinite that are 4 to 12 feet thick and rich enough and sufficiently shallow to justify extensive mining operations have been found and are now being mined commercially. Smith (1933) has reported that the proved area of commercial potash salts in New Mexico covers 33 square miles and contains a reserve of more than 100,000,000 tons of high grade salts, that is, a minimum of 14 per cent potash. More recent estimates made by Dolbear (1946) would expand the estimates of reserves of the New Mexico field to 400,000,000 tons. In addition it is said that probably as many as 70 counties of Texas are underlain with beds of potash salts that have been adequately explored. Much of this deposit appears to be made up of polyhalite and sylvinite.

Mining of potash salts was inaugurated on land leased from the Government in 1931, at Carlsbad, Eddy county, New Mexico, by the United States Potash Company, orginally the American Potash Company, and the first commercial shipment of potash from Carlsbad took place that year. In 1935 mining was also begun at Carlsbad by the Potash Company of America, and in 1940 by the Union Potash and Chemical Company which has recently been amalgamated with the International Minerals and Chemical Corporation. The deposits at Carlsbad are in the nature of a vein 10 to 14 feet thick and lie at a depth of about 1000 feet. They are com-

Table 76

PERCENTAGE COMPOSITION OF POTASH SALTS FROM NEW MEXICO

	Muriate	*Run of Mine Manure Salts*
Potash	62.49	27.05
Potassium chloride (KCl)	98.92	42.82
Sodium chloride (NaCl)	0.825	53.42
Sodium oxide (Na_2O)	..	28.33
Sulfur trioxide (SO_3)	0.024	..
Chlorine (Cl)	..	53.52
Insoluble matter	0.025	..
Moisture (loss at 130° C.)	0.10	0.45

posed largely of sylvinite. The crude salts as mined analyze 20 to 30 per cent of potash and are sold as such on the American market as manure salts. The crude salts are also refined and the resulting muriate of potash is the principal source of potash on the American market. American producers

FIG. 80. Influence of potash on the growth of apple trees. (*Center*) Tree received potash. (*Right and left*) Trees received no potash. (Courtesy, Florida Agricultural Experiment Station.)

have not yet attempted to supply the American farmer with potassium phosphates or nitrates, but they are producing potassium sulfate and sulfate of potash-magnesia from langbeinite. Over 80 per cent of the potash production of this country comes from the Carlsbad region.

It is unfortunate that the principal sources of soluble potash in the United States are so remote from the agricultural areas of maximum use.

As already stated 20 years ago potash beds were discovered in Utah. In 1944, the Defence Plant Corporation put down some deep holes and found three beds of potash at depths ranging from 3400 to 4210 feet. The thickest bed was 90 feet thick. Thus there appears to be in Grand County, Utah, a potash area potentially as important as that of New Mexico. The potash salts found in these deposits occur chiefly as the highly soluble minerals sylvinite and carnallite and appear to be 10 to 20 times as thick as the sylvinite beds of New Mexico.

Small beds of potash salts are also known to be located in Kansas, Ohio, New York, and Michigan, and in Westmoreland County, New Brunswick, Canada. Nevertheless, all known reserves of soluble potash in America are less than 2 per cent of the world reserves.

German and French Sources of Potash

The first boring by which potassium minerals were discovered in Germany was begun in 1839, near the town of Stassfurt in Prussian Saxony, and completed in 1843. The potash mineral carnallite was discovered in 1856.

The German deposits of potash salts underlie an area extending roughly from Thuringia on the south, and Hanover on the west, to Mecklenburg on the north. The bulk of the potash-bearing strata lies at a depth of 1200 to 3500 feet.

In 1904 the potash salt, sylvinite, was discovered as a result of a boring made during the prospecting for oil and coal at Whittelsheim, near Mulhouse in Alsace, but exploration of the deposits was not begun until 1908, and the deposits were not exploited extensively until after World War I when this area was returned to France. The French deposits extend over 85 square miles in Alsace largely between the old Maginot and Siegfried lines. Although these deposits produce a crude salt containing 5 to 25 per cent of potash, there is a complete absence of potassium and magnesium in the sulfate form, and the operators are, therefore, unable to manufacture potassium sulfate or potassium magnesium sulfate by methods used in Germany. The crude salts mined from these deposits contain on an average about 17 per cent of potash, compared with about 13 per cent of potash in the mined German salts. These deposits, which are of Tertiary formation, are found 1600 to 3000 feet below the surface and are tilted at an angle of about 2 to 3 degrees. The deposits of potash salts occur largely in two beds, one at an average depth of about 2050 feet and another at a depth of about 2150 feet. The upper layer is about 4 feet thick and will analyze 20 to 25 per cent of potash. The lower layer is larger in area and is said to average about 9 feet in thickness and to analyze 15 to 21 per cent of potash.

FIG. 81. Cross section of European potash mine showing strata of potash salts. The wide dark areas within the white salt deposit are potash salts. (Courtesy, N. V. Potash Export My., Inc.)

Origin and Nature of the Subterranean Soluble Potash Salts

Both the American and the European deposits of soluble potash minerals were laid down in ancient geologic times and had their origin in the evaporation of sea water. It is thought that lakes of sea water were cut off from the ocean and became saline to an extent comparable to the water of the Great Salt Lake and the Dead Sea of today, and that finally through further evaporation of water their salts were precipitated. It is also supposed that these lakes were connected with the ocean by a channel, or channels, through which further incursions of sea water were at times forced into the lakes by storms. Following each incursion of sea water evaporation continued until as the saturation point was reached the salts were precipitated in the reverse order of their solubility—i.e., calcium sulfate on the bottom, then calcium chloride, magnesium sulfate, potassium chloride, and magnesium chloride, in the order named. This order is not invariable throughout the deposits, for in many places strata of mixed salts can be found.

Over 30 different minerals, mostly potassium salts, have been discovered in the American and European deposits. The most common of the minerals are anhydrite, carnallite, gypsum, langbeinite, sylvite, kainite, kieserite, halite, polyhalite, and krugite. These deposits of soluble salts have been largely protected from the action of rainwater by a covering of "salt clay" impervious to water. The beds of crude salts are of enormous extent. The position of these deposits of salts follows well understood physical and chemical laws, and the apparent heterogeneity of the deposits can be explained by such laws when local geologic conditions are taken into consideration. In some localities, where the deposits have been exposed to the action of water, the salts have been dissolved and then redeposited so that secondary deposits of kainite and sylvinite have arisen. Estimates as to the duration of the period over which the disposition of the German and French salts took place vary from 1500 to 13,000 years.

Mining of Potash Salts

The deposits of potash-bearing minerals are usually found from 650 to 5200 feet below the earth's surface in Germany and France and at about the 1000 foot depth in the United States. They are reached by mine shafts.

The Carlsbad deposits of the United States are mined by what is known as the Room and Pillar method of mining. After a shaft is sunk to a point slightly below the level of the ore body to permit a gravity ore loading bin, a large tunnel is driven on both sides of the shaft. At right angles to this tunnel the potash ore is mined in large rooms about 50 feet wide and 300

FIG. 82. In potash mine. Drilling equipment and undercutter at the face. (Courtesy, U.S. Potash Co.)

feet long. A pillar or wall about 50 feet thick is left between the rooms to support the roof. When the potash minerals have been removed from an area in this manner another tunnel is driven and the operations continued. The ratio of ore extracted to ore left in the pillars is about 60 to 40. No mined areas have been robbed, for there is water in the overburden and water must be kept out of potash mines at all costs.

The line of demarcation between the potash ore and other salt deposits is rather clear. Undercutting machines are used to cut a slot to a depth of 8 feet across the lower part of the potash bed, and into the face of potash ore are drilled a number of holes in which dynamite is placed. The dynamite is then exploded and the ore is broken down clearly and loaded into cars by a dragline scraper. Electric locomotives haul the loaded cars to the shaft where the ore is carried up by elevators to bins in the head frame. American Mines are dry, well ventilated and the colored crystals give an effect of clearness seldom associated with mines. The potash ore is taken from the bins to the refinery, or it is crushed and screened and prepared for the trade as "run of mine" potash salts.

The characteristics of some of the most important potash minerals and salt mixtures now taken from the European and American deposits are given below.

Langbeinite

Langbeinite is mined at one of the mines at Carlsbad and the ore is refined for the production of potassium sulfate, sulfate of potash magnesia, and by-product magnesium chloride. It is harder than sylvinite or halite and is about as hard as limestone. It is brittle, shatters easily, and is for this reason more easily blasted than sylvinite. It has the formula ($K_2SO_4.2MgSO_4.XNaCl$).

Carnallite

Carnallite is the principal potash mineral found in the German deposits but at present is economically of only secondary importance. It is a mixture of magnesium potassium chloride, $KCl.MgCl_2.6H_2O$, together with impurities such as common salt and magnesium sulfate. As mined, it contains only about 9 to 12 per cent of potash. This, together with its hygroscopic character, makes it unsuited for long distance shipping for fertilizer use. It is commonly used for fertilizer in European localities within easy transport distance of the mines.

Kainite

The mineral kainite is found in the American, German, and French deposits. The color of the salt may vary, and all shades of white, gray, yellow and red may be found. Kainite deposits are often intersected with rock salt and other minerals. In fact, very little pure kainite, $KCl.MgSO_4.3H_2O$, occurs as such.

The commercial product, kainite, consists largely of a mixture of potassium chloride, magnesium sulfate, and magnesium and sodium chlorides. Any low-grade potash salt of subterranean origin, carrying 14 to 22 per cent of potash, may now be referred to in the American trade as kainite. In recent years a kainite analyzing 20 per cent of potash has replaced much of the lower grades on the American market. Kainite usually carries about 46 per cent of chlorine. It is alkaline to litmus. When kept in storage kainite, like most of the potash salts, often "sets up" and has to be reground.

Sylvinite

Sylvinite is the principal potash salt mixture in the American and French deposits. It is thought to be a decomposition product of carnallite and is composed largely of sylvite (KCL) and sodium chloride, together with small amounts of magnesium chloride and calcium sulfate. The name sylvinite, like kainite, has now a commercial usage and in France is applied to various fertilizer salts having a potash content of 20 to 42 per cent, regardless of composition. The mine-run material taken from the American deposits is marketed as manure salts.

Hardsalz

Hardsalz is not a definite mineral, but a natural mixture of sylvite, kieserite ($MgSO_4.7H_2O$), and rock salt. In 1937 82 per cent of the production of crude potash salts in Germany was composed of the more valuable and readily worked hardsalz, kainite, and sylvinite bearing mixtures. Like carnallite, hardsalz is sometimes ground and used in Europe for fertilizer purposes, and is usually sold under a guarantee to contain 16 per cent of potash.

Methods Employed in the Manufacture of Muriate of Potash

Muriate of potash, or potassium chloride, occurs in the American, German, and French deposits only to a small extent in its mineral form,

FIG. 83. Mine site of the U.S. Potash Co.—headframe, ore bins, crushing plant, and tram cars. (Courtesy, U.S. Potash Co.)

sylvite. The product that is sold on the market as muriate of potash is a refined product that has been secured by removing sodium chloride from sylvinite of the American and French deposits or by removing magnesium chloride or sulfate from carnallite and hardsalz of the German deposits. The German muriate carries some magnesium sulfate and chloride, whereas the French muriate does not. While the methods of refining employed at the different refineries vary somewhat, the general procedure at most American and European plants is about the same.

The method employed in separating the potassium chloride from the other salts is largely dependent upon lixiviation, usually in hot solutions, followed by cooling and crystallization. Owing to their difference in solubility in water at different temperatures, potassium, sodium, and magnesium chlorides and sulfates disassociate themselves when the crude salts of which they are constituents are dissolved in water. Thus, the mineral carnallite readily breaks down when it is dissolved in water. Potassium chloride, when mixed with sodium chloride, is much more soluble in hot water than in cold water; whereas sodium chloride is somewhat more soluble in cold water than in hot water. This may be due to the fact that potassium and chlorine are isosteric, whereas sodium and chlorine are not.

The plant of the Potash Company of America concentrates the crude salts by the process of flotation. The mineral is ground to separate the crystals of KCl and NaCl, filmed with a reagent, a selective with respect to NaCl, and passed through a series of flotation cells. In this way fine particles of halite may be separated from fine particles of sylvite, although these two salts differ only 0.16 in specific gravity.

Muriate of potash is manufactured more extensively than any other potash fertilizer. It has been estimated that at least 90 per cent of the potassium content of potash salts consumed in the world is in the form of potassium chloride. Muriate of potash may contain 48 to 62 per cent of potash and about 47 per cent of chlorine. The American preferred product contains 62 per cent of potash and this grade makes up over three-fourths of the American production. Some of this muriate is known as "red muriate." The impurities are composed largely of common salt. The name muriate is derived from the term muriatic acid, which is the common name for hydrochloric acid. Muriate of potash is usually the cheapest carrier of potash found on the market.

Methods Employed in the Manufacture of Potassium Magnesium Sulfate

Potassium magnesium sulfate is sold on the market as sulfate of potash magnesia. At one time this material was called "double manure salts." It is useful where a low chlorine content fertilizer or a magnesium carrying fertilizer is desired. The American Association of Official Agricultural Chemists has recommended that potash salts sold in the United States as sulfate of potash magnesia contain not less than 25 per cent of potash nor less than 25 per cent of magnesium sulfate, and not more than 2.5 per cent of chlorine. The German material sold in the United States in prewar years carried about 48 to 53 per cent of potassium sulfate.

Potassium magnesium sulfate is manufactured in Germany and is prepared by dissolving kieserite in hot water until a solution having a specific gravity of 1.32 is secured. The solution is then allowed to cool to 30° C., and potassium chloride is added until the magnesium chloride content amounts to 10 to 20 per cent, the amount of potassium chloride added depending upon the composition of the kieserite. At this point the double sulfate of potash magnesia precipitates as fine crystals. The reaction is about as follows:

$$2KCl + 2MgSO_4.7H_2O = K_2SO_4.MgSO_4.6H_2O + MgCl_2.H_2O$$

In the United States potassium magnesium sulfate is produced from brines at Searles Lake, California, and from langbeinite which is mined at one of the operations at Carlsbad, New Mexico. Washed langbeinite came on the market in 1941.

Methods Employed in the Manufacture of Potassium Sulfate

Potassium sulfate was formerly manufactured in Germany from kainite but it is now made by dissolving any double sulfate of potassium and magnesium in water and then adding a concentrated solution of potassium chloride. Potassium sulfate precipitates itself and is separated from the solution by simple decantation. The sulfate is then dried, screened, ground, and sacked for shipment. One American plant produces its potassium sulfate through the interaction of potassium chloride and sodium sulfate, another through the action of caustic potash and potassium chlorate, and still another produces it from langbeinite. The French and some American operators have produced a little potassium sulfate by treating potassium chloride with sulfuric acid. The method is similar to that used in the manufacture of HCl from NaCl. High grade potassium sulfate contains at least 48 per cent of potash and not more than 2.5 per cent of chlorine. The product sold in the United States contains 90 to 95 per cent of potassium sulfate. As there is less demand for potassium sulfate than for potassium chloride, smaller quantities of potassium sulfate are manufactured. Potassium sulfate has a better physical condition than most of the potash salts and is much more finely divided. Potassium sulfate would be more generally used were it not for its high market value which results from the expense involved in its production.

Potash Salts

Potash salts that have been only partially refined are manufactured in both Europe and America. They vary in composition from 9 to 42 per cent of potash.

The German producers export 5 grades, as follows: (1) Carnallite, analyzing 9 to 12 per cent of potash; (2) crude salts, analyzing 12 to 15 per cent of potash; (3) fertilizer salts analyzing 18 to 22 per cent of potash; (4) fertilizer salts analyzing 28 to 32 per cent of potash; (5) fertilizer salts analyzing 38 to 42 per cent of potash.

The French producers export 3 grades, as follows: (1) Sylvinite analyzing 12 to 16 per cent of potash; (2) sylvinite analyzing 20 to 22 per cent of potash; and (3) sylvinite analyzing 30 to 40 per cent of potash.

The American producers are selling crude salts that analyze 20 to 30 per cent of potash under the name of *manure salts*.

Potassium Magnesium Carbonate

Small amounts of a double carbonate of potassium and magnesium have been imported into this country from Germany, and at some of the New England experiment stations this product has proved a satisfactory potash fertilizer. It contains about 20 per cent of both potash and magnesia. It is free from chlorides and can therefore be applied safely to crops which are injured by chlorides. It has a neutralizing effect on soil acids.

Potassium Silicate

Formerly potassium silicate containing about 25 per cent of potash was manufactured in Germany. It has been tested in this country by the Massachusetts Experiment Station and found to be a high-grade product comparable with potassium chloride and potassium sulfate.

By-products of the Potash Industry

In recent years the increased competition which European and American potash has had to meet has resulted in attempts to reduce the cost of potash to the final consumer. As a result high-grade potash salts are now found on the market in the place of the crude salts formerly sold. Refining of the potash minerals is accomplished by rejecting the nonpotash constituents and concentrating the potash in a small bulk. In this way the freight on long distance shipments of useless material is saved. Such by-products as rock salt, magnesium sulfate, sodium sulfate, Glauber's salt, bromine, boric acid, bromides, lithium salts, and others are now produced from the European and Californian deposits, and potassium hydroxide and magnesium chloride from the Carlsbad deposits so that potash need not carry the entire expense of the industry. Research work leading to industrial efficiency is constantly carried on.

European Control of the American Potash Market

After World War I, the Versailles Treaty gave the Galician deposits of potash to the recreated state of Poland and the Alsatian deposits to France. With the return of Alsace-Lorraine to France, the French made a successful effort to exploit the Alsatian deposits. The exportation of their product in competition with potash from German sources soon became a factor in the world trade of potash. This competition resulted in an agreement between the French and the Germans, in August 1924, which put an end to the competition in the United States and in other parts of the world.

This agreement was the beginning of the international cartel which took over the functions of the prewar German monopoly in controlling overproduction and allocating orders. This agreement fixed the proportion of sales allowed to each nation in the American market. According to this agreement France was allotted 37.5 per cent and Germany 62.5 per cent of the American sales. These figures were based on the percentages of the deliveries in 1923. In April, 1926, the agreement was revised so that sales in all foreign markets were divided between the French and German producers in the ratio of 30 and 70, respectively. In 1927 the sales offices for both the German and French products in the United States were combined under the supervision of a Dutch corporation the N.V. Potash Export My., so that foreign propaganda relative to the use of potash salts was directed from one office until the beginning of World War II in 1939 when these offices were again separated. In 1935 the Spanish potash producers became affiliated with the European potash cartel with an allocation of 15 per cent of the export market served by the group. In like manner Poland was allocated 4 per cent and it is reported that agreements were reached with producers in Russia and Palestine. Because of legal restrictions the producers in the United States were never included in the cartel.

The advent of World War II in 1939 brought an end to importations of the European potashes into the United States. In 1941 only 13 short tons of potash salts were imported into the United States and in 1944 only 5

FIG. 84. Influence of potash on tomatoes. (1) Tray from standard fertilizer plat. (2) Tray from check plat. (3) Tray from extra potash plat. (Courtesy, Better Crops and Extension Department of the Ontario Agricultural College.)

tons. It is doubtful that foreign potashes will ever again monopolize the American Market, but no doubt the American producers can look for severe competition. World War II also put an end to the German cartel but it is not at all unlikely that an Anglo-Russo-French cartel will now take its place, for the potash industry is government owned and controlled in Russia, and to a large extent in France. England controls 22 per cent of the potash resources of Germany and also the potash resources of the waters of the Dead Sea. The end of World War II brought about the occupation of Germany by Russia, Britain, and the United States so that the potash resources of Germany are controlled to the extent of 61 per cent by Russia, 22 per cent by Britain, 14 per cent by the United States, and 3 per cent by France.

Other Foreign Sources of Potash

During World War I, deposits of potash, largely sylvinite and carnallite, were found near the town of Suria in northeastern Spain. Mining operations were started a short time later. The workable crude sylvinite contains 18 to 34 per cent of potash, and the crude carnallite-bearing salts about 12 per cent of potash. By 1935 Spain had become one of the world's leading producers of potash. During the Spanish Civil War the mines were damaged and production was brought to a standstill from the latter part of 1936 through 1939. Normal output was again attained in 1940 and by 1944 Spanish production aggregated 128,000 tons of potash.

The mining of potash in Poland dates from about 1862. Very little development of the industry took place, however, until after World War I. One authority has estimated Poland's resources of potash salts in the Galician region to amount to nearly 250,000,000 metric tons. The crude salts of these deposits contain on an average about 16 per cent of potash. The potash produced in Poland, like that produced in the United States, was intended primarily for domestic consumption. In 1939 the greater part of Galicia, including the potash mines, was ceded to Russia. Production ceased, however, with the advent of World War II and many mines were flooded. A considerable time will have to elapse before they can be put into full production.

In Italy, potash has been extracted with some success from leucite, but it is very improbable that potash from this source will ever enter international trade.

The potash industry of Japan is largely based upon kelp. The kelp is harvested by manual labor, usually by women divers, and then sun-dried and shipped to the factory. Iodine and many other materials are produced as by-products of potash manufacture. Both potash and iodine were produced just prior to World War II in such quantities as to provide an excess for export.

The water of the Dead Sea, like that of our Great Salt Lake, contains a high percentage of salt. The Dead Sea lies 1300 feet below sea level, is about 47 miles long and 9 miles wide and represents the remnant of a much larger lake of the Quarternary Era. Methods for the extraction of the potash salts, on a commercial scale by salar evaporation, has received considerable attention, but many difficulties remain to be overcome before these salts can be profitably extracted. It has been estimated that the Dead Sea contains 1200 million tons of potash. In 1939 Palestine produced 80,000 tons of potash, and this was not materially increased during the period of World War II for production emphasis was laid on bromine.

Russian deposits of potash discovered in 1925 are said to be quite extensive underlying the province of Perm. Production of sylvinite and carnallite has expanded quite rapidly since 1932. Most of the production is at Ssolikamsk in the northern Urals. Little is known as to what new discoveries were made during the period of World War II. Recently Soviet engineers have urged the exploitation of the potash deposits of Central Asia. It would seem that because of their inland location the Russian deposits are not a promising source of potash for international trade.

Potash deposits are also found and worked to some extent in Abyssinia, Brazil, Canada, Norway, Peru, and Sweden. Economically, they have yet to prove of commercial importance.

World's Reserve of Soluble Potash

Mansfield (1926), of the United States Geological Survey, has estimated the world's reserve of potash in the form of soluble salts to be 2,338,000,000 metric tons. This reserve was estimated to be sufficient at the then rate of production to last nearly 2,000 years. The reserves of Germany alone were estimated at 2,500,000,000 tons. Two mines of Poland have a total estimated reserve of 10,000,000 tons. Considerable unexplored territory of Poland is believed to be underlaid by workable beds. The potash content of Spanish deposits has not been fully determined, but it is estimated that the Syneline at Suria contains 10,000,000 tons of potash. The potash deposits of Russia are of unknown extent. An estimated tonnage of 1,250,000 tons has been found in Abyssinia. As has already been stated, rather recent evidence has been secured which would indicate extensive deposits, probably amounting to millions of tons, in southeastern New Mexico, southwestern Texas, Utah, as well as in other Rocky Mountain States of the United States. Dolbear (1946) estimates these gross reserves of potash to be 106,963,000 tons.

In Table 77 is given an estimated world reserve of potash, as reported by Turrentine, (1943).

Table 77

ESTIMATED WORLD POTASH RESERVES

	Tons
United States	85,000,000
Russia	700,000,000
Poland	10,000,000
Germany	2,500,000,000
France	300,000,000
Spain	270,000,000
Palestine	1,200,000,000

Soils to Which Potash Is Applied

It is to be expected that crops grown on soils that are deficient in potassium, such as muck, peat, and sandy soils, should be particularly responsive to the use of potash. The results of many experiments have shown this to be true. Crops grown on clay soils, especially those with a high organic matter content in the younger agricultural areas of the country and of those particular soil series that have the capacity to release potassium in greater quantities than normal from the nonexchangeable forms, seldom require large applications of potash fertilizers, whereas sandy soils nearly always produce their greatest yields of crops where liberal applications of potash fertilizers have been made. In fact, in the United States potash is applied to the greatest extent to crops grown on the light gray, sandy soils of the southeastern states.

The five southeastern states—North Carolina, South Carolina, Florida, Georgia, and Alabama—consume about 40 per cent of the potash used in agriculture in the United States. In Table 78 is given the estimated consumption of potash in the United States, by states, in 1944. It will be noted from Table 78 that potash is not extensively used in our western states that have a low rainfall.

Table 78

THE ESTIMATED CONSUMPTION OF POTASH IN THE UNITED STATES BY STATES IN 1944

Rank	*State*	*Tons*	*Rank*	*State*	*Tons*
1	North Carolina	69,913	25	Massachusetts	5,388
2	Florida	58,563	26	Connecticut	4,702
3	Georgia	49,404	27	Iowa	4,053
4	South Carolina	40,560	28	Minnesota	3,940
5	Indiana	39,695	29	Delaware	3,509
6	Ohio	38,150	30	West Virginia	2,482
7	Alabama	28,442	31	Washington	1,750
8	Maine	27,460	32	New Hampshire	1,342
9	Virginia	25,542	33	Rhode Island	1,214
10	Pennsylvania	24,012	34	Vermont	1,195
11	New York	23,764	35	Oregon	1,080
12	New Jersey	19,774	36	Oklahoma	606
13	Michigan	19,661	37	Kansas	536
14	Illinois	17,889	38	Arizona	472
15	Wisconsin	16,544	39	North Dakota	182
16	Maryland	15,271	40	Idaho	180
17	Mississippi	12,416	41	Colorado	60
18	Tennessee	11,859	42	New Mexico	22
19	California	9,778	43	South Dakota	18
20	Arkansas	9,145	44	Nevada	6
21	Kentucky	8,689	45	Nebraska	4
22	Louisiana	7,301	46	Montana	3
23	Texas	7,100	47	Utah	2
24	Missouri	6,200	48	Wyoming	2

Collings (1929) has called attention to the profitableness of th European practice of making large applications of potash fertilizers t crops grown on all classes of soils.

Influence of Potash Salts On Soil Acidity

The available evidence as to the influence of the potash salts on so acidity is somewhat conflicting, but it does show that the influence c quantities generally applied is not very great and is apparently masked b the influence of other soil, cropping, and fertilizer practices. MacInti *et al.*, (1941) found that successive annual additions of potassium sulfa exerted no direct influence upon the ultimate pH of the soil. Skinner an Beattie (1917) applied potassium sulfate at the rate of 100 pounds p acre annually, and after five years determined the lime requirements of th surface soils of treated and untreated plats on which 7 different crops ha been grown. Their results are given in Table 79. These results show th potassium sulfate, when applied to the soil under test, at the rate of 100 pounds per acre annually increased the lime requirement of the soil in each case.

Table 79

INFLUENCE OF THE CONTINUED USE OF POTASSIUM SULFATE ON THE LIME REQUIREMENT OF THE SOIL, ARLINGTON, VIRGINIA, EXPERIMENTAL FARM

(Lime requirement, pounds $CaCO_3$ per acre)

Treatment	*Wheat*	*Rye*	*Clover*	*Timothy*	*Corn*	*Cowpeas*	*Potatoes*	*Average*
Soil untreated	1782	1566	1386	1691	1654	1780	1876	1677
Soil and potassium sulfate	2136	1780	1958	3104	2136	2136	2241	2212

On the other hand, Burgess (1922), of the Rhode Island Experiment Station, concluded that all potash salts reduced slightly the acidity of the soil. The lime requirements of all plats which had received equivalent quantities of potash, but in different carriers, for 10 years are given in Table 80.

Table 80

RELATIVE INFLUENCE OF THE CONTINUED USE OF VARIOUS CARRIERS OF POTASH ON THE LIME REQUIREMENT OF THE SOIL, RHODE ISLAND EXPERIMENT STATION

Carrier of Potash	*Total Crop Grown During Last Five Years (tons)*	*Lime Requirement, Jones Method (lbs. CaO per acre)*
None	11.35	1800
Potassium sulfate	16.59	1530
Potassium chloride	17.92	1350
Potassium magnesium sulfate	14.11	1260
Kainite	17.72	1080

Russell (1932), of the Rothamsted Experiment Station, England, has been unable to find evidence at that station that potassium sulfate tends to increase the acidity of the soil. Theoretically, this is what would be expected, for both KCl and K_2SO_4 are neutral salts.

Crops to Which Potash Is Applied

In this country a large percentage of the potash consumed by field crops is applied to cotton, corn, and potatoes. This is not due necessarily

FIG. 85. (*Left*) Healthy corn plant. (*Right*) Corn plant showing marginal leaf burn and other symptoms of potash deficiency. (Courtesy, South Carolina Agricultural Experiment Station.)

to a greater demand for potash by these crops than by other crops, but because these crops are grown on soils deficient in potash. Furthermore, in the case of tobacco and potatoes, potash requirements are greater owing to the intensive system of cultivation under which they are grown. Farmers use potash for these crops because it is a profitable practice and they would do likewise with any other crop grown under similar conditions.

The potato farmers of Aroostock County, Maine, commonly apply 200 pounds of soluble potash per acre, and this is also a common practice of tobacco growers in the Connecticut Valley.

Potash is also applied in large amounts to truck and vegetable crops. This may be accounted for by the fact that most truck crops are grown on sands and mucks and the potash content of these soils is very low. Furthermore, sands and mucks are subject to heavy losses by leaching and for this reason a reserve supply of potash may not be built up as a result of fertilization over a period of years. It often happens that the colloidal complex of the soil becomes so devoid of potassium that when small applications of potash fertilizers are added the soil will compete with the crop for the added potassium to such an extent that very little influence on the crop will be observed from the fertilization.

The application of potash fertilizers to pastures tends to bring about an increase in the clover content of the flora. This is very noticeable in pastures in the Southeast, and it has been observed by the author on the Park at the Rothamsted Experiment Station, England. Similar observations have been made by Bear (1927), of the Ohio Experiment Station, and by Van Alstine, of the Vermont Experiment Station.

Potassium a Radioactive Element

Because potassium is the only radioactive element among the essential elements, it has been suggested that this property may allow potassium to function in some vital manner to supplement the energy from sunlight. This subject should be investigated further.

When Potash Salts Should Be Applied

Potash salts are generally applied before the crop is planted and this appears to be a well-founded practice. Patterson (1903), of the Maryland Experiment Station, has reported that little difference was secured in crop yields when potash salts were applied at the time of planting and when they were applied 3 to 6 months before planting. These results are interesting in that they show that potash is not readily lost from the soil In most cases, however, the application of potash salts some months previous to planting would not be practical.

In recent years the practice of making side applications of potash fertilizers to cultivated crops, especially to cotton, has been advocated and some favorable results have been reported. Where large quantities of potash salts are to be applied it appears best to make a part of the application as a side dressing in order to avoid injury to germination and a loss by leaching from sandy soils. When applied, however, in amounts commonly used, side applications of potash are not generally advised. This was the conclusion of Cooper and Wallace (1937). Their results are given in Table 81.

Gourley and Wander (1939) concluded that in applying potash to orchards it might be best to drop the fertilizer behind a deep tillage tool. This, they concluded, would put the potash within the active absorbing zone of the tree roots and below the soil zone in which potassium is fixed by alternate wetting and drying. This practice has much to commend it, not only in the fertilization of tree crops, but with field and horticultural crops as well.

According to Bartholomew and Janssen (1931), potash may be

Table 81

AVERAGE YIELD IN POUNDS OF SEED COTTON PER ACRE FROM EXPERIMENTS ON RATE AND TIME OF APPLYING POTASH FERTILIZER TO COTTON, 1931–1936, SANDHILL EXPERIMENT STATION, SOUTH CAROLINA

Pounds of Potash Applied	*Time of Application*	*Yield of Seed Cotton in Pounds per Acre*							*Average for Each Rate*
		1931	*1932*	*1933*	*1934*	*1935*	*1936*	*Average*	
None		..	788	657	696	552	412	621	621
15	All before planting	1238	966	871	937	801	809	937	..
	Half before planting and half at chopping	1230	1014	867	998	839	790	956	947
30	All before planting	1317	1267	1127	1161	1071	1153	1183	..
	Half before planting and half at chopping	1469	1316	1175	1184	1137	1196	1246	1218
	All at chopping	1449	1278	1174	1142	1123	1181	1224	..
45	All before planting	1525	1424	1357	1329	1213	1163	1335	..
	Half before planting and half at chopping	1506	1429	1425	1313	1245	1177	1349	1338
	All at chopping	1505	1393	1399	1291	1209	1176	1329	..
60	All before planting	1633	1548	1538	1451	1271	1385	1471	..
	Half before planting and half at chopping	1726	1577	1517	1521	1316	1393	1508	1490
	All at chopping	1618	1590	1503	1515	1324	1389	1490	..

absorbed by plants during all stages of development and at night as well as during the day.

Relative Value of the Potash Fertilizers

In view of the fact that the test for availability of potash fertilizers is their solubility in water, there is no problem of availability as is the case with the nitrogen and phosphoric acid fertilizers.

On the whole, experimental evidence with most truck and fruit crops shows that muriate of potash gives results as satisfactory as does potassium sulfate, but farmers of some sections are often prejudiced against some particular fertilizer material. Thus, truck farmers of Florida are prejudiced against the use of the muriate and will buy the sulfate although they have to pay more per unit for the potash in the sulfate form. Several ears ago Newell, of the Florida Experiment Station, estimated that Florida armers would save $160,000 a year by substituting the muriate for the ulfate. Because low-grade kainite and other crude potash salts contain high percentage of impurities, it often becomes necessary to apply large uantities where average, or greater than average, applications of potash re to be made. As the impurities are largely soluble salts injury to the ermination of seeds and to the growth of seedlings has often been noted, specially in dry seasons, from the application of low-analysis potash salts.

With a few notable exceptions, most field crops also appear to respond s well to potash salts carrying chlorine as they do to potassium sulfate. n fact, some crops, such as bright tobacco, show an additional response to hlorine although they may be injured by excessive quantities. Moss 1927), of the North Carolina Experiment Station, has shown that muriate f potash, which carries chlorine, gives somewhat higher yields of bright obacco than potassium sulfate, but when applied as the sole source of otash, in quantities sufficient to furnish more than 20 to 25 pounds of hlorine, it produces tobacco of poorer burning quality. The same observation has been reported by many other investigators. It has been observed lso, that plants receiving chlorides appear to have a better color than those eceiving sulfates.

Garner and McMurtrey (1930) have called attention to the ease with which the chlorides enter the plant, which is in sharp contrast with the behavior of sulfates and phosphates. This observation has also been made by Pettinger (1932). Garner and McMurtrey (1930) think that potassium

FIG. 86. Influence of potash on the growth of cotton. (*Left*) Plants defoliated due to potash deficiency. (*Right*) Healthy plants which had received an abundance of potash. (Courtesy, N. V. Potash Export My., Inc.)

chloride enters the plant as such, while the potassium furnished by the sulfate enters the plant in some other form.

MacIntire *et al.* (1938, 1943) concluded from lysimeter studies conducted at the Tennessee Experiment Station that the addition of potassium sulfate to acidic soils enhances the quantities of both calcium and magnesium in the soil solution.

Beets and Irish potatoes also have been reported to be sensitive to large quantities of chlorides. Chlorides, when present in large amounts, appear to lower the crystallizable sugar content of beets and to make the starch of Irish potatoes somewhat waxy. The Massachusetts Experiment Station has reported that the yield of potatoes remained about the same over a series of years where equivalent amounts of potash were applied, either in the form of potassium chloride or potassium sulfate. But the potatoes grown on the potassium sulfate plats were earlier and of a better quality than those grown on the potassium chloride plats. Also, during some years the potato vines on the chloride plats died earlier and were more affected by disease than the vines on the sulfate plats. Results secured at the Rothamsted Experiment Station (1924) show that higher yields of potatoes are secured with potassium sulfate than with potassium chloride when the crop is grown in dry seasons, but that the reverse is true during wet seasons.

In the Mobile, Alabama, potato growing section, and in Florida, the practice is to derive practically all of the potash for potatoes from potassium sulfate. On the other hand, Pierre and Bower (1945) point out that the potassium absorption by plants is usually decreased by the presence of high concentrations of other cations in solution. In the Beaufort section of South Carolina, and in the Wilmington section of North Carolina, the growers use both potassium chloride and potassium sulfate. In Aroostook County, Maine, practically all of the potash for potatoes is derived from muriate. Potassium sulfate is demanded by the citrus growers of southern California and the tobacco growers of the East.

Use of Potash Salts as a Herbicide

The value of sodium chloride as a herbicide for weeds has been known for years, but recent experimental work in France has shown that finely ground sylvinite is even more effective for this purpose. It can be used for the destruction of tares and other broad-leaved weeds of European grain crops. If the weeds are to be completely destroyed, it is important that the salt be applied when the weeds are quite young.

FIG. 87. On the dark spots of this field in Germany the charlock was destroyed by an application of kainite. (Courtesy, N. V. Potash Export My., Inc.)

Changes Potash Salts Undergo in the Soil

The reactions between potash and the soil are just as interesting as are those between phosphoric acid and the soil, and in many cases just as baffling.

When soluble potassium salts like potassium chloride and potassium sulfate are added to the soil, they normally dissolve and undergo ionization with a part of the potassium being absorbed by the colloidal complex of the soil in an exchangeable form. Potassium which has been so absorbed is not lost from the soil as readily as sodium, magnesium, and calcium, but it is not generally fixed in as unavailable a form as is the case with phosphorus. Under some conditions, however, varying with the soil type, potassium may be fixed to a limited extent in a nonexchangeable form which DeTurk *et al.* (1943) think is recoverable in part in a few weeks, and perhaps completely in a longer time. Kolodny and Robbins (1940), however, have shown that this nonexchangeable form of potash is only slightly available to tomato plants. Joffe and Levine (1939) have demonstrated that this fixation is peculiar to potassium and that other cations, such as calcium, magnesium, and barium, do not appear to be fixed in this way. Volk (1938) ascribes this fixation to several different processes and does not think it is due to any definite soil mineral. Truog and Jones (1938) explained the fixation by drying as due to the strong attraction offered by absorbed potassium which prevents a reexpansion of the crystal lattice and thus reëntrance of water. Several investigators have shown that if the soil is prevented from drying out the tendency to fix potash is much reduced and that the greatest fixation of potash, in an unavailable state, is in those layers of the soil that are subjected to alternate wetting and drying.

Because potassium may leach back to the soil from the above-ground parts of plants, and may even go back to the soil from plant roots, and because it dominates all cations in plant composition, it must be looked upon as being very nomadic, a fact which certainly hinders our understanding of its behavior. Potassium leaches from humid soils more rapidly than does phosphorus. This has been shown in the study of drainage water by Voelcker (1874) at the Rothamsted Experiment Station, by Lyon and Bizzell (1918) at the Cornell Experiment Station, and by Collison and Walker (1916) at the Florida Experiment Station. In its movement potassium may be considered to be somewhat intermediate between nitrogen and phosphorus. Unlike phosphoric acid, after potash salts go into solution they may be carried down into the soil several inches before the potassium becomes absorbed by the soil.

In the sandy soils of the Atlantic and Gulf coastal plain potash is subject to leaching in rather large quantities. This is due to the low base-exchange capacity of these soils.

12

Fertilizers Carrying the Secondary Essential Elements—Sulfur, Calcium, and Magnesium

Recent intensive studies of plant nutrient deficiencies in soils, particularly in the United States, has led to the revelation that many soils may be almost as deficient in one or more of those elements sometimes labelled "secondary essential"—sulfur, calcium and magnesium—as they are in nitrogen, phosphorus, and potassium.

Sulfur an Essential Element for Plant Growth

Sulfur has been known since remote antiquity. According to Clarke, the earth's crust contains 0.06 per cent of sulfur which is found in its elemental state and as sulfides and sulfates.

Sulfur was determined to be an essential element for plant growth

F.G. 88. Plant of the Texas Gulf Sulphur Company at Newgulf, Texas. (C esy, Texas Gulf Sulphur Co.)

about the year 1860. This element is a constituent of the amino acid cystine, and cystine is a constituent of protein. Sulfur is also required by the plant in the synthesizing of the essential oils. It is a constituent of mustard oil, which is found in mustard, and of allyl sulfid and vinyl sulfid which are found in onion and garlic. Members of the plant families Resedaceae, Capparidaceae, Umbelliferae, and others, synthesize special organic sulfur compounds. In the absence of sulfur, plants often are characterized by yellowing of leaves and generally show symptoms that resemble those of nitrogen starvation. This has been shown to be true in

Table 82

POUNDS OF SULFUR AND PHOSPHORUS REMOVED FROM THE SOIL PER ACRE BY SOME COMMONLY GROWN CROPS

Crop	*Dry weight, pounds*	*Sulfur, pounds*	*Phosphorus, pounds*
Wheat, grain, 30 bu	1530	2.6	6.2
Wheat, straw	2653	3.7	3.0
Total crop	4183	6.3	9.2
Barley, grain, 40 bu	1747	2.6	7.0
Barley, straw	2080	3.1	2.0
Total crop	3827	5.7	9.0
Oat, grain, 30 bu	1625	3.0	5.7
Oat, straw	2353	4.9	2.8
Total crop	3978	7.9	8.5
Corn, grain, 30 bu	1500	2.6	4.4
Corn, stalks	1877	2.2	3.5
Total crop	3377	4.8	7.9
Tobacco, leaf	1800	6.4	3.5
Tobacco, stalk	3200	2.0	3.5
Total crop	5000	8.4	7.0
Meadow hay	2822	4.5	5.4
[illegible]d clover hay	3763	6.2	10.9
[illegible]lfa hay	9000	25.9	17.4
[illegible]toes	3360	4.6	9.4
[illegible]ge	4800	39.2	26.6

tea by Storey and Leach (1933), in tobacco by McMurtrey (1938), in tomatoes by Fisher (1935), in sugar cane by Martin (1934), in citrus trees by Haas (1936), and in the navel orange by Chapman and Brown (1941).

In 1940 Alway reported a soil belt through Minnesota where sulfur was deficient for the maximum production of alfalfa. As a general rule, the veins of a sulfur-deficient leaf are lighter in color than the rest of the leaf. Chlorosis resulting from sulfur deficiency is similar to that produced by nitrogen, but sulfur-deficient plants do not readily lose their lower leaves.

The sulfur content of many plants is equal to or greater than the phosphorus content. In fact, in many legumes and members of the Cruciferae family the sulfur content exceeds that of phosphorus. In Table 82 is given the sulfur content of some of the common field crops, as reported by Hart and Peterson (1911), of the Wisconsin Experiment Station and recalculated by the author. The phosphorus content also is given so that a comparison can be made between the sulfur content and the phosphorus content in the crops listed. Plants frequently contain more sulfur than phosphorus, calcium or magnesium.

Sulfur Lost from the Soil in Drainage Water

The annual loss of sulfur from the soil in drainage water varies with the sulfur content of the soil, and the precipitation under which the soil exists. At the Rothamsted Experiment Station, the annual loss of sulfur in drainage water has been estimated by Hart and Peterson (1911) to be about 20 pounds per acre. In this country, Erdman and Bollen (1925) have reported an annual loss through leaching of 50 pounds of sulfur per acre,

FIG. 89. Harvesting a field of fertilized potatoes in New Jersey.

from Iowa soils, and Hart and Peterson (1911) have estimated that unde Wisconsion conditions the annual loss through drainage is 15 to 2 pounds per acre.

Sulfur Added to the Soil in Rain Water

Sulfur gases are present in the atmosphere and are evolved throug the decomposition or the burning of organic matter. These gases may b absorbed and removed from the atmosphere by rain water. As would b expected, the amount of sulfur added to the soil annually by rain water i variable. Soils located near large cities, where immense quantities o hydrogen sulfide and sulfur oxides are continuously being discharged int the atmosphere, receive more sulfur than those located at greater distance from centers of population. Hart and Peterson (1911) have estimated tha the soils of the Rothamsted Experiment Station, England, receive fron the atmosphere on an average the equivalent of 7 pounds of sulfur pe acre annually. Stewart (1920), working at the University of Illinois, ha reported an annual gain of sulfur per acre, due to precipitation, of 45. pounds, and Collison and Mensching (1932) report a gain of 41 pound at Geneva, New York. Alway *et al.* (1937) reported an annual fall of sulfu of 100 pounds per acre in Minneapolis, Minnesota, and of 5 pounds pe acre in the northern part of the state of Minnesota, but in 1940 he ques-tioned his results secured in Minneapolis. Harper (1943) has reported fron Stillwater, Oklahoma, the annual addition by rain water of 8.7 pounds o sulfur per acre of soil.

The annual addition to the soil of sulfur in rain water has a direc bearing on the problem of sulfur fertilization. Hall (1915), Lyon an Bizzell (1918), and others, have presented results showing that the soi annually loses through drainage and cropping much more sulfur than i added in precipitation. Lysimeter studies at Cornell University, as reporte by Bizzell and Lyon (1927), have shown an annual addition to the soil o 30.1 pounds of sulfur, whereas the annual loss due to cropping and leach-ing from plats on which a rotation containing legumes was followed amounted to about 44.1 pounds. Where such conditions prevail it woul appear that sulfur must be applied as a fertilizer, or this element will even-tually become a limiting factor in crop production.

Value of Elemental Sulfur as a Fertilizer

It was not until 1912 that the work of several foreign investigator demonstrated the value of elemental sulfur as a fertilizer. In this countr

FIG. 90. Storage pile of raw sulfur. (Courtesy, Texas Gulf Sulphur Co.)

Shedd (1914), of the Kentucky Experiment Station, reported increased tobacco yields due to its influence in 1914, and in 1919 very marked influences of sulfur on alfalfa yields were reported by Reimer and Tartar of the Oregon Experiment Station. In 1941, Tolman and Stoker, of the Bureau of Plant Industry, reported increased yields of sugar beets following applications of sulfur.

Many soils of the Pacific Northwest produce higher yields of crops when sulfur-carrying fertilizers are applied. Alway (1940) has reported a similar area of soils in Minnesota. Cases have been reported where the sulfur hunger of legumes, especially alfalfa, has been so great that the application of sulfur has increased the yields 500 to 1000 per cent, and increases of 50 to 500 per cent are common. The sulfur-treated alfalfa experimental plats at the Oregon and Washington Experiment Stations are said to show the most striking response to fertilizer ever seen, nitrogen fertilizers not excepted. Harris *et al.* (1945), of the Florida Experiment Station, has recently reported sulfur to be deficient on three soil types of Florida, and they conclude that for the sandy soils of Florida sulfur nutrition merits greater consideration.

Necessity for Applying Sulfur

The necessary for using sulfur or sulfur compounds as fertilizer depends upon the amount and availability of the sulfur in the soil, the amount that is added in precipitation during the growing season, and upon the crop requirements. According to Erdman and Bollen (1925) the sulfur content of the surface soils of the United States varies from 200 to 3,000 pounds per acre.

Crocker (1923) has summarized the results of 14 different analyses or groups of analyses involving 395 soils of the United States and 21 of England. He states that the sulfur content of the soils of Oregon, which are generally looked upon as being very deficient in sulfur, might be considered as average, being surpassed in lowness of their sulfur content by many other soils of the United States. He believes that sulfur availability, which involves sulfofication and decomposition of organic matter, is of more importance in determining yields of crops than the total amount of sulfur in the soil. The sulfur content of most soils is generally less than that of phosphorus.

In the past there has been little need for applying sulfur in any form as a fertilizer. This has been due not only to the fact that soils contain organic matter and many soils carry large quantities of sulfur-containing minerals, but also to the fact that in the fertilizer-consuming areas of the United States calcium or other sulfates have been unintentionally applied in superphosphate, kainite, and other materials.

Table 83 gives an estimation, according to Smalley (1931), of the quantities of sulfur and phosphorus applied to the soils of the United States in 1930 in some of our leading fertilizer materials.

It should be mentioned, however, that MacIntire (1941) found in his

Table 83

SULFUR AND PHOSPHORUS SUPPLIED IN FERTILIZER MATERIALS USED IN THE UNITED STATES IN 1930

Material	*Sulfur Trioxide (tons)*	*Phosphorus Pentoxide (tons)*
Superphosphate	1,277,000	788,000
Bone	..	11,000
Ammonium phosphate	..	10,500
Miscellaneous organics	..	14,000
Ammonium sulfate	374,000	..
Potassium sulfate	41,000	..
Miscellaneous potash salts	40,000	..
Ammonium-sulfate nitrate	3,500	..
Totals	1,735,500	823,500

lysimeter studies in Tennessee, that in humid regions rainfall substantially affects the rate of removal from the soil of ordinary applications of sulfates.

Use of Elemental Sulfur

Elemental sulfur is employed to produce acidity in the soil, to reclaim alkaline land, and as a fertilizer for crops, particularly the legumes. When elemental sulfur is added to the soil it readily undergoes a change to the sulfate form. Lipman (1911) has named this transformation process sulfofication. Sulfur oxidation may be brought about by many forms of soil microörganisms. The chemical reactions involved are probably as follows:

$$S + H_2O + O_2 = H_2SO_3$$

$$H_2SO_3 + CaH_2(CO_3)_2 = CaSO_3 + 2H_2O + 2CO_2$$

$$2CaSO_3 + O_2 = 2CaSO_4$$

Where large quantities of elemental sulfur are applied to a soil deficient in sufficient lime to correct the resulting acidity, an acid condition of the soil is produced. Potato growers of the middle Atlantic and New England states have taken advantage of this in their attempt to control the potato scab (*Actinomyces scabies*). For these farmers the limiting factor for high yields of marketable potatoes is often the presence of scab. As scab growth is encouraged by alkalinity, and hindered by acidity, the application of free sulfur helps to control the disease. Martin (1920), of the New Jersey Experiment Station, found that potato scab could be controlled when sulfur was applied at the rate of 300 to 500 pounds per acre when the pH of the soil was 5.8 or less, and 700 to 1200 pounds per acre when it was 6.0 or greater. Martin found the proper soil reaction for potatoes to be a pH value between 5.0 and 5.4 but the calcium content of the soil may be more important than the pH value. Likewise, the control of root rot of bright tobacco has been obtained at a pH value of 5.4. Because of leaching, the acidity of leached soils will gradually decrease so that further applications must be made in the following years.

Sulfur has also been used on the burnt-over peat and muck soils of Florida for the purpose of reducing the alkalinity of the soil and in this way to increase the quantity of manganese available to plants of succeeding crops.

Fraps and Fudge (1936) found that induced acidity resulting from the application of sulfur or sulfuric acid decreased and disappeared after a few years. They also found that the induced acidity did not penetrate the

FIG. 91. Influence of sulfur on the growth of alfalfa. (Courtesy, Oregon Agricultural Experiment Station.)

soil for more than 2 or 3 inches unless the soil became so acid that no plant growth could occur.

Inoculated Sulfur

Inoculated sulfur consists of free sulfur to which has been added small amounts of compost that contain numerous sulfur-oxidizing organisms. It has been shown by Lipman and his coworkers of the New Jersey Experiment Station that sulfofication takes place in the soil more rapidly when inoculated sulfur is added than when the uninoculated sulfur is applied.

Influence of Sulfur on Plant Growth

Pitz (1915), of the Wisconsin Experiment Station, has found that sulfates increase nodular development in alfalfa, and Miller (1919) has shown the same to occur in red clover. Hart and Tottingham (1915), and Reimer and Tartar (1919) have reported that sulfur increases root development. In fact, Crocker (1923) thinks that the increased root development that is often observed after an application of superphosphate may be due in part to the sulfur added as calcium sulfate. It appears also that sulfur may have a part in the development of chlorophyll. Both the Washington and the Virginia Experiment Stations have called attention to the deep green color of alfalfa which was grown on sulfur-treated plats. Other investigators have noted this influence with corn and legumes. A tobacco committee of the southern agronomists in 1933 reported that, as experiments had shown that sulfur delays maturity of bright flue-cured tobacco and produces a red coloring of cured leaves, fertilizer mixtures for bright

Table 84

INFLUENCE OF SULFUR ON THE YIELD OF ALFALFA WHEN GROWN ON SALEM CLAY LOAM, OREGON*

Treatment	Pounds of Fertilizer Applied per Plat	Pounds of Sulfur Applied per Plat	Yields in Pounds per Plat (plat 2 × 8 rods)			
			1915	1916	1917	Totals
Check	0	0.0	96	114	184	394
Gypsum	59.5	10.0	492	752	948	2192
Monocalcium phosphate	41.0	0.0	178	120	228	526
Superphosphate	82.3	10.0	553	768	888	2209
Check	0	0.0	286	388	344	1018
Sulfur	10	10.0	435	930	952	2317
Sulfur	30	30.0	618	1126	1100	2844
Iron sulfate	84.0	10.0	899	1060	1016	2975
Rock phosphate	56.0	10.0	368	306	576	1250
Check	0	0.0	273	296	368	937

* The fertilizer was applied to 2-year-old alfalfa March 8, 1915, and another application was made March 27, 1917.

tobacco should be formulated so as to contain a minimum quantity of sulfur. That sulfur is used by plants in the production of protein is shown by the work of Tottingham (1918) with red clover. Skinner (1941) has reported that sulfur-deficient tomato plants have a remarkable capacity for elongation of the stem, not natural in a healthy plant.

FIG. 92. Showing the value of sulfur in reclaiming alkali land. Alfalfa planted on untreated and treated plats. Plat B in the foreground, untreated; pH of soil 9.8. Plat C in the background, received 1500 pounds of sulfur and 10 tons of manure per acre; pH of soil 7.8. (Courtesy, Oregon Agricultural Experiment Station.)

In Table 84 are given the results secured by Reimer and Tartar (1919), of the Oregon Experiment Station, with alfalfa that had been fertilized with sulfur-carrying fertilizers.

World's Reserve of Native Sulfur

The United States Geological Survey has estimated the world's reserve of native sulfur to be between 56,000,000 and 121,000,000 metric tons. Italy alone has an estimated reserve ranging from 25,500,000 to 84,200,-000, metric tons. In the United States the Texas-Louisiana field, which was discovered in 1865, is said to contain about 30,000,000 tons of sulfur without allowance for new discoveries. These are the largest known sulfur deposits in the world. Chile has a reserve estimated to be from 525,000 to 1,750,000 metric tons. Japan has a reserve but figures as to their extent are not available. The principal known sources of sulfur in the world today are in the United States, Italy and Japan. The deposits of Texas, Louisiana, Sicily, Japan, and most of Spain are readily accessible to cheap water transportation. The United States leads the world in sulfur production.

Practically all of the sulfur used in the United States comes from Louisiana and Texas. These deposits occur in the cap rock overlying certain salt domes at depths ranging from 500 to 1500 feet below the surface. Although about 200 of these domes are known, only a few are now producing commercial sulfur.

Use of Gypsum as a Fertilizer

Gypsum ($CaSO_4.2H_2O$) has been used as a fertilizer from the early Greek and Roman periods, and it was used extensively in Europe a century and a half ago. Its use in this country was probably introduced by German immigrants, for in the early days it was extensively employed in Pennsylvania and Ohio.

Deposits of gypsum and anhydrite ($CaSO_4$) are found in this country in the northern districts of Ohio along the lake shore and in the northwestern part of New York State. The mineral is usually pure white in color, although it is sometimes colored by impurities, especially iron. It is soft and can easily be scratched with a knife. Gysum is commonly called "land plaster" and is a hydrated sulfate, a neutral salt of calcium. It is approximately the same chemical compound as that formed in the manufacture of superphosphate when sulfuric acid is added to the ground phosphate rock. Gypsum is largely used in the industries in the manufacture of Portland cement. It is not very soluble in water and does not readily decompose in the soil. Cameron and Bell (1906) have shown that if lime is applied to the soil at the same time as gypsum the solubility of the gypsum will be decreased.

Ames and Blotz (1916) have called attention to the 20-year average yields of crops grown on the soils of the Wooster, Ohio, Experiment Station, which show that phosphorus carriers containing gypsum, such as superphosphate and dissolved bone black, produce more corn, oats, and wheat per acre than do bonemeal and basic slag. They think it probable the gypsum content of the added superphosphates accounts in part for the increased yields.

Erdman and Bollen (1925) have shown that on some Iowa soils application of gypsum increased the yields of oats, red clover, and especially alfalfa. In Table 85 are given the results secured with alfalfa by these investigators on O'Neill sandy loam.

Table 85

THE EFFECT OF GYPSUM ON THE YIELD OF ALFALFA GROWN ON O'NEIL SANDY LOAM IN IOWA

Treatment (lbs. per Acre)	*First Crop (lbs.)*	*Increase or Decrease (lbs.)*	*Second Crop (lbs.)*	*Increase or Decrease (lbs.)*	*Total Yield (lbs.)*	*Increase or Decrease (lbs.)*
100 lbs. sulfur	5,400	1,100	1,700	600	7,100	1,700
500 lbs. gypsum	5,100	800	1,350	250	6,450	1,050
300 lbs. gypsum	5,100	800	1,450	350	6,550	1,150
Check	4,300	..	1,100	..	5,400	..
150 lbs. gypsum	5,400	1,100	950	−150	6,350	950
50 lbs. gypsum	5,600	1,300	1,050	− 50	6,650	1,250

Willis and Rankin (1930) demonstrated the value of gypsum in superphosphate in counteracting the toxic free ammonia produced in soils as a result of the decomposition of organic matter. Willis and Piland (1931) have called attention to the practical possibilities of gypsum as a supplement to all fertilizers or other soil treatments supplying free ammonia to soils in toxic concentrations. They found calcium sulfate to be more effective in this regard than calcium carbonate or phosphate. It has long been recognized that gypsum and other calcium salts may be used advantageously in correcting abnormal nutrient media.

Skinner and Beattie (1917), working at the Arlington Farm of the

United States Department of Agriculture, concluded that calcium sulfate had no power to correct soil acidity. After five years of treatment they found the lime requirement of the soil, as determined by the Veitch method, to be slightly greater in those plats which received the calcium sulfate.

Gypsum has been used extensively in the western states to dealkalize Solenetz, or black alkali, soils. It will react with black alkali, which is largely sodium carbonate, and convert it into white alkali, which is largely sodium sulfate. As black alkali is more toxic to plant growth, it is advised that the application of gypsum, lime or other calcium salts be made before irrigation.

Gypsum is sometimes used also as a preservative to be applied to farm manure. Generally, an application of about 100 pounds per ton of manure is recommended. The value of gypsum as a preservative is supposed to be due to its ability to convert the volatile ammonium carbonate into ammonium sulfate. Ammonium carbonate is produced in manure piles as an end-product of ammonification. As a preservative of farm manure gypsum has the advantage over many other preservatives in that it is safe material to use where livestock are allowed to tramp over the manure after treat-

FIG. 93. Influence of gypsum in correcting toxicity arising from the use of concentrated fertilizers. Representative cotton plants from 3 plats: (1) Concentrated fertilizer alone; (2) concentrated fertilizer with 20 pounds of gypsum per acre; (3) concentrated fertilizer with 40 pounds of gypsum per acre. (Courtesy, North Carolina Agricultural Experiment Station and *The Fertilizer Review*.)

ment. So far, however, the efficacy of the practice has not been satisfactorily determined.

Influence of Gypsum on the Solubility of Soil Constituents

Advocates of "plastering land" have often made claims as to the influence of gypsum on the release of unavailable plant nutrients for plant absorption. Scott and Roberston (1926) have reported that the use of gypsum in Victoria, Australia, has aided in the liberation of potassium, phosphorus, magnesium, and silicon. They reported also that the rate of ammonification was increased and that the finer soil particles were flocculated following its use. Likewise, Andre (1913) has reported that the solubility of potash in microline is greatly increased in the presence of calcium carbonate and gypsum. On the other hand, Briggs and Breazeale (1927) were unable to show that either lime or gypsum had any power to increase the availability of potash in soils derived from orthoclase-bearing rocks. In fact, they found that in some cases there was a marked depression of the solubility of the potash in the presence of gypsum. Likewise, it has been reported that large applications of gypsum applied annually for 40 years at the Pennsylvania State Agricultural College have exhibited no tendency to increase the solubility of the soil constituents.

How Gypsum Should Be Applied

When an application of gypsum is to be made it is best to apply it in such a manner that it becomes thoroughly mixed with the soil. Broadcasting the gypsum, and then harrowing it into the soil before the planting of crops, is probably the most satisfactory manner of application. The most commonly recommended application is 200 pounds per acre but 500 pounds is sometimes advised. Peanut growers of the Southeast commonly apply 400 to 500 pounds of gypsum directly to the vines at blossoming time, in order to increase the soluble calcium content of the soil around the peanut pegs and thus insure a low percentage of pops. Caldwell and Brody (1945), at the North Carolina Experiment Station, found that the use of gypsum increased the average weight of peanut kernels.

Gypsum as a Source of Sulfuric Acid

Sulfuric acid is produced as a by-product of various smelting operations, but the greater part of the world's production is manufactured from sulfur, brimstone, and pyrites. At the present time, however, gypsum is being substituted to a limited extent for brimstone in the manufacture of sulfuric acid in Germany and in England.

World's Reserve of Pyrites

Pyrite ores, which include pyrite, marcasite, and pyrrhotite are sold in large quantities in the world markets to be manufactured into sulfuric acid. The world's reserve of pyrites, as estimated by the International Geological Congress, ranges from 465,000,000 to 907,000,000 metric tons. The sulfur content of pyrites normally ranges between 40 and 50 per cent. Spain has the most extensive reserve; it has been estimated to be 132,000,000 to 264,000,000 metric tons. Russian reserves rank second and have been estimated at 17,300,000 to 62,000,000 tons. Other deposits of considerable size are found in Germany, Portugal, Italy, Sweden, and France. Many deposits are found scattered over the United States.

The largest body of pryites located in the Eastern part of the United States is in Virginia. At one time this deposit produced 200,000 tons of ore a year. The use of pyrites by sulfuric acid manufacturers is rapidly being replaced by the use of brimstone.

Calcium an Element Essential for Plant Growth

It cannot be said definitely just what essential function, or functions, calcium performs within the plant. It is generally believed that calcium has an influence on the translocation of the carbohydrates. Plants suffering from a deficiency of calcium often show poor development and drying of feeding roots, and in many cases a deficiency of calcium results in the death of the terminal bud. Calcium appears to be essential for the construction of healthy cell walls, and there is evidence that it assists in collaboration with potassium, magnesium, and possibly boron in neutralizing organic acids and maintaining a delicate physiological balance within the plant. In many vegetable crops a lack of calcium results in retarded vegetative growth and the production of thick, woody stems. Roots of calcium-deficient trees are often short and stubby. Certainly, calcium is one of the most important elements insofar as crop fertilization is concerned. Albrecht (1943) has pointed out that calcium and potassium are seemingly reciprocals in plant composition, calcium supporting the proteinaceous and potassium the carbonaceous properties. It is a common observation that an application of a calcium salt will often greatly help plants suffering from abnormal nutrition. As pointed out by McMurtrey (1938) a chief function of calcium appears to be its ability to antagonize other ions. There seems to be little or no transfer of calcium from older plant tissues to the growing points. Young plants suffering from calcium deficiency readily succumb to winter-killing and summer drouth when grown on acid soils. Although calcium-deficiency symptoms can be produced in controlled cultures they are rarely seen and recognized under field conditions.

FIG. 94. Influence of calcium on the growth of tobacco. (*Left*) Plat receiving no calcium. (*Right*) Plat receiving calcium. (Courtesy, Dr. W. W. Garner.)

The quantity of calcium present in the average soil is so large that rarely except in highly leached soils do we find a soil that does not contain a sufficient amount to supply the nutritional needs of ordinary yields of nonleguminous plants, although the supply is often insufficient for the legumes. Legumes generally remove from the soil larger quantities of calcium than do the nonleguminous plants. A crop of the commonly grown nonleguminous plants seldom takes from the soil over 25 pounds of calcium, whereas a crop of legumes often removes 100 pounds or more. Calcium is usually found in the soil as the carbonate, phosphate, sulfate, or silicate, but the silicate is the most widely distributed. Russell (1932) has presented data which show that the amount of calcium in the soil influences the total amount of calcium absorbed by plants. As plant leaves mature, calcium, unlike phosphorus and nitrogen, is not translocated to the stem or seed but remains in the leaf. If present in the seed it would demobilize phosphorus. The amount of calcium in a plant reaches its maximum at maturity.

Forms in Which Calcium May Be Applied to the Soil

Calcium compounds are commonly applied to the soil as constituents of fertilizing materials or soil amendments. Calcium is most commonly applied in the form of agricultural lime. The term agricultural lime is used indiscriminately to include a great variety of materials which are usually composed of either the oxide, hydroxide, or carbonate of calcium.

FIG. 95. Limestone plant of the American Limestone Company, Knoxville, Tennessee. (Courtesy, American Limestone Company.)

or calcium and magnesium. The most commonly used forms of agricultural lime are ground limestone, hydrated lime, marl, burnt lime, and oyster shells. Materials like limestone, shell, and marl should be ground to such a fineness that about 60 per cent of the material will pass through a 100-mesh sieve. Finer grinding will make the material more soluble but this generally is not necessary. MacIntire, Sanders, and Shaw (1933) found that limestone of 100- to 200-mesh fineness was equal in its influence on nitrification to an equivalent of hydrated lime.

Calcium is most commonly applied to the soil in the form of pulverized limestone, either calcitic or dolomitic, but usually dolomitic. It is often applied, especially in the New England States, as burnt or hydrated lime. Calcium is applied also in such fertilizer materials as superphosphates, gypsum, calcium cyanamide, and calcium nitrate. Calcium silicate is now available in some states and may be used as a carrier of calcium when a change in the pH of the soil is not desired. A substantial quantity of a relatively new type of quenched slag is now becoming available from rock phosphate reduction furnaces. This by-product is made up largely of calcium silicate and contains most of the component elements of phosphate rock, except the volatilized phosphorus together with the added silica. From laboratory and pot studies made at the Tennessee Experiment Station, MacIntire (1940, 1943) concludes, that a 2- to 3-ton $CaCO_3$ equivalent of the material passing a 20-mesh sieve, constitutes a safe and effective liming application for ordinary soils. The solubility of the ground quenched slag is comparable to ground limestone of similar fineness.

In addition, calcium silicate is now produced as a by-product in South Carolina from the smelting of aluminum ores. Experimental evidence secured by the author appears to indicate that calcium silicate is a good source of calcium for pasture plants.

FIG. 96. Quenching a flow of calcium silicate slag showing the dispersal resultant from impingement of the 8 jets of water against the underside of the arc of the slag as it drops from the trough. (Courtesy, Dr. W. H. MacIntire, the Tennessee Agricultural Experiment Station, and the Tennessee Valley Authority.)

Because the soils of the southern and southeastern states are inadequately buffered against acidity, many fertilizer specialists are now advising that fertilizer mixtures which are to be used upon these soils should contain enough liming material to correct the acidity effects of the ammoniacal or organic forms of fertilizer nitrogen usually used in the mixtures.

Use of Lime as a Soil Amendment

The application of lime is an old practice. According to Pliny, lime was used 200 to 300 years before Christ. The Virginia farmer, Edmund Ruffin (1794–1865), was probably the first man in this country to recognize the

FIG. 97. Red clover growing on Fullerton silt loam that has been slagged with 4 tons of quenched calcium silicate slag from Wilson Dam. In addition, the soil has been fertilized with 200 pounds of K_2O and 80 pounds of P_2O_5 per acre. The source of the P_2O_5 was varied as follows: Pot 518, no phosphate; pot 520, concentrated superphosphate; pot 522, dicalcium phosphate; pot 524, fused tricalcium phosphate; pot 526, calcium metaphosphate; pot 528, potassium metaphosphate. An uniformity of response is to be noted from the different types of phosphates. (Courtesy, Dr. W. H. MacIntire and the Tennessee Agricultural Experiment Station.)

general prevalence of soil acidity. He conducted field tests with lime and wrote a treatise on calcareous manures. Ruffin has been called the father of soil chemistry in America.

The major purposes of applying liming materials to the soil are, to neutralize acids, to precipitate soluble compounds that are toxic to plant growth, and to introduce calcium into the soil colloidal complex. For

FIG. 98. Influence of lime on the growth of young cotton plants at the South Carolina Sandhill Experiment Station. (*Left*) Unlimed. (*Right*) 1500 pounds of ground dolomitic limestone per acre. (Courtesy, South Carolina Agricultural Experiment Station.)

these reasons liming materials are generally classed as soil amendments and not as fertilizers. Nevertheless, liming materials should be looked upon as fertilizers when applied to soils that are low in their calcium content.

Some agronomists feel that the development of a system of permanent agriculture in a community should not be attempted unless the farmers of the community are willing and able to make the necessary applications of lime required to produce conditions favorable for plant growth. Initial applications of from 1 to 5 tons of ground limestone per acre appear to be necessary on tens of thousands of acres in our Eastern states. The quantity of limestone to be applied as an initial application depends upon the acidity of the soil, the buffer capacity of the soil, and the crop or crops to be grown. The amount generally considered necessary to maintain a favorable soil reaction is from 2,000 to 4,000 pounds of ground limestone, or its equivalent, per acre, once in a rotation cycle or from 400 to 600 pounds per acre in the drill row when crops are planted. This rate of application has been found sufficient for most crops except where the soil is very acid or where the promise of increased crop yields is exceptionally good. If it were not for the excessive leaching to which calcium is subjected in most soils of humid regions, smaller applications made at longer intervals would suffice. As stated above the rate of application should be based upon the pH of the soil that is to be limed and the crop that is to be grown. In addition, seasonal variations in pH, which in exceptional cases have been shown by Hester and Shelton (1933) to be as much as 2.0 pH, must be taken into consideration. Under usual soil conditions about 1 ton of limestone per acre is required to change the soil reaction 1 pH in a strongly acid sandy loam, and about 1.5 tons in a heavy loam.

There is much evidence to indicate that on sandy soils which are low in their content of magnesium, manganese, zinc, and iron, quantities of lime smaller than those generally considered necessary should be used. Excessive quantities of lime will bring about the conversion of manganese, zinc, boron, and iron into unavailable forms. Pierre and Browning (1935) concluded that the temporary overliming injury noted in West Virginia was due to a disturbed phosphate nutrition. Cooper (1930) thinks that heavy applications of lime may result in the formation of the hydroxides of magnesium, manganese, and iron. As these hydroxides are of low solubility, they are thus rendered less available for plants. Albrecht (1946) has proposed the concept that since the saturation of the soil clay with calcium would displace other essential elements the greatest benefit might be derived from small applications of lime, or from the application of the less soluble forms, or from the application of the granular form of soluble materials.

Ten years' results from the Cornell University lysimeters, secured by Bizzell and Lyon (1927), show an average annual loss of calcium carbonate per acre equivalent to 995 pounds from bare soil, and 535 pounds from plats that grew crops and where a rotation including legumes was followed. In the North Central States the annual loss of lime probably ranges between 100 and 500 pounds, so that an application of 2 tons of limestone should take care of this loss for a period of from 8 to 20 years.

Changes that Lime Undergoes in the Soil

Burnt and hydrated limes are converted largely into the carbonate form within a short time after they are incorporated in the soil. As the carbonate is insoluble in water it must be converted into the bicarbonate, or into some other soluble calcium salt, before it can be leached from the soil or before it can become available for plant use. These changes readily take place and in all probability have resulted in more calcium than sodium being carried to the ocean in river water. The following equations may parallel the complex changes through which burnt lime passes in the soil in its conversion to the soluble bicarbonate form. The number of reactions involved will depend upon the form of lime that is added to the soil.

$$CaO + H_2O = Ca(OH)_2$$
$$Ca(OH)_2 + CO_2 = CaCO_3 + H_2O$$
$$CaCO_3 + H_2CO_3 = CaH_2(CO_3)_2$$

FIG. 99. Experimental pasture plats near Wardensville, W. Va. (*Left*) No lime or fertilizer. (*Right*) Lime and fertilizer. Note the presence of daisies in the unfertilized plat. (Courtesy, West Virginia Experiment Station.)

FIG. 100. Cylinders for soil investigations installed at New Jersey Agricultural Experiment Station in 1898. Crop and Fertilizer records have been kept to date. Series A shows crop growth when lime was omitted during last 20 years. Series B was limed throughout. Both series received identical annual fertilizer applications from the beginning. (Courtesy, New Jersey Agricultural Experiment Station.)

The rate of solubility of dolomite is only about 50 per cent as rapid as that of calcite.

Methods Employed in the Application of Lime

Liming materials are generally broadcast and then worked into the soil. Unless they are mixed well with the soil their efficiency in correcting acidity may be very low. Lime is effective only in correcting soil acidity in that portion of the soil in the immediate vicinity of the lime particle. This fact has been substantiated by Weidemann (1936) and others. Blair and Prince (1934), however, found that the application of limestone at the rate of 2,000 to 4,000 pounds per acre at 5-year intervals had, after 25 years, changed the reaction of the subsoil (6⅔–13 inches) of Sassafras loam to about the same degree that it changed the reaction of the surface soil. The mechanical downward movement of the limestone through the porous Sassafras loam might account for these results. Nevertheless, agronomists usually recommend that lime be thoroughly mixed with the soil. It is most successfully applied with a lime distributor.

It is often desirable to apply lime to pastures for the purpose of encouraging the growth of clovers and the most palatable and nutritious pasture grasses. Where possible, the lime should be broadcast on pastures during the late fall in order that the alternate freezing and thawing during the winter months will tend to work it into the soil. Scarification of pastures following an application of lime should give quicker results, and is to be advocated in those instances where the grass can be scarified without material injury and where erosion will not be increased.

Generally, when applications of lime are made to field crops they are commonly applied in such large quantities that it has become the custom to apply them separately and at a different time than when applications of commercial fertilizers are made. Small amounts of dolomitic limestone in the place of filler are commonly used now in fertilizer mixtures for the purpose of producing a nonacid-forming mixture, as well as for the purpose of supplementing the mixture with calcium and magnesium. This practice is to be recommended. Parker (1932) estimates that the addition of 5 pounds of limestone per pound of nitrogen derived from ammonium sulfate or chloride, and 2.5 pounds of limestone per pound of nitrogen derived from free ammonia, urea, or ammonium nitrate will produce a nonacid-forming fertilizer. Bartholomew (1933) was able to find no serious decrease in the availability of soluble phosphorus when agricultural lime was added with superphosphate to an acid soil.

FIG. 101. Tobacco plant showing symptoms of calcium deficiency. (Courtesy, Dr. W. W. Garner.)

It appears that when light to medium applications of lime are made to an acid soil the processes of nitrification are hastened. Inasmuch as biological activities are stimulated, the immediate need for fertilizers is often reduced since nitrogen and mineral nutrients may be released from organic matter and minerals as a result of these activities. The nitric acid produced as a result of nitrification is one of the most potent agents in making available the unavailable mineral plant nutrients in the soil. On the other hand, heavy applications of lime may result in a potash deficiency for succeeding crops. Apparently, the intake of potassium may be insufficient for normal plant growth if the available calcium and magnesium is disproportionately high. MacIntire and Sanders (1930) have reported that lime will fix the potassium in the soil in a nonreplaceable state, and MacIntire *et al.* (1936) concluded after a 12-year test that all forms of lime repressed the out-go of potassium from the soil in drainage water. Lagatu

Nitrogen deficiency in blueberries. A healthy leaf is shown at the left in contrast with three leaves showing various stages of nitrogen starvation. (Courtesy, J. B. Hester and the Chilean Nitrate Educational Bureau, Inc.)

Calcium deficiency in cauliflower. The leaves gradually lose their dark green color and turn yellow beginning first around the margins. (Courtesy, Chilean Nitrate Educational Bureau, Inc.)

Magnesium deficiency in cotton. These leaves show various stages of the deficiency, the bright leaves and yellow pattern being the latter phase. (Courtesy, Chilean Nitrate Educational Bureau, Inc.)

FIG. 102. Influence of limestone on the growth of sweet clover. (*Left*) 6 tons of limestone per acre. (*Right*) No limestone. (Courtesy, Ohio Agricultural Experiment Station.)

and Maume (1928) found that plants grown on limed soil contain less potassium than those receiving the same amount of potassium and grown on unlimed soil. Following heavy applications of lime, rust in cotton and similar manifestations of potash deficiency in other plants are commonly seen in crops grown on the gray soils of the Southeast.

Crop Response to Liming

Plant response to lime depends upon a number of factors. The legumes in general respond to liming to a greater extent than do the nonleguminous plants. Recent studies have shown that the fixation of nitrogen from the atmosphere by legume bacteria is much more effective where high levels of available calcium are present. The legumes are known to be able to precipitate calcium within the plant as the oxalate, and may in this way accumulate within their bodies large quantities of calcium. Of the legumes commonly grown on the acid soils of Eastern America red and Dutch clover and alfalfa appear to respond to lime most markedly. Garden and field peas and soybeans respond to liming, but not to the extent observed with red or Dutch clover. Likewise, cowpeas, vetch, Austrian winter peas, and lespedeza appear to be benefited to a degree by lime. Davis and Brewer (1940), working in Louisiana, found that liming increased the calcium, phosphorus, and nitrogen content of both Austrian winter peas and common vetch. On the other hand, serradella and the lupines may be injured by liming. Contrary to the belief of many, all legumes do not respond favorably to lime.

The reported experimental evidence of response among the non-leguminous plants is confusing. This is to be expected when we consider the wide range of soil and climatic conditions under which many of our lime experiments have been conducted. Asparagus, barley, corn, Kentucky blue grass, oats, sorghum, timothy, and wheat appear to be benefited by liming. Irish potatoes, redtop, and strawberries appear to be little influenced, while watermelons and blueberries may be markedly injured by liming. The response of cotton to lime is much greater than formerly thought. Data reported by Paden (1942) and given in Table 86 suggest that liming of most cotton soils would be a beneficial practice.

Table 86

YIELDS OF SEED COTTON, OATS, AND LESPEDEZA HAY SOWN ON CECIL SANDY LOAM AT VARYING pH LEVELS

pH Value	*Seed Cotton, lbs. per Acre Average (5 yrs.)*	*Oats, bushels per Acre (1 yr.)*	*Lespedeza Hay, lbs. per Acre (1 yr.)*
5.0	1093	46.2	2119
5.5	1263	49.9	2524
6.0	1346	49.0	3115
6.5	1382	46.6	3273

Magnesium an Element Essential for Plant Growth

Magnesium deficiency has affected a larger area of the United States than any other secondary essential element. This may be explained in part by the fact that superphosphate—a carrier of calcium—and ammonium sulfate—a carrier of sulfur—have been used so extensively in former years that crop needs for calcium and sulfur have been largely taken care of.

The essential functions which magnesium performs within the plant are not definitely known. At one time magnesium was thought to play a part in protein synthesis, but this view is no longer generally held. It is now thought by some plant physiologists that magnesium may play a role similar to that of potassium in the translocation of starch. It also appears to be essential for the formation of fats and oils, and to function as a carrier of phosphorus. Magnesium is a key constituent of chlorophyll, and to this element we owe the green world of vegetation. Cooper (1945) has called attention to the fact that magnesium is one of the strongest metallic cations that could be readily assimilated by plants in connection with such relatively light-stable anions as carbonate and phosphate. As plant leaves

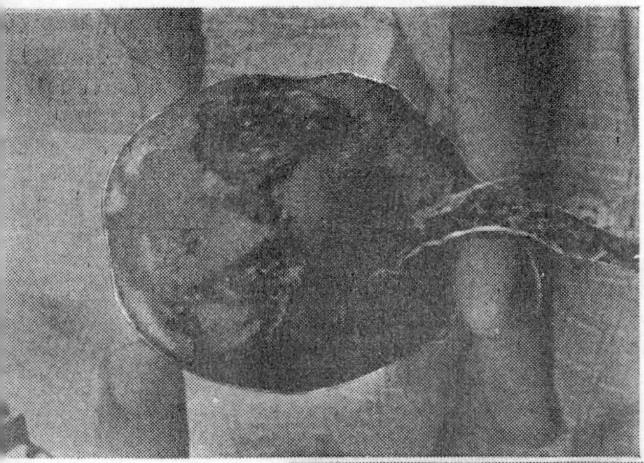

Boron deficiency in beets. The black corky scars on the surface of the beet are symptoms of the disease. When cut across, the beet is corky and blackened around the growth rings. (Courtesy, Chilean Educational Bureau, Inc.)

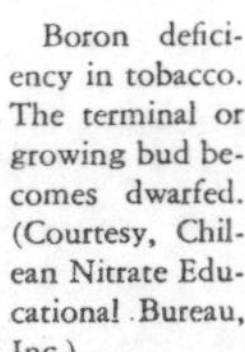

Boron deficiency in tobacco. The terminal or growing bud becomes dwarfed. (Courtesy, Chilean Nitrate Educational Bureau, Inc.)

Manganese deficiency in soybeans. The yellow area in the middle of the field was found deficient in manganese. (Courtesy, L. G. Willis and the Chilean Nitrate Educational Bureau, Inc.)

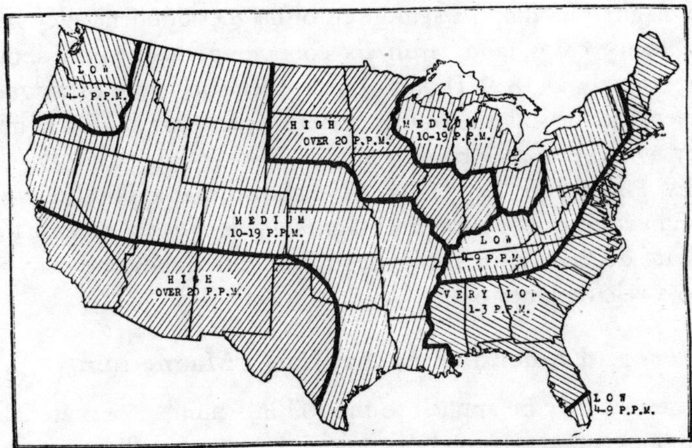

FIG. 103. Magnesium content of drainage waters of U.S. (Courtesy, S. B. Detwiler Soil Conservation Service, and *American Hortigraphs and Agronomic Review.*)

mature the magnesium is translocated to the seed and magnesium-deficiency symptoms may appear unless sufficient magnesium is present in the soil. In some of the vegetable crops only the lower leaves become chlorotic. Davidson (1941) has reported that magnesium deficiency greatly reduces the resistance of seedling fruit trees to winter injury and produces in strawberries abnormally thin leaves of a bright green color. Analyses of the ash of most plants will show a magnesium oxide content of about 5 to 10 per cent. The average yield of most field crops removes from the soil at least 10 pounds of magnesium per acre.

Bartholomew (1933) has secured evidence which indicates that soluble magnesium salts increase the absorption of phosphorus. Magnesium also may play a part in allowing the translocation of phosphorus in plants. Cooper (1930) calls attention to the fact that when the soil colloids are relatively high in H- ions and Mg- ions, conditions are often favorable for the dominance of resinous plants, such as coniferous and gum trees.

Magnesium Content of Soils

Magnesium is a constituent of some of the most common soil-forming minerals, such as biotite mica, augite, olivine, serpentine, talc, and chlorite. It is also found as an impurity in all limestone, and in the potash minerals carnallite and kieserite. Many soils contain sufficient magnesium to supply the normal demands of field crops, but in the Coastal Plain section of eastern South Carolina and other southeastern states, and in the gray sandy areas of the Piedmont Plateau section, many soils are so deficient in magnesium that the symptoms of magnesium deficiency are common in many crops, and fertilizers containing magnesium should be applied. As early as 1859 Thomas Green Clemson wrote of magnesium: "Its presence in the soil is a requisite to fertility and its addition of manifest necessity wherever it may be wanting."

Willis, Piland and Gay (1934) have shown that applications of non-magnesium lime to soils low in magnesium invariably accentuated the magnesium deficiency. In such cases a dolomitic limestone is to be preferred to a calcitic limestone.

Fertilizers and Amendments that Carry Magnesium

Magnesium may be applied to the soil in a number of materials. The principal ones commonly used are: dolomitic limestone (limestone with 20 per cent or more of magnesium carbonate), potassium magnesium sulfate, magnesium limestone, magnesium sulfate, and magnesite together with its by-products magnesium oxide and hydroxide. Manure salts and kainite also sometimes carry magnesium salts.

The mineral magnesite is a carbonate of magnesium and the largest

FIG. 104. Tomato plants that were shifted from a complete nutrient solution, when 8 inches high, to one having one of the elements left out as indicated. (*Right to left*) Receiving complete nutrient solution, no magnesium, no phosphorus, no potassium, no nitrogen, no calcium. (Courtesy, New Jersey Agricultural Experiment Station.)

Manganese deficiency in peanuts. Note the dabs of yellowish-white at the tips of the leaves (Courtesy, Chilean Nitrate Educational Bureau, Inc.)

Iron deficiency in citrus. In severe cases the leaves are quite small, yellow, and are quickly shed. (Courtesy, Chilean Nitrate Educational Bureau, Inc.)

Copper deficiency in citrus. A section of a healthy orange (*right*) is shown in comparison with one suffering from copper deficiency. The diseased fruit is dry and pulpy. (Courtesy, Chilean Nitrate Educational Bureau, Inc.)

deposits in this country are found in California. The mineral dolomite is a double carbonate of calcium and magnesium. Deposits of dolomite are widely distributed in the United States, and the mineral is commonly employed as building stone, and for the manufacture of agricultural lime. The ground dolomite commonly sold for agricultural lime contains the equivalent of about 20 per cent magnesium oxide and 25 per cent calcium oxide. Nearly all calcium limestones carry varying amounts of magnesium. The amounts of magnesium contained in calcitic limestones may vary from a mere trace to quantities equivalent to those found in the dolomitic limestones. Although the magnesium in limestone is water-insoluble it becomes slowly available in the soil, and when incorporated with complete fertilizers a small portion is made water-soluble as a result of reactions that take place with superphosphate.

MacIntire *et al.* (1938) have proposed a selectively calcined dolomite ($CaCO_3.MgO$) as a conditioner and a superior compatable component of phosphatic compounds where quickly available magnesium is desired.

As magnesium salts are common impurities in the subterranean potash deposits of Europe and America, magnesium is nearly always present in the unrefined or partially refined potash salts of subterranean origin.

Bartholomew (1933) has reported that the presence in the soil of applied magnesium sulfate slightly increased the availability of ground phosphate rock.

Calcium-magnesium Ratio in Soils and Plants

About the beginning of the present century the injurious results on plants noted from the application of magnesium compounds alone, and the beneficial results noted from the application of calcium compounds alone led some investigators to suggest a balance between the two elements. Loew (1901) proposed that for the best development of field crops a certain ratio should exist between the quantities of calcium and magnesium in the soil. He reported that soils in Japan varied widely as to this ratio and similar variations in the calcium-magnesium ratio of soils were found in this country. Loew concluded that in soils the best ratio of calcium to magnesium on which the common field crops are to be grown appeared to be about two parts of calcium to one part of magnesium. Following the report of Loew many experimenters in various parts of the world studied the problem and drew varying conclusions. In this country it is now generally agreed that there is little correlation between the ratio of calcium and magnesium in the soil and crop yields. C. B. Lipman (1916) concluded

FIG. 105. (*Left*) Normal cotton leaf. (*Right*) Leaf showing magnesium deficiency symptoms—green veins, purplish-red between veins. (Courtesy, South Carolina Agricultural Experiment Station.)

that there is little evidence available to support the hypothesis that a specific calcium-magnesium ratio exists for any plant or group of plants.

The analyses of crops also have shown a very wide range in the calcium-magnesium ratio within plants. As a general rule most crops contain more calcium than magnesium. Results secured by Wheeler (1921) and Hartwell indicate a wide range in the ratio of calcium and magnesium in plants without any correlation existing between the ratio and the yield. Nevertheless, some investigators hold the view that calcium and magnesium, as well as nitrate and potassium, potassium and magnesium, and iron and manganese are associated in nutritional effects. McMurtrey (1941) has reported that quantities of magnesium which produced normal growth in the presence of calcium appear to be toxic in its absence.

MacIntire and Shaw (1930) concluded that in humid climates toxic or excessive amounts of magnesium cannot result from the use of dolomitic limestone in view of the fact that the loss of added magnesium by leaching exceeds that of the added calcium.

Symptoms of Magnesium Deficiency

Because most magnesium compounds are very soluble and because magnesium is weakly absorbed by soil colloids, its deficiency in soils is probably greater than now thought.

Zinc deficiency in corn. The plants are stunted and flat-topped. The older leaves are yellow and marked with a vivid green striping. (Courtesy, L. G. Willis and the Chilean Nitrate Educational Bureau, Inc.)

Magnesium deficiency. "Sand Drown," in tobacco. The leaves gradually turn yellow, remaining green along the veins until the entire leaf loses its normal color. (Courtesy, Chilean Nitrate Educational Bureau, Inc.)

Boron deficiency, "alfalfa yellows," in alfalfa. Plants are stunted and terminal leaves are yellow. (Courtesy, J. B. Hester and the Chilean Nitrate Educational Bureau, Inc.)

Magnesium is a component of chlorophyll, and magnesium deficiency shows itself in crops by a characteristic coloration of the leaves and in serious cases in some crops in a premature defoliation of the plant. Plant symptoms of magnesium deficiency were first noted in tobacco by Garner and Brown (1919), of the Bureau of Plant Industry. They showed that the characteristic chlorosis of tobacco, known as "sand drown" could be cured by supplying magnesium salts. Cotton plants suffering from an insufficiency of magnesium produce purplish-red leaves with green veins. Corn and sorghum leaves show the deficiency by becoming striped, the veins remain green but that portion of the lamina between the veins becomes yellow in corn and purple in sorghum. In the legumes the deficiency is evidenced by chlorotic leaves with green veins. In Irish potatoes the deficiency is referred to as "potato sickness," in citrus as "bronzing," and in apples as "blotch." Fudge (1938) has shown that magnesium-deficiency symptoms in citrus result from the translocation of magnesium from the leaves to the fruit.

During wet seasons more magnesium is leached from the soil than during dry seasons and because of this the plant symptoms of the deficiency are more pronounced during wet years. Collison, Beattie and Harlan (1933) found from a study of lysimeter investigations in New York that 80 per cent of the total magnesium lost from the soil was in the drainage water.

According to Cooper (1932), magnesium deficiency symptoms are more pronounced in crops grown on soils of a low magnesium content, following applications of fertilizer salts. The aggravation of the trouble appears to be related to the solubility of the magnesium salts formed in the soil as a result of the addition of the fertilizer salts. The addition of chlorides, nitrates, sulfates and phosphates would result in the formation and loss in drainage water of magnesium chloride, nitrate, sulfate and phosphates. The molar solubility of magnesium chloride is 5.76, of magnesium nitrate 5.00, of magnesium sulfate 2.88, and of magnesium phosphate 0.002. It is to be expected, therefore, that the loss of magnesium from the soil would be greater where chlorides are applied and least where phosphates are applied. This is borne out, in part, by Cooper's work. Cooper (1930) has suggested also that the deficiency of magnesium, as well as manganese and iron, may be due in part to the fact that the stronger ions, such as potassium and calcium, may exclude or limit the absorption by the plant of weaker ions. Magnesium appears to be less available in soils containing high levels of available calcium and potassium.

Magnesium deficiency may be corrected by the addition to the soil of materials carrying magnesium such as magnesium sulfate, calcined kieserite, potassium-magnesium sulfate, magnesium-ammonium phosphate or magnesium carbonate. Calcined kieserite contains 19 per cent water-soluble magnesium. In recent years fertilizer manufacturers have been incorporating *Emjeo*, a magnesium sulfate, in their fertilizer mixtures to supply soluble magnesium. *Emjeo* analyses 90 to 92 per cent magnesium sulfate. Fifteen hundred pounds of dolomitic limestone, or 200 pounds of kainite, has been found to correct the deficiency for general field crops at the South Carolina Sand Hill Experiment Station. When the pH of the soil is above 6.0 magnesium sulfate may be preferable to dolomitic limestone. Chucka (1934) found that soluble carriers of magnesium were more effective than dolomitic limestone in preventing magnesium deficiency symptoms in potatoes. Recent tests have shown that the magnesium of dolomite is more readily available to crops after the dolomite has been mixed for several weeks with fertilizer mixtures, even though the mixtures may be neutral in reaction. This probably results in a reaction of the dolomite with superphosphate with the resulting formation of dimagnesium phosphate. It is well to note here that apparently magnesium phosphate does not combine with the free fluorides of superphosphate and revert to the magnesium fluorphosphate. For a number of years a tobacco committee of the southern agronomists has recommended that all complete fertilizers for tobacco carry 2 per cent of magnesia and that at least 1 per cent be derived from water-soluble materials.

13

Fertilizers Carrying the Rarer Essential Elements

Six essential elements of plant growth compose a group that is now commonly referred to as the rarer elements. These elements are iron, manganese, boron, copper, zinc and molybdenum. In addition, Hutchinson (1945) believes that aluminum should be accepted as an essential element for ferns and other vascular cryptograms. Unlike the three so-called secondary essential elements, the rarer elements appear to be helpful in most cases only at concentrations of the order of one-fourth part per million, while often two parts per million in solution may be toxic.

The response of plants to the presence of the rarer elements varies with different plants and with different soil environments. Powers (1940), of the Oregon Experiment Station, thinks that the functions of the rarer elements are as follows:

1. They may be essential nutrients for life and growth.
2. They may serve as substitutes for other elements.
3. They may help form enzymes or vitamins.
4. They may antidote toxicity.
5. One element may aid in keeping another reduced or oxidized.
6. One element may liberate another element.
7. They may increase resistance to disease.
8. The addition or presence of one element may precipitate ions of another element.
9. An indirect effect may occur from effect on reaction or action of an element upon microörganisms.
10. They may have stimulating effects.
11. They may have possible colloidal effects.

In general, farmers should not attempt to supply the rarer elements to their crops without first consulting specialists of their state experiment station or extension service.

Iron an Element Essential for Plant Growth

Iron has long been accepted as an element essential for plant growth. Iron appears to catalyze the production of chlorophyll. If plants are grown in a medium free of iron they develop a pathological chlorosis which can be relieved only by the application of small quantities of iron. The chlorosis appears first on the young parts of the plant and although it is somewhat similar to that produced by a deficiency of nitrogen and magnesium it develops much faster when produced by a deficiency or an inactivation of iron. The part that iron plays in plant metabolism is not known but Bennett (1945) believes that the chlorosis is a disturbance of both nitrogen and iron metabolism and that the two are intimately related. Unlike many other elements, once iron becomes fixed in the plant it is not readily transported to another part. More iron appears to be needed by the plant in light of high intensity than in light of low intensity and more iron is found in leaves of plants growing on noncalcareous soil than in leaves of plants growing on calcareous soil.

Nearly all soils contain an abundance of iron and some soils may contain 20 tons or more of iron oxide per acre six inches. Nevertheless, there are a few acid soils and many alkaline soils that have been reported deficient in this element. Iron deficiency has been recognized in California for many years. Aston (1929) has reported that forage plants grown on certain soils in New Zealand contain insufficient iron for the proper nutrition of

FIG. 106. Influence of an iron spray on the growth of pineapples produced on the manganese soils of Hawaii. (*Left*) Plants receiving a spray of iron sulfate. (*Right*) Plants unsprayed. (Courtesy, Hawaiian Agricultural Experiment Station.)

animals. Unreported analyses of plants made by the South Carolina Experiment Station would indicate that iron may be a limiting factor of plant growth for truck crops on some of the poorly drained sandy soils of the Coastal Plains section. The presence of decomposing organic matter under poor drainage conditions tends to deplete the iron content of soil colloids. This has been demonstrated by Holmes, Hearn, and Byers (1938). The chlorosis of plants due to a deficiency of iron is usually found on soils rich in calcium and manganese. These elements appear to immobilize the iron in the soil and in the plant and to bring about a condition of deficiency especially in the growing plant. Iron is comparatively immobile in growing plants and for this reason it is not moved from older leaves to younger tissues when a deficiency sets in. In fact, pathological accumulation of iron and aluminum in diseased corn stalks has been described by Hoffer and Carr (1923). The high content of iron in potassium-deficient plants is due to the higher pH values of the sap, a condition that brings about the precipitation and accumulation of iron in abnormal amounts. Iron deficiency may develop in sandy soils of low iron content where heavy applications of lime have been made. Lime-induced chlorosis of grapes has been noted in France, and of sugar cane in Puerto Rico.

The fertilizer material that should be applied to a soil deficient in iron may depend upon what other nutrients are required. If the soil is also deficient in phosphorus, both the phosphorus and iron may be supplied in basic slag. When iron deficiency is lime-induced it may be corrected by the use of acid-forming fertilizers.

The iron chlorosis of pineapples in Hawaii appears to be manganese-induced and can be cured by applications of iron compounds in the form of sprays. The excess of manganese dioxide in the soil, resulting from an acid condition, keeps the iron oxidized to the ferric form and thus depresses its absorption. The ferric form of iron is not readily available to plants. Pineapple growers who farm the acid manganiferous soils of Hawaii, often find it profitable to apply as a spray a 5 per cent solution of ferrous sulfate at the rate of 50 to 100 gallons per acre every week or 10 days during the growing period. Johnson (1916) has reported the cost of this iron spray to be about 60 cents per acre.

The iron chlorosis of fruit and shade trees can be overcome by driving iron nails into the trunks, or by boring small holes into the trunks and filling with a soluble iron salt such as iron citrate or tartrate, as has been described by Bennett (1931).

Toxic Compounds of Iron in the Soil

Often deoxidation processes are set up by anaerobic organisms in soils that are not thoroughly drained, and such oxides as ferric oxide may be reduced to the lower oxide, ferrous oxide. These lower oxides of iron, and probably also other similarly derived compounds of iron, are toxic to many plants. When such toxic conditions exist they may be corrected by drainage and the employment of soil management practices which will aerate the soil.

Manganese an Element Essential for Plant Growth

Manganese was first shown to be present in plants by Scheele in 1774. Since that time its presence in many plants has been noted and in recent years it has been generally accepted as an essential element for crops. McHargue (1926), of the Kentucky Experiment Station, has reported results which demonstrate conclusively that manganese is essential for plant growth. He has shown that manganese functions with the aid of iron in the synthesis of chlorophyll and in the assimilation of carbon and that legumes appear to be more sensitive to a lack of manganese than the nonlegumes. McHargue (1945) states, however, that leguminous plants usually contain the least amount of manganese and the grasses the greatest amount. Manganese also appears to be needed for the activation of oxidases. Many investigators concur in the belief that manganese tends to accumulate in the leaves, but chemical analyses show that

FIG. 107. Effect of manganese on potatoes growing on marl (calcareous) land in Florida. (*Left*) Fertilizer without manganese. (*Right*) Fertilizer with manganese. (Courtesy, Carus Chemical Company, Inc.)

FIG. 108. Showing symptoms of magnesium deficiency in turnips. (*Left*) Deficiency symptoms. Note ragged outline and chlorotic condition. (*Right*) Normal leaf. (Courtesy, *American Hortigraphs and Agronomic Review.*)

the manganese content of plants of the same crop will vary widely and that the element is probably the most variable of the mineral nutrients. Burstrom (1939) is of the opinion that the reduction of nitrates in plants is dependent on an enzyme system in which manganese has an essential role. Rigg and Askew (1929) found pasture plants in New Zealand to contain more manganese than iron. The soils upon which these plants were grown were not deficient in iron, so it would appear that manganese is capable of replacing iron to some extent, and of being selectively absorbed by plants ahead of iron.

Manganese Content of Soils

Manganese exists in most soils as the dioxide or the hydrated oxide in the mineral pyrolusite, manganite, and psilomelane. The manganese content of most soils is rather low, but manganese occurs in available amounts sufficient to meet crop needs in most American soils, with the exception of the calcareous soils of the glades of Florida, and some of the sandy soils of the Atlantic Coastal Plain. Like iron, manganese exists in the soil in an oxidation-reduction equilibrium. This equilibrium exists between the available manganese (manganous) and a continuous series of manganese of a higher state of oxidation (manganic) representing a range from the most active to the most stable form. In certain black Hawaiian soils the manganese oxide content may be as high as 8 to 10 per cent. Schreiner (1930) reports that owing to the presence of excessive quantities of manganese in these soils great difficulty is experienced in growing pineapples. Sherman *et al.* (1942) concluded that with the gray-brown podsolic soils the active manganese is concentrated in the A horizon and that erosion exposes subsoils that are deficient in manganese.

Schreiner and Dawson (1927) directed attention to the soils occurring in the glades of Dade County, Florida, which fail to produce a crop unless fertilized with stable manure or manganese sulfate in addition to a balanced inorganic fertilizer. Stable manure usually contains small quantities of manganese. Hester (1941) found that soils with a low manganese content produced tomatoes with a low vitamin C content.

Powers (1940), of the Oregon Experiment Station, reported an increase in the yields of many crops following the application of manganese sulfate. On certain acid peat soils he secured an increase of 82 per cent in the yield of tomatoes. Further, Marsh and Powers (1945) found that manganese was needed on the basic peats of Oregon.

Influence of Manganese on Plant Growth

Excess quantities of soluble manganese, such as may be present in very acid soils, may be toxic to plants and may explain in part the ill effects of high soil acidity. As the clay crystal acts as an acid, clays tend to reduce manganic manganese and to retard the oxidation of manganous manganese. It seems reasonable to believe that manganese toxicity on the highly acid soils of the Southeast is more prevalent than is commonly realized.

Malnutrition in plants due to a deficiency of manganese often occurs on soils which contain fair amounts of manganese but which are neutral or slightly alkaline in reaction. This deficiency is commonly noted where the soil has been heavily limed. Alkalinity resulting from liming reduces the solubility of manganese. It appears that the manganese ion is only absorbed by plants in the bivalent form which is readily oxidized to a higher valency when the soil is alkaline. The fixation of manganese in soils above pH 6.0 is due to the oxidation of the manganous manganese to the hydrated manganic oxide and to inert manganic oxides. Likewise, in strongly acid soils the oxidation-reduction conditions favor the reduction of the manganic manganese to the manganous form. Manganese can be readily leached from a strongly acid soil. Jacobson and Swanback (1932) and Sherman, McHargue and Hodgkiss (1942), as well as other investigators, have shown that liming greatly reduces the active manganese content of the soil. The liming of strongly acid soils to an alkaline reaction is the most common cause of manganese deficiency. However, this condition is hard to obtain on soils having a high clay content. Willis and Mann (1930) reported that manganese deficiency is more pronounced when liming materials have been added in amounts sufficient to increase the pH of the soil solution to 6.5 and above. McVickar (1942) found that the manganese uptake by bluegrass in parts per million on soils below pH 5.0 averaged 7.68 as compared with 1.84 on soils of pH 5.0 or above. Spots showing manganese deficiency are often noted in fields of sandy soils of the Coastal Plain where brush has been burned, for wood ashes are composed largely of lime. When manganese deficiency occurs in plants growing on acid soils it is thought to result from the leaching of soluble manganese.

Manganese chlorosis of plant leaves first appears between the veins as a pale green, yellow or red area, depending upon the crop. These spots gradually become white surrounded by the green of the veins.

Gilbert, McLean, and Hardin (1926), of the Rhode Island Experiment Station, have reported instances where the chlorosis or frenching of crops that had been induced by liming could not be corrected by an application of iron, but was corrected by the application of manganese. Samuels and Piper (1928) have shown that the "gray speck" or "white wilt" disease of oats which occurs in certain parts of Australia and Europe is due to an insufficient supply of manganese. Sherman and Horner (1941) have been able to control gray speck disease by any soil treatment which will increase the exchangeable manganese in the soil to 3 p.p.m. or more. Manganese toxicity in tobacco has been reported from Connecticut by Jacobson and Swanback (1932) and from Kentucky by Bortner (1935).

Piper (1931) has reported that unavailable manganese in the soil may be made available by subjecting the soil to such reducing conditions as those brought about by temporary water-logging prior to seeding. Such indirect methods of supplying a crop with more manganese is based on the reversion of the manganous-manganic equilibrium. The same thing can be accomplished by changing the soil reaction with sulfur and other acid-forming materials.

Application of Manganese Fertilizer

The most common carrier of manganese used as a fertilizer and the most economical is manganese sulfate. Where manganese deficiency occurs 50 to 100 pounds of manganese sulfate per acre annually is generally recommended, although 200 pounds is recommended for the marl soils of Florida. The more alkaline the soil the greater appears to be the need for the salt, and on alkaline organic soils Sherman and Harmer (1941) recom-

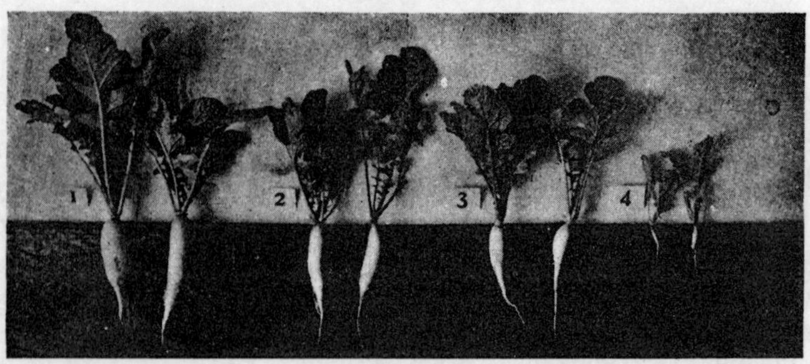

FIG. 109. Response of radish to treatment with copper, manganese, and manure on raw saw-grass soil of the Florida Everglades. (1) Copper; (2) manganese; (3) manure; (4) check. (Courtesy, Florida Agricultural Experiment Station.)

mended 100 to 400 pounds of manganese sulfate per acre. However, the application of manganese directly to a soil of high fixation capacity is a very inefficient method for the correction of manganese-deficient condition. In such cases, either an application of manganese sulfate should be made with equal application of sulfur or preferably the manganese sulfate should be applied as a spray to the foliage of the crop. A too liberal supply of manganese is not economical and may cause harm. Manganese toxicity in a soil may be corrected by applying lime or phosphates. Certain manufacturers of mixed fertilizers in Florida make it a practice to add manganese sulfate to all their complete fertilizers.

Sherman (1943) concludes that manganese deficiency in soils has been caused by man, and due to his carelessness in the management of the soil and to his ignorance of the fundamentals of soil chemistry.

Because of its manganese content barnyard manure has been shown to alleviate manganese deficiency. It is probably true that the increasing scarcity of manure in America explains in part the increase in manganese deficiency noted during recent years.

Boron an Element Essential for Plant Growth

Boron is widely distributed in nature. It was first discovered to occur naturally in plants by Wittstein and Apoiger in 1857. It is found in rocks and soils and in plant and animal tissues. The extensive distribution of boron in the soils of America was demonstrated by Whetstone *et al.* (1942). Garman (1946) recently determined the boron content of 20 soil types of South Carolina. His results are given in Table 87. Boron deficiency is not as widespread in acid soils as in those on the alkaline side. McHargue *et al.*

Table 87

Available Boron in Some Soils of South Carolina

Soil Type	*Depth (inches)*	*County*	*pH*	*Boron (p.p.m.)*
Charleston fine sandy loam	0–10	Beaufort	4.8	0.30
	10–22		4.7	0.30
Edisto sandy loam	0– 6	Beaufort	5.7	0.33
	6–20		5.7	0.09
Dunbar fine sandy loam	0– 6	Jasper	5.0	0.54
	6–18		4.7	0.30
Edisto fine sandy loam	0– 9	Beaufort	4.7	1.03
	9–20		4.7	0.56
Stono fine sandy loam	0–12	Charleston	5.5	0.23
	12–24		4.5	0.34
High Marsh	0– 8	Charleston	7.1	1.87
	9–24		6.5	2.00
St. Johns fine sandy loam	0– 9	Charleston	4.0	0.58
	9–20		4.4	0.34
Nosbig fine sandy loam	0– 8	Charleston	5.7	0.23
	8–22		4.8	0.11
Dupont fine sandy loam	0–10	Charleston	5.0	0.21
	10–18		4.6	0.15
Bladen fine sandy loam	0– 6	Horry	4.6	0.76
	6–17		4.2	0.54
Dunbar, very fine sandy loam	0– 6	Horry	4.5	0.53
	6–12		4.6	0.12
Norfolk fine sandy loam	0–10	Georgetown	5.7	0.16
	10–24		4.1	0.12
Onslow fine sandy loam	0–10	Georgetown	4.2	0.24
	10–15		4.2	0.16
Portsmouth fine sandy loam	0– 8	Charleston	5.3	0.70
Lynchburg fine sandy loam	0– 6	Georgetown	5.0	0.22
	6–12		6.2	0.08
Fairhope sandy loam	0– 6	Colleton	4.9	0.14
	6–17		4.5	0.17
Marlboro, silty loam	0– 7	Florence	5.5	0.25
	7–13		5.6	0.32
Iredell silty clay loam	0– 7	Abbeville	6.1	0.33
	7–14		6.4	0.20
Cecil clay loam	0– 8	Abbeville	5.6	0.21
Davidson clay loam	0– 8	Abbeville	5.7	0.16

(1940) have reported its widespread occurrence in plant and animal materials. It occurs in most plants in quantities up to about 200 parts per million and Munsell and Brown (1943) report a higher content in legumes than in nonlegumes. According to Robinson and Edgington (1945) boron does not exceed 70 p.p.m. in normal naturally grown plant leaves. McHargue *et al.* (1939) have reported from .4 to 143 p.p.m. in various plant parts. Some of the data of McHargue *et al.*, is given in Table 88. Boron makes up about 10 per cent of the mineral tourmaline, and tourmaline is almost universally present in soils, although generally in very

Table 88

BORON CONTENT OF SOME FIELD CROPS

Material	*Boron (p.p.m.)*
CEREALS:	
Corn, white, Ky	1.17 (2)*
Corn, yellow, Ky	2.30 (3)
Wheat, Ky	0.59 (3)
Oats, ?	1.02
Barley, Ky	0.64
Rye, Ky	1.50
Average	1.19
PASTURE AND HAY CROPS:	
Bluegrass, Ky	11.80
Bluegrass, Ky	6.00
Orchard grass, Ky	5.20
Timothy and clover mixed, Ky	3.30
Average	6.57
LEGUMINOUS HAYS:	
Red clover, Ky	31.30
Alfalfa, normal, Ky	33.30
Alfalfa, abnormal, Ky	47.00
Lespedeza, Ky	18.10
Lespedeza, Ky	20.30
Average	30.00

* Average of number of samples indicated.

small quantities. Its persistence in soils would indicate that the mineral is almost insoluble. Whetstone, Robinson, and Byers (1942) have made a comprehensive survey of the distribution of boron in the soils of America. Three regions of low boron content are recognized—the Atlantic Coastal Plain, a region across Michigan, Wisconsin, and Minnesota, and the Pacific coastal region of the Northwest. Many soil types have been reported to be deficient in boron, including soils from many foreign countries and from at least 20 of the states in the United States. Schuster and Stephenson (1940) have found that soils that have long been clean cultivated and that are low in organic matter contain less available boron than uncultivated soils of the same type. Also, deficiency symptoms appear to be more prevalent during dry seasons. Boron, unlike some other rarer elements, such as copper and zinc, is not absorbed by soil colloids.

Influence of Boron on Plant Growth

Although it was demonstrated by Agulhon as early as 1910 that boron is required by some plants, its essential nature was not established until after World War I. Boron appears to be closely related to some function that calcium performs in the plant. As in the case of calcium, there appears to be little transfer of boron from old tissues to younger growing points. Warington(1923) and Brenchley (1927), of the Rothamsted Experiment Station, were among the first to demonstrate the essential nature of boron and this was soon concurred in by Sommer and Lipman (1926), Johnson and Dore (1928), and others in this country. Brenchley and Warington believe that boron enables the plant to absorb or utilize calcium. Recently Parks, Lyon and Hood (1944) have secured results at the Plant, Soil, and Nutrition Laboratory of the U.S. Department of Agriculture at Ithaca, N. Y., which indicate that an increased absorption of boron by plants may bring about an increase in the absorption of all other mineral nutrients to an extent of several hundred per cent. Reeve and Shive (1944) report that the rate of boron absorption and accumulation by plants increases as potassium in the growth medium increases. Altogether some 20 or more physiological disorders affecting cultivated plants are attributed to boron deficiency.

Very small quantities of boron are required by plants and, unlike most other essential elements, if present in concentrations much above that required for normal growth it becomes toxic. This sometimes occurs naturally in arid regions. Eaton (1944) has presented data showing that there is considerable overlapping between the injurious and the beneficial effects of boron within plants. Collings (1927), working at the New Jersey Experiment Station in 1923, was among the first to describe boron toxicity. In the same study Collings also found that 2.5 parts per million of boron, either as boric acid or borax, stimulated the growth of plants that were grown to maturity. During World War I certain American potashes which contained borax were put on the American market. Damaging results to crop growth were secured in nearly all sections where these potashes were used, and especially where they were used on the sandy types of soil. Since World War I the methods of refining these American potashes have been improved to such an extent that the borax, when present, is removed almost entirely. The irrigation water of some sections of California and other western states often carry over one-half part per million of boron. This concentration of boron is sufficient to produce toxicity symptoms in citrus trees. Natural boron toxicity is not likely to occur except under arid conditions.

Many physiological diseases of plants have now been shown to be due to a deficiency of boron. Davidson (1941) thinks that in recent years deficiencies of boron in apple orchards have occurred more frequently than deficiencies of any other nutrient except nitrogen. Plants appear to need a continuous supply of boron for, unlike some other elements, boron is not stored in the plant. Also, as pointed out by Walker (1944) the first symptoms of boron deficiency are not externally visible. It appears that in the absence of boron the development of meristematic tissue of plants and the nodulation of legumes does not take place properly. Scripture and McHargue (1945) concluded that boron is functional in the metabolism and translocation of carbohydrates and Hoagland (1944) suggests that it has a role in the formation of pectin compounds and forms compounds with sugar or organic acids through adjacent OH groups. McMurtrey (1935) in 1935 showed the effect of boron deficiency on the growth of tobacco plants. He found that in the absence of boron the bud turned a light green color and then died. Dennis (1937) has presented evidence to indicate that cauliflower "browning," "brown heart" of turnips, and "roan" of swedes and rutabagas, as well as "internal cork" of apples, "cracked stem" of celery, and "yellows" of alfalfa are due to a deficiency of boron. Jones and Brown (1941) have reported a deficiency of boron to bring about sloughing in Irish potatoes. Boron deficiency in the legumes usually produces a red coloration, often with a purplish tint, of the margins or tip half of leaves. Willis and Piland (1928) and later Piland, Ireland and Reisenauer (1944) showed that in North Carolina alfalfa yellows was due

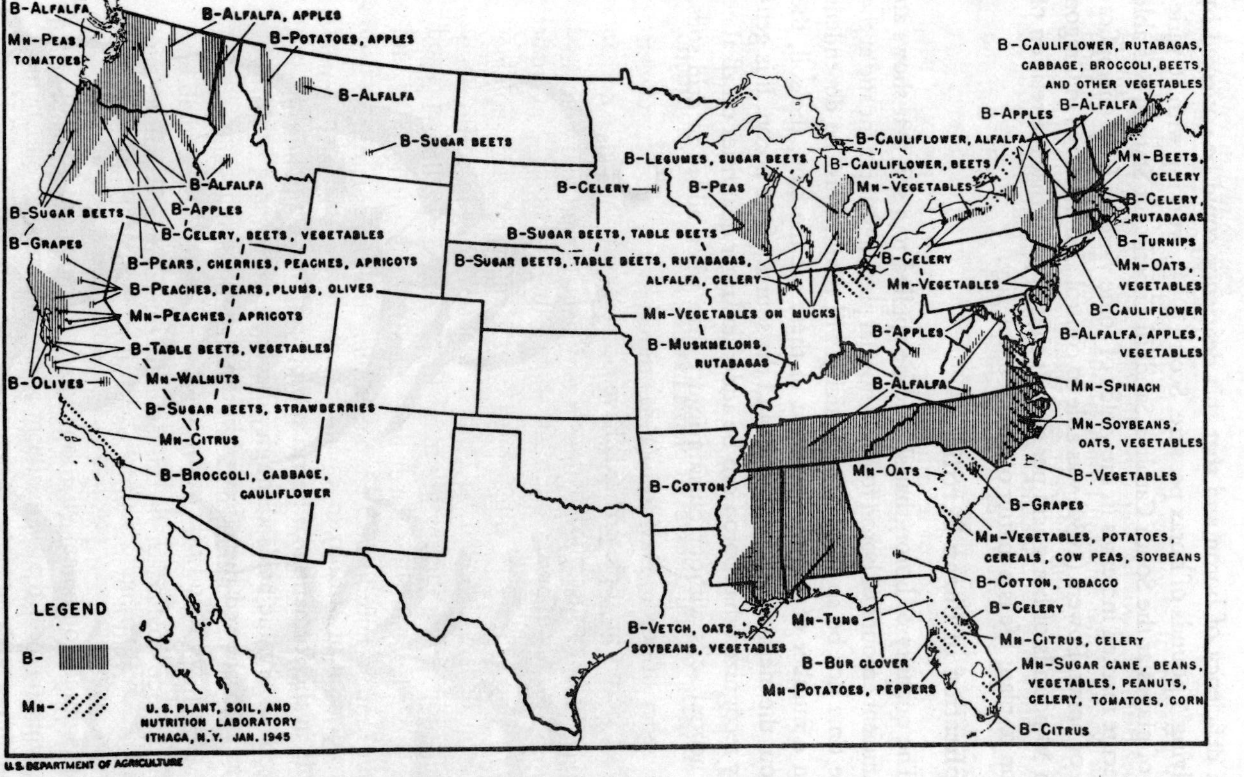

FIG. 110. Generalized map of the occurrence of boron and manganese deficiencies in plants in the United States. (Courtesy, Kenneth C. Beeson and the United States Plant, Soil, and Nutrition Laboratory.)

to a deficiency of boron and that this deficiency could be corrected by applying 5 pounds of borax per acre. Scott (1944), working with 50 varieties of grapes at the South Carolina Sandhill Experiment Station, was able to secure marked increases in the yields of some varieties from an application of borax. Powers (1939) was able to control alfalfa yellows in Oregon with 30 pounds of boric acid per acre. It thus appears that the relation of boron to plant life is a matter of great complexity.

Application of Boron Fertilizers

The quantity of boron that should be applied to a soil that shows an insufficient amount of boron to allow normal plant development and at the same time not be so much that toxicity will result, must be dependent upon a number of factors. Included in these are the crop, the soil, the season, the method of application, and the source of the boron. It is generally recognized that boron can be added in larger quantities on alkaline and neutral soils without causing injury than when added to acid soils.

FIG. 111. Influence of zinc sulfate on the growth of peach trees. (*Left*) Shoot from check tree grown in zinc deficient soil. (*Right*) Shoot from tree which received one-half pound of zinc sulfate. (Courtesy, Florida Agricultural Experiment Station.)

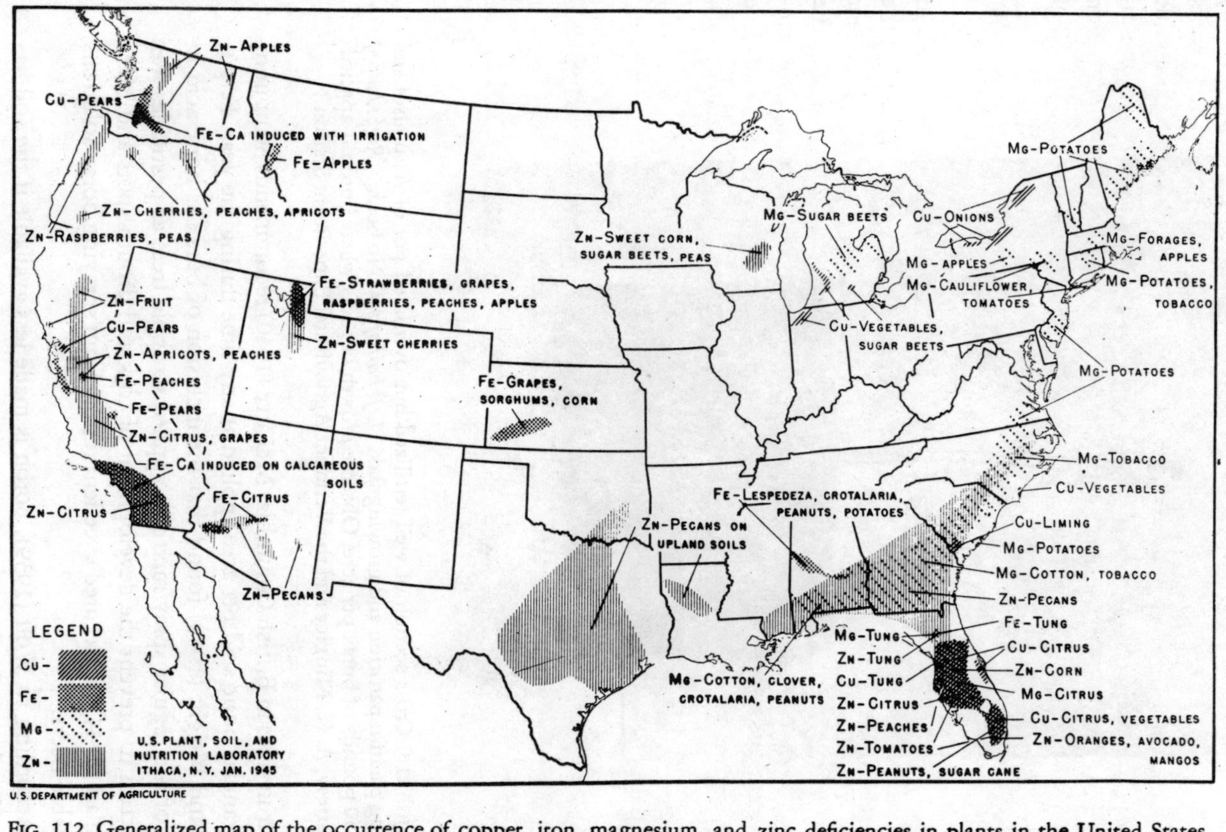

FIG. 112. Generalized map of the occurrence of copper, iron, magnesium, and zinc deficiencies in plants in the United States. (Courtesy, Kenneth C. Beeson and the United States Plant, Soil, and Nutrition Laboratory.)

FIG. 113. Cross section of well fertilized but overlimed pot of A_0 podsol soil having paraffin partition and growing flax. (*Left half of pot*) No borax. (*Right half of pot*) 50 pounds of borax per acre. Observe growth of roots where boron is applied. (Courtesy, A. K. Midgley and the Vermont Agricultural Experiment Station.)

For fruit trees in British Columbia, McLarty (1940) recommends 30 pounds of granular boric acid per acre applied at any time during the year. Ferguson and Wright (1940) found that the inclusion of borax as boric acid in two of the regular spray mixtures applied to apple trees as pesticides was sufficient to prevent the development of cork and drouth spots and MacLachlan (1944) has secured excellent results from spraying borax solutions on table beets.

According to Wolf (1939), boron is made less available if the soil has been limed to a pH above 7.0. He also demonstrated that plants grown on soils treated with borax showed an increase in boron content as the amount of borax applied to the soil was increased. It appears, therefore, that boron deficiency of crops is correlated not only with the amount of boron in the soil but, in addition, is also influenced by the soil reaction and the fertilizing practices followed. In 1940, Wolf demonstrated that the percentage and total amount of boron in plants decrease as the pH of the soil increases. Midgley and Dunklee (1939) concluded that the limited degree of fixation of soluble boron was chemical rather than biological, and that the ability of soils to fix boron depends on the degree of acidity and the extent to which they are subsequently limed. Harmer and Benne (1945) found that crops having a high sodium requirement were among the first to exhibit boron deficiency.

Borax (sodium tetraborate) is the most commonly used fertilizer material carrying boron. In experimental work potassium borate, boric acid, and finely ground borosilicates, have been found satisfactory. They may be applied alone or mixed with fertilizers. For most crops and for most soils 10 to 50 pounds of borax is recommended for use with complete fertilizer, but 1 pound per acre is sufficient for tobacco and 2 to 5 pounds may be satisfactory for some truck crops. Killinger *et al.* (1943) recommend 5 to 10 pounds of borax per acre for pastures. As soluble compounds of boron are not held as such in the soil annual applications of borax may be necessary. In Table 89 is given according to Reeve *et al.* (1944) the amount of borax leached from four soil types by an amount of water equivalent to one-fourth of the annual rainfall following an application of

FIG. 114. Response of sugar cane to an application of copper sulfate. (*Left*) No copper sulfate. (*Right*) Copper sulfate. (Courtesy, Florida Agricultural Experiment Station.)

Table 89

BORON LOSSES RESULTING FROM LEACHING SOILS WITH WATER

Soil Type	B Added to Soil*	B Leached from Soil†
	mg.	*mg.*
Penn silt loam	8.49	6.07
Lansdale silt loam	8.49	6.30
Sassafras loam	8.49	6.52
Lakewood sand	8.49	7.20

* Equivalent to an application of 20 lbs. of borax per acre.
† By an amount of water equivalent to a depth of 11 inches.

20 pounds per acre. In Table 90 is given, according to Purvis (1939), the plants which have been reported as showing boron deficiency in the United States. It shows also the range of field recommendations made in various states for boron deficient soils. Purvis (1945) has also classified plants into four groups: very sensitive, sensitive, tolerant, and very tolerant. The very sensitive group cannot stand over 5 pounds of borax per acre. In this group are included cowpeas, cucumbers, snap beans, and strawberries. The sensitive group can stand up to 20 pounds of borax per

Table 90

PLANTS SHOWING BORON DEFICIENCY IN THE UNITED STATES AND RANGE OF FIELD RECOMMENDATIONS MADE IN VARIOUS STATES FOR DEFICIENT SOILS

Plant	Recommendation, lbs. borax per acre	Plant	Recommendation, lbs. borax per acre
Alfalfa	10–40	Mangels	10
Apples*	5–30	Narcissus	None
	¼ to 1 lb. per tree	Pears*	½ lb. per tree
Beets	10–50	Potatoes	25
Broccoli	10	Prunes	None
Cabbage	10–12	Radishes	10–20
Carrots	10–20	Rutabagas	10–25
Cauliflower	10–25	Strawberries	10–12
Celery	10–100	Sugar beets	10–40
Corn	10	Tobacco	½–10
Cotton	½–10	Tomatoes	10–20
Eggplant	10–20	Tung trees	None
Lettuce	10	Turnips	10–30

* Boric acid used for apples and pears in Washington and for apples in Montana.

acre. This group includes celery, muskmelon, peas, potatoes, squash, and watermelon. The tolerant group can stand up to 30 pounds of borax per acre and includes such crops as cabbage, kale, lettuce, lima beans, okra, onions, peppers, radish, spinach, and sweet potatoes. The very tolerant group appears not to be injured by 50 pounds of borax per acre and includes beets, cauliflower, mustard, tomatoes, and turnips. It should be kept in mind that for a given crop borax can be applied in heavier applications to soils which have a high organic matter content, a high exchange capacity, or a high pH, than can safely be applied to light acid soils.

Copper an Element Essential for Plant Growth

Experimental evidence is now sufficient to justify the acceptance of copper as an essential element of plant growth. In the United States copper deficiency in vegetables is largely confined to the Portsmouth soil types and to peat and muck but in Florida copper deficiency frequently occurs with magnesium deficiency on acid sands. Copper is present in most plants in quantities up to 100 parts per million, and in many agricultural soils up to 50 parts per million. It is present in high concentrations in mushrooms and in the seeds of certain plants. It is readily absorbed by soil colloids but because it is a weak ion it is easily replaced by most other cations. Jamison (1942) found that copper is retained by the soil in slowly soluble or slowly replaceable forms which prevents ordinary applications of copper sulfate from becoming toxic to plants. In Table 91 is given the copper contents of some soils of the Atlantic Coastal Plain as determined by Holmes (1943). Copper is assimilated by animals and is present in rather large quantities in certain forms of marine life.

The function of copper in plant nutrition is not known. Hoagland (1944) suggests that copper plays a role in the synthesis of chlorophyll. Jacks and Scherbatoff (1934) think that the chief function of copper, both in the plant and in the soil, is that of an oxidizing agent. In recent years growth abnormalities of many plants produced on peat and muck soil, especially those that contain appreciable quantities of ferrous iron, have been corrected by the application of copper compounds. McMurtrey and Robinson (1938) think that some of the beneficial effects derived from the application of copper compounds may be due to the influence of copper in precipitating the toxic sulfide ion of peat soils, and Willis and Piland (1934–36) suggest that copper sulfate may serve not only as a nutrient but as a soil amendment which decreases the availability of iron and possibly of manganese. According to Sommer (1945) much more copper is neces-

Table 91

COPPER AND ZINC CONTENTS AND OTHER DATA FOR PROFILE SAMPLES OF SOILS OF THE ATLANTIC COASTAL PLAIN

Soil Series	*Depth (Inches)*	*Organic Matter (Per Cent)*	*Clay (Per Cent)*	*pH*	*Copper (p.p.m.)*	*Zinc (p.p.m.)*
Orangeburg	0– 6	0.55	5.0	5.1	18	25
	16– 30	0.32	47.1	4.8	22	34
	80–100	0.08	27.7	4.6	20	25
Ruston	0–10	0.88	7.9	6.0	25	26
	16–30	0.31	29.4	5.0	27	28
	40–54	0.17	30.8	4.8	26	27
Norfolk	0–12	0.89	8.9	5.1	22	31
	12–34	0.40	28.8	4.6	17	22
	36–80	0.24	31.3	4.6	17	22
Dunbar	0– 5	3.47	8.0	4.9	18	21
	5–12	0.72	11.6	4.5	19	13
	16–28	0.23	21.8	4.4	18	17
Coxville	0– 9	4.39	21.1	4.3	9	25
	26–34	0.52	40.8	4.4	8	29
	44–66	0.42	44.9	4.2	6	30
Bladen	0– 8	2.74	25.5	4.4	10	23
	10–38	0.64	30.8	4.3	5	23
Portsmouth	0–15	7.00	6.7	4.1	9	31
	15–35	0.19	15.2	4.3	10	16
	50–60	0.34	15.3	4.4	10	20
Average	..	..	..	..	16	24

sary for the correction of copper deficiency in soils of high humus content than in soils of low humus content. The addition of copper sulfate to some muck soils of New York has prevented the premature dying of onions, and in Florida and Holland it has been very helpful in preventing so-called "reclamation disease." Crop response to copper sulfate has been secured in Holland when crops were planted on newly reclaimed moor soils and on the peat soils of Michigan and Delaware in this country.

McMurtrey and Robinson (1938) showed that the upper leaves of copper-deficient tobacco plants are unable to maintain their turgor and consequently wilt badly, and that growth is reduced in proportion to the shortage of the element. In some plants a chlorosis develops. Allison, Bryan, and Hunter (1927) secured responses in growth of many plants from the use of antimony, barium, copper, nickel, tin, and zinc, but exceedingly favorable results were secured from the use of copper. In 1937

FIG. 115. Chlorotic dieback in flax near Fargo, N. Dakota. (Courtesy, A. C. Dillman, H. H. Flor, and the U.S. Bureau of Plant Industry.)

Hill and Bryan (1937) reported that copper is deficient for optimum growth of mustard on many soils of Florida.

Application of Copper Fertilizers

When used with the proper fertilizers very encouraging results have been secured from applying copper sulfate at the rate of 75 to 200 pounds per acre to the sugar cane lands of the Florida Everglades. Farmers of Florida and California commonly apply 50 to 75 pounds per acre of copper sulfate or blue vitriol to citrus groves, with their fertilizers, as a remedy for dieback. Dieback (Exanthema) is now considered to be a physiological disease resulting from an insufficient supply of copper. Some truck growers in New York, Florida, and Michigan have found it is necessary, in order to grow lettuce successfully on certain soils which are high in organic matter, to apply copper at the rate of 25 to 50 pounds of copper sulfate per acre.

Toxic effects on plants from excessive copper might be expected in special cases where the acidity of the soil is high. McKaig, Carns, and Bowen (1937) have reported a toxic effect secured from the application of 75 pounds of copper sulfate per acre to soybeans grown on Norfolk sand at the Sandhill Station of the South Carolina Experiment Station, and Lipman and MacKinney (1931) found one-fourth part per million of copper to be distinctly toxic to flax and barley that were grown in solution cultures.

It may be found inadvisable to mix copper salts with complete fertilizers for cupric phosphate may be formed and this phosphate is relatively unavailable. It may prove best to apply copper in a spray directly on the plants.

Zinc an Element Essential for Plant Growth

Although a quarter of a century ago Bertrand suggested that zinc was necessary for plant growth, only during recent years has zinc been commonly accepted as essential. Its present importance to the agriculture of Florida has been shown by a number of investigators. The quantity of the zinc required by plants, like that of boron and copper, is not great and some plants either require less than others or are better able than others to obtain their needs from the same environment. As zinc does not undergo reversible valance changes, its functions in oxidation-reduction systems must be less direct than those of iron and manganese.

Zinc is widely distributed in small quantities in soils, plants, and animals. Soils usually show from 5 to 50 parts of zinc per million, but the zinc content of sandy soils is usually very low. In the Southeast, zinc deficiency is more often noted in pecans grown on upland soils than in pecans grown on the flood plains. The availability of zinc declines as the pH of the soil rises. The critical point is between pH 5.5 and 6.5. It appears that zinc may again become slightly more available above pH 7.85. This may be due to the inclusion of zinc in a soluble form in the zincote anion. Hibbard (1940) analyzed 140 samples of soils and rocks of central California, and concluded that zinc and other rarer metals are accumulated in the surface soil by the action of vegetation. Plant parts usually show 1 to 50 parts per million of zinc, but 100 and 200 parts and more have been reported. Rogers *et al.* (1939) have secured data which indicate that weeds and volunteer grasses are able to secure more zinc from soils that are low in their zinc content than are the plants of most cultivated crops. Oysters and animal livers are usually very high in their zinc content.

Zinc, unlike boron and like copper, is absorbed by the clay acids of the soil and at least a part of it is held in an exchangeable form. Zinc, like copper may be precipitated in insoluble forms, such as the phosphate, when the reaction of the soil is neutral or alkaline. Alben and Boggs (1936) have reported zinc deficiency to be more prevalent on basic soils than on acid soils. Nevertheless, zinc deficiency has been reported in citrus, tung, and pecan trees growing on sandy soils of Florida which are slightly to highly acid in reaction. Camp (1945) says that it may be safely concluded that zinc deficiency occurs to some degree wherever citrus is grown. Hoagland (1944) calls attention to the fact that zinc-deficiency diseases are generally found in regions of high summer light and temperature, and that the sunny side of a citrus tree is usually much more mottled than the shady side. Hoagland also cites work with tomatoes by Strout, showing better zinc assimilation when plants were grown under conditions of reduced light. Zinc appears to be necessary for normal chlorophyll production and growth.

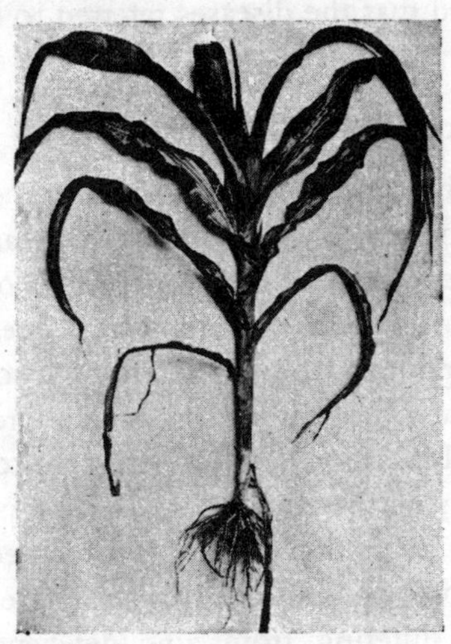

FIG. 116. Corn plant showing symptom of zinc deficiency. (Courtesy, Florida Agricultural Experiment Station.)

FIG. 117. One method of applying zinc sulfate to pecan trees. (Courtesy, *American Hortigraphs and Agronomic Review.*)

Beginning with the early nineteen thirties, a number of investigators reported beneficial effects on plant growth resulting from an application of zinc. In fact, during the period of 1931–1936 zinc became established as a plant nutrient and as a valuable fertilizer and spray material. Alben, Cole, and Lewis (1932), Finch and Kinnison (1934), and Alben and Boggs (1936), showed that "rosette" of pecans results from a deficiency of zinc. Barnette and Warner (1935) likewise reported that the "white bud" of corn is due to a deficiency of zinc. Thorne, Laws, and Wallace (1942) have reported widespread symptoms of zinc deficiencies in Utah on sweet cherries, apricots, peaches, apples, and plums. Other investigators have reported that the bronzing of tung oil trees, the yellows of walnut trees, the mottle leaf of citrus, the rosette of apple, abnormal leaves of pineapple plants, and the little leaf of the stone fruits and grapes, appeared to be due to a deficiency of zinc, and often appear on soils that have been overlimed or where there has been a reduction in the use of proteid organic commercial fertilizers. Zinc deficiency in tobacco is usually manifested by the development of dead spots all over the leaf, and not specifically at the tips and margins. Early (1943) has reported a varietal difference in susceptibility of soybeans to zinc toxicity. Chandler (1937) after reviewing the extensive literature on the subject, suggested that the diseases referred to above be called "zinc deficiency."

Application of Zinc Fertilizers

The application of zinc, usually in the form of zinc sulfate, has been recommended to correct zinc deficiency. For the control of pecan rosette Alben and Boggs (1936) recommend three or four applications of a spray made of 1 pound of zinc sulfate in 50 gallons of water. Citrus trees may be sprayed with a spray made of 5 pounds of zinc sulfate, 2.5 pounds of hydrated lime and 50 gallons of water. In general, sprays are more reliable than soil treatment. Dickey and Blackmon (1940) were able to correct little leaf of peaches either by applying the zinc sulfate as a spray or as an application to the soil. Beneficial results have also been secured by orchardists through the application of ½ pound of dry zinc sulfate in small concentrated areas around trees suffering from zinc deficiency. This amount may seem large but zinc appears to be subject to several types of soil fixation. Hoagland (1944) thinks that it may be fixed in the crystal structure of soil colloids. The driving of zinc wedges into the trunks of trees near the foliage has proved a favorable influence when the trees showed signs of zinc deficiency, a fact which would indicate that the trees were using the element as a nutrient. Reed (1941) has demonstrated that zinc deficiency has a special effect in inhibiting seed formation and development.

Toxic influences of zinc may be expected, especially on acid soils, from a too large application of a soluble zinc salt. Toxic effects on plants grown in water cultures have been abundantly proved. Zinc leaf injury, or chlorosis, would suggest an iron deficiency, and it would seem that the various metals do not act entirely independently in the plant. McKaig, Carns, and Bowen (1937) have reported a toxic effect on soybeans from an application of 75 pounds of zinc sulfate per acre when grown on Norfolk sand. Dickey and Blackmon (1940) secured a toxic injury to peach trees resulting from an application of 5 pounds of zinc sulfate per tree. There is no evidence of immobilized or inactive zinc within the plant.

Just why it is necessary to make such heavy applications of zinc sulfate as have been reported by many investigators is not thoroughly understood, but as explained above it is probably due to clay crystal fixation. Cooper (1945) thinks that zinc compounds applied to some soils, as Norfolk, Blanton and Eustis, combine with organic compounds and become unavailable. Gall and Barnette (1940) have been able to show that the application of lime to a soil definitely increases the concentration at which replaceable zinc becomes injurious to crops.

Molybdenum an Element Essential for Plant Growth

The essential character of molybdenum for barley was demonstrated by Arnon in 1937, and for tomatoes by Arnon and Stout in 1939.

14

Fertilizers Carrying Elements Not Accepted as Essential for Plant Growth

Among the so-called nonessential elements that have been reported by investigators to have a beneficial influence on plant growth and arranged in ascending order of atomic weights are lithium, beryllium, boron, fluorine, sodium, aluminum, silicon, chlorine, titanium, vanadium, chromium, manganese, iron, cobalt, nickel, copper, zinc, gallium, germanium, arsenic, selenium, bromine, rubidium, strontium, yttrium, silver, tin, iodine, cesium, barium, lanthanum, cerium, praseodymium, neodymium, samarium, scandium, europium, gadolinium, terbium, dysprosium, erbium, ytterbium, gold, mercury, thallium, lead, radium, thorium, and uranium. Research data are available which would indicate that many of these elements may be beneficial to plant growth. A few of them are absorbed in large amounts by many plants, for example, aluminum by the sweetleaf, barium by the brazilnut, and the rare earths by the hickory, while others appear to be absorbed only in minute quantities. Robinson (1943) found exchangeable rare earths in many soils and plants. It would be expected that the rare earths would remain in the soil more tenaciously than the common elements, as they lie at the end of the exchange series and will readily replace even barium in the soil colloidal complex. Aluminum, arsenic, barium, chromium, fluorine, lead, molybdenum, selenium, and thallium have been shown to be toxic to plants or animals at relatively low concentrations. Trelease and Trelease (1938), however, suggest that selenium may be essential to plants. Slater, Holmes and Byers (1937) concluded from an extensive study of soils secured from various erosion experiment stations that the distribution of the trace elements—those elements usually present in very small quantities in the average soil including the essential elements boron, copper and zinc—was very extensive. They concluded also that the total absence of these elements from any soil was improbable. Likewise, Gaddum and Rogers (1936) found as a result of spectrographic analyses that trace elements are present in many fertilizer materials. The phosphatic materials, hardwood ashes and the insoluble organic nitrogen sources contained the greatest variety of trace elements. It appears not altogether improbable that further research will show that some of these elements are essential for field and horticultural crops. Many may be incorporated purposely in the complete fertilizers of the future which are designed for specific crops and soils. The elements that have attracted the greatest attention of plant physiologists are sodium, chlorine, and silicon.

Influence of Sodium on Plant Growth

While sodium has not yet received complete acceptance as an element essential for plant growth, it is known to have an influence on cell turgor and to influence plant metabolism. It is essential to animal life and perhaps to marine plant life.

There is rather general agreement among agricultural authorities that sodium is valuable: (1) as an essential element for certain plants; (2) as a substitute for potassium in certain functions in all plants; (3) as an alkaline agent to reduce acidity in humid climates; and (4) as an agent to maintain a high degree of availability of phosphate applied to the soil as superphosphate.

Many European and American investigators have reported an increase in the yields of crops resulting from an application of sodium salts. Results secured by Hartwell and Wessels (1913) and by Hartwell and Damon (1919), at the Rhode Island Experiment Station, would indicate that sodium may, to a limited extent, replace potassium in certain plants.

The partial substitution of one nutrient element for another in plant metabolism has been noted by many investigators. Cooper (1930) believes that one element may be substituted for another in nutrition in several ways: (1) by substituting as an actual constituent of a specific organic compound; (2) in oxidation-reduction reactions; and (3) as catalytic agents. Cooper thinks that the possibilities of substitution in the two latter cases are probably much greater than in the first; also, that the extent of the possible substitution will probably vary with the plant and properties of the elements involved. As pointed out by Collander (1941), some ions closely related chemically act almost like isotopes in their absorption-interrelations.

The function and value of sodium in plant nutrition has received increasing attention in recent years. Volk (1945) reports that greenhouse experiments conducted with cotton, oats, wheat, sugar beets, vetch, Austrian winter peas, and turnips definitely show that sodium is necessary as a plant food element in order to obtain maximum production. He found the response to sodium was most pronounced when potash was deficient but that high amounts of potash did not eliminate the beneficial effects of the sodium.

Russell (1945), reporting on the famous Broadbalk wheat experiments at the Rothamsted Station, England, says the tests showed quite clearly that the sodium had a good effect on wheat. Analysis revealed that the sodium nitrate had increased the uptake of potash of the plant, which might be taken to mean that it facilitated the absorption of potash by the roots. Further evidence that sodium is essential was provided by Holt and Volk (1945), who reached the conclusion that "the substitution of other forms of nitrogen in place of sodium nitrate as a source of nitrogen for sandy soils, may in time necessitate the addition of sodium in some form to obtain maximum yields of crops." They further concluded that plants vary in their need for, and capacity to utilize, sodium, and that sodium can be substituted for calcium to some extent.

Presence of Sodium in Soils and Plants

Sodium is a constituent of nearly all soils and especially of soils near the seacoast. The common soil-forming minerals, orthoclase, plagioclase, diorite, and the plagioclase feldspars, carry sodium to the extent of 1.5 to 6.0 per cent. Most minerals that carry sodium are readily eroded, and in humid climates most of the sodium that is converted into soluble compounds is carried to the sea in drainage water. As sodium salts are readily leached from the soil, humid soils contain a smaller percentage of sodium than semiarid or arid soils. In alkali soils sodium carbonate, so-called black alkali, is very prevalent, and in some soils the percentage of this salt is so high that crops cannot be grown.

Chemical determinations have shown that most plants contain sodium. As might be expected, the amount of sodium in plants varies, not only with the kind of plant, but also with the amount of sodium in the soil in which the plant is grown. Plants grown on soils that have a high sodium content always contain a higher percentage of sodium than similar plants grown on soils that have a low sodium content. As a general rule plants that are grown near the sea contain a higher percentage of sodium than plants that are grown farther inland. This may be explained by the fact that small amounts of salt sea spray are blown inland by sea breezes.

FIG. 118. Influence of sodium applied to a potash deficient soil. (*Left*) Sodium itrate. (*Right*) Calcium nitrate. (Courtesy, South Carolina Agricultural Experiient Station.)

'lant Response to Sodium Salts

When sodium is applied purposely as a fertilizer, it is generally applied the form of common salt or sodium chloride. For many years common lt has been used as a fertilizer in some sections of Europe. Where used, is customary to make top dressings to such crops as mangels, potatoes, id barley at the rate of 200 to 300 pounds of salt per acre. Mangolds, igar beets, and asparagus respond markedly to dressings of sodium iloride, and Lehr (1941) thinks that sodium is an almost indispensable ement for beets, approaching potassium in importance. Nevertheless, it iould not be overlooked that in those cases where increased yields ave been reported from the application of sodium chloride, that chlorine ay have been responsible in part for the increase.

At the Vermont Experiment Station, Jones and Orton (1897) observed marked increase in the growth of meadow fescue following an application f 3000 pounds of sodium chloride per acre that had been applied to kill awk weed, and Shelton (1889), of the Kansas Experiment Station, has ported a 4-bushel increase of wheat following an application of 300 ounds per acre. Likewise, Gardner (1931), of the Michigan Experiment ation, has reported an increase in the yield of sugar beets due to the plication of 500 pounds of sodium chloride per acre. Barnette (1925),

Table 92

INFLUENCE OF SODIUM CHLORIDE ON THE YIELD OF IRISH POTATOES. AVERAGE OF 3 YRS. AT ORLEY AND 1 YR. AT NORFOLK, VIRGINIA

Fertilizer Treatment	*Sodium Chloride per Acre, (lbs.)*	*Early Harvest Marketable Potatoes per Acre (bu.)*	*Late Harvest Marketable Potatoes per Acre (bu.)*
Complete fertilizer...........	2000	304.8	333.3
Complete fertilizer...........	1500	285.8	304.2
Complete fertilizer...........	1000	300.0	338.2
Complete fertilizer...........	500	270.1	294.2
Complete fertilizer...........	None	231.9	266.6

working at the New Jersey Experiment Station, conducted both pot and field tests with sodium chloride and concluded that while under certain conditions sodium chloride may be a proper amendment to certain fertilizers yet the cost and time necessary for the application of the salt would offset any small increases to be expected from its application. At the Virginia Truck Experiment Station, which is located in the Atlantic coastal plain, Johnson (1924) secured marked increases in the yield of Irish potatoes when sodium chloride was used in addition to a complete fertilizer. These results are given in Table 92.

Harmer and Benne (1945) classified crops, with reference to their response to sodium, in the presence of a deficiency and in the presence of a sufficiency of potash. Their results are given in Table 93.

Loew (1901), of the Bureau of Plant Industry, concluded that common salt has a depressive effect on photosynthetic processes, and as a result may decrease the percentage of sugar in sugar beets and the percentage of starch in potatoes. This influence which Loew noted is probably due to the chlorine and not to the sodium. Harmer and Benne (1945) concluded that sodium appears to have definite functions which it can best fulfill in the presence of ample potash. This is evident in improved vigor, disease resistance, and decreased wilting in hot, dry weather.

It is generally overlooked that sodium, in the form of sodium nitrate, as well as in the form of impurities in low-grade potash salts, is added extensively to the soils of the fertilizer using sections of the United States. Certainly, the good showing often made by sodium nitrate in fertilizer tests is partly due to the sodium which is added to the soil in this way. An excellent example of this influence on the growth of cotton at the South

Table 93

EFFECT OF SODIUM APPLIED AS A NUTRIENT ON SEVERAL CROPS

Number	*Degree of Benefit in Deficiency of Potash*		*Degree of Benefit in Sufficiency of Potash*	
	1. None to very slight	*2. Slight to medium*	*3. Slight to medium*	*4. Large*
1	Buckwheat	Asparagus	Cabbage	Celery
2	Corn	Barley	Celeriac	Mangel
3	Lettuce	Broccoli	Horseradish	Sugar beet
4	Onion	Brussels sprouts	Kale	Swiss chard
5	Parsley	Caraway	Kohlrabi	Table beet
6	Parsnip	Carrot	Mustard	Turnip
7	Peppermint	Chicory	Radish	..
8	Potato	Cotton	Rape	..
9	Rye	Flax	..	..
10	Soybean	Millet	..	..
11	Spinach	Oat	..	..
12	Squash	Pea	..	..
13	Strawberry	Rutabaga	..	..
14	Sunflowers	Tomato	..	..
15	White bean	Vetch	..	..
16		Wheat	..	..

Carolina Sand Hill Experiment Station is shown in the illustration on p. 329.

The yields of wheat on the sodium sulfate plats of Broadbalk field at the Rothamsted Experiment Station, for the first 50 years of the experiment, were about as great as the yields on the potassium sulfate plats. During more recent years, however, the yields of the sodium plats have decreased more than the yields of the potassium plats. This is due, probably, to the fact that all the absorbed potassium which was held by the soil colloids has been substituted by sodium during the earlier years, and that crop demands in recent years for potassium could not continue to be met.

Cooper (1941) concluded that after the addition of about 20 to 25 pounds of potash to the soil, sodium may be almost as effective as additional potash in increasing the yields of cotton on many cotton soils. Ten-year results reported by Cooper and Garman in 1942 show that at all levels of potash fertilization the plats receiving sodium produced more seed cotton per acre than those not receiving sodium. Similar results with seven other crops have been reported by Harmer and Benne (1945) and their results are given in Table 94. Harmer and Benne (1945) concluded that in sugar beets and celery the total milliquivalent content of the four

Table 94

EFFECT OF ORDINARY SALT (SODIUM CHLORIDE), WHEN APPLIED WITH DIFFERENT AMOUNTS OF POTASH,* ON YIELDS OF SEVERAL CROPS GROWN ON MICHIGAN ORGANIC SOIL

Fertilizer Analysis, 800 lbs. per Acre Applied Annually*	*Annual Salt Application per Acre*	*Yield per Acre*						
		Swiss Chard (Above Ground) Ave. 2 yrs.	*Table Beets (Roots)*	*Sugar Beets (Roots) Ave. 3 yrs.*	*Cabbage (Solid Heads) Ave. 3 yrs.*	*Parsnips (Roots) Ave. 2 yrs.*	*Carrots (Roots) Ave. 2 yrs.*	*Potatoes (Tubers) Ave. 3 yrs.*
	lbs.	*tons*	*tons*	*tons*	*tons*	*tons*	*tons*	*bu.*
0–12–0	0	8.5	2.6	3.3	5.6	4.9	16.0	109
0–12–0	600	20.1	5.3	5.6	7.0	4.8	15.2	110
0–12–24	600	30.9	13.7	11.5	15.5	10.1	22.6	175
0–12–12	600	30.0	12.3	11.2	14.4	9.2	22.0	175
0–12–6	0	25.5	9.9	9.0	14.2	10.0	23.7	182

* German potash used in the mixtures was KCl, containing 50 per çent K_2O, and approximately 10 to 12 per cent NaCl as an impurity.

bases, potassium, sodium, calcium, and magnesium remains fairly uniform whether or not sodium is applied as an additional nutrient.

Influence of Chlorine on Plant Growth

Analyses of most crops will reveal that the chlorine content of plants is higher than that of either phosphorus or sulfur. Although it has long been known that under some conditions chlorides may have a beneficial influence on crop yield, positive proof that chlorine is an essential element for plant growth has not been secured. Beneficial effects of chlorine have been noted on asparagus, buckwheat, cotton, potatoes, tobacco, and tomatoes. Symptoms of chlorine deficiency lack distinctiveness, and Haas (1945) concluded that there appears to be no threshold concentration below which chlorine-sensitive crops are uninjured by sodium chloride.

Chlorine appears to cause earlier death of plant leaves and in general to hasten maturation. Russell (1932) thinks that the effect of chlorine is to reduce transpiration and to increase the temperature of the leaf on hot days. He believes that since mangolds, sugar beets, and barley are tolerant of high temperatures, this may explain the benefit they derive from the use of chlorine as a fertilizer. On the other hand, as potatoes grow better during cool periods, they seem to be injured both in yield and quality by applications of chlorine. According to Garner *et al.* (1930), chlorine appears to have an important influence on water relations in the plant tissue. He states that under conditions of comparative drouth, chlorine appears to prevent destructive desiccation of the leaf tissue. When large quantities are absorbed the chlorine may induce such an increase in the water content of the tissue that carbohydrate metabolism may be disturbed and growth retarded. Gradient distribution of chlorine has been observed in tobacco and in avocados. There is some evidence to indicate that the presence of chlorine in the soil may bring about an increase in the absorption of magnesium by the plant.

Garner and many other investigators have shown that excessive amounts of chlorine may affect adversely the quality of the cured tobacco leaf, for chlorine tends to retard combustion. A tobacco committee of southern agronomists recommends that tobacco be fertilized with not more than 20 pounds of chlorine per acre but this amount seems necessary for the best development of the crop and the production of the highest yields. Karraker (1932) has shown that 25 pounds of chlorine in the form of muriate of potash had no injurious effect upon white Burley tobacco grown at the Kentucky Experiment Station. Skinner (1931) secured results showing that when fertilizer mixtures carrying as much as 112 pounds of chlorine per acre were applied, the yields of cotton were reduced when grown on a number of soil types in North Carolina. C. B. Lipman (1938) reported that chlorine definitely benefited the growth of buckwheat and peas.

Collison and Mensching (1932) found that at Geneva, New York, 16 pounds of chlorine is added to the soil annually through precipitation. Russell (1932) has reported an annual addition in precipitation of 10 to 24 pounds at the Rothamsted Experiment Station.

Influence of Silicon on Plant Growth

Many plants contain silicon and in some plants the percentage has been reported to be as high as 40 per cent ash in cereal straws to 70 per cent ash in the scouring rushes. This is not surprising when it is recalled that with the exception of oxygen, soils contain a larger amount of silicon than of any other element. Silicon is not accepted as an element essential for plant growth for apparently plants may be grown to maturity in media that are free of this element. Just what functions, if any, silicon may perform in the plant is not known. Silicon is known to alter the structure of certain plants. It was formerly thought that silicon added stiffness to the culms of grasses, but this view is not held by Russell (1932). C. B. Lipman (1938) found that the seed production of sunflowers and barley was definitely benefited by the presence of silicon in the culture medium, and similar results have been secured by Toth (1939) with rape, barley, and sudan grass. Cooper (1930) found a negative correlation between the content of phosphorus and silicon in pasture plants. These results would appear to indicate that silicic acid, or its salts, play a role in nutrition.

Hall and Morison (1906), summarizing the 40-year results secured from plats at the Rothamsted Experiment Station, which had been treated with soluble silicates, concluded that the presence of silicates increased the assimilation of phosphoric acid but that there was no evidence that the functions of phosphorus in the plant could be replaced by silicon. On the other hand, Brenchley (1927), of the same station, thinks that silicon may serve in the plant as a partial substitute for phosphorus. Scarseth (1935) has reported that sodium silicate produced a growth of plants without a phosphatic fertilizer almost equal to that produced on a phosphate-fertilized soil. The soil was a very acid, heavy clay with a silica-sesquioxide ratio of 2.3. Raleigh (1945) secured better growth of beets when they were supplied with silicon.

There is a physical influence of silicon in soils that should not be overlooked. The capacity of the mineral fraction of a soil to hold fertility is largely determined by its clay minerals. The higher the percentage of silicon in the clay, the greater the ionic exchange capacity and the maximum fertility level of the soil. Montmorillonite contains twice the silicon content of kaolinite, and has an ionic exchange capacity ten times the value of kaolinite. For this reason the saturated fertility level of the soils of the North and West, that contain a high montmorillonite content, is much higher than the soils of the South whose percentage clay fraction of kaolinite is higher.

Influence of Aluminum on Plant Growth

Aluminum has been reported in plants to be as high as 70 per cent of the ash. Its absorption appears to be a regional phenomenon, dependent to a large extent on climatically determined soil types. The native vegetation of soil types that are high in aluminum consists of aluminum tolerant species.

Some botanists have concluded that there is a gradient in the aluminum content of plants, the amount falling off from root to leaf. This is probably true for many plant species that grow on heavy acid soils.

Hutchinson (1945) believes that aluminum is an essential element in the production of ferns. He also believes that the relative unimportance of aluminum in organisms is due to its low solubility in neutral solutions. Aluminum in feeds is excreted in the feces of animals, although a part may be concentrated in the bile.

Soluble aluminum is often present in acid soils. It is toxic to plants and may be precipitated by applying lime or soluble phosphates.

Use of Radium as a Fertilizer

Because it was early shown that radium might affect the physiological processes of plants, it was thought by many that radium salts, as well as ore tailings containing minute quantities of radium, might be used advantageously as a fertilizer. However, Hopkins and Sachs (1915) showed that radium applied as radium barium chloride and radium barium sulfate at a cost of $1, $10, or $100 per acre produced no effect upon crop yields either the first or second season. Similar results have been reported since that time by other invesigators.

Use of Cobalt as a Fertilizer

Mitchell (1945) believes that a cobalt content of 0.08 p.p.m. in herbage is necessary to insure animal health. According to Askey (1939) cobalt pining in animals can be prevented by top-dressing pastures with cobaltized superphosphate at the rate of 2 pounds of the cobalt salt per acre.

15

Adjusting Soil Reaction and Fertilizer Practice to Crop Requirement

The availability of soil nutrients is influenced to a very great extent by the reaction of the soil and by the soil management practices followed by the farmer. The acidity of the soil governs in part the intake of ions and as a result the concentration of a nutrient which might be sufficient at one degree of acidity may be insufficient at another. This has been shown by the work of Albrecht and Schroeder (1942) in Missouri. The increased growth of crops, secured from the employment of good soil management practices, is, in most cases, due to an increase of available soil nutrients resulting from these practices. In other words, the problems of soil management are largely those which center around the problems of nutrient availability.

How Soil Reaction is Expressed

Soils may be spoken of as being acid, neutral (sweet), or basic (alkaline) in reaction. Soils containing an excess of acid materials are said to be acid in reaction, while soils containing an excess of basic materials are said to be basic or alkaline in reaction. If the quantities of acidic and basic materials are equal the soil is said to be neutral. Farmers often refer to a soil as being "sweet" when the soil has a reaction near the neutral point. The degree of acidity or basicity depends upon the excess quantity of basic or acidic materials present.

Briefly, the method commonly used in measuring and expressing degrees of acidity or alkalinity in soils is in terms of pH values. The term pH which is now in common use was adopted without explanation of its derivation by a Danish scientist, S. P. L. Sørensen. The scale for measuring soil reaction is made up of 14 divisions or values. These values represent the logarithms of the reciprocals of the hydrogen-ion concentrations of the soil solution at different soil reactions. A pH value may be expressed mathematically as:

$$pH = Log. \frac{1}{(H+)}$$

The scale is centered on pH 7.0 which represents neutrality. Values from 7 to 1 represent the acid range and values from 7 to 14 the alkaline range. Because pH values are logarithmic there are some disadvantages to their use by the farmer. In addition, pH values express alkalinity as well as acidity in terms of H-ion concentration, and this may be confusing to persons who are not chemists. Again, the change in the H-ion concentration as represented by pH values is geometric and not arithmetic, which is the common method of thinking. For instance, there are nine acid units between pH 6 and 7, 90 between pH 5 and 6, 900 between pH 4 and 5, etc., and this fact is often overlooked even by many who commonly use pH values to express degrees of soil acidity.

FIG. 119. Edmund Ruffin—successful Virginia farmer. Father of soil chemistry in America. First to contend that upland mineral soils in humid regions are often acid; strong advocate of liming of acid soils; brilliant writer; a great southern patriot. (Courtesy, National Fertilizer Association.)

When pH values are used it must be kept in mind also that they represent only the amount of H-ions or free acidity in the soil solution, and that they are not an index of the quantity of potential or combined acidity that may be in the soil. In other words, pH values represent the intensity of acidity rather than the quantity, and therefore soils with the same pH value may require different amounts of lime. Clays, silts, and sands having the same pH value will vary as to the amounts of their potential acidity, and as a consequence must be limed with different amounts of lime if comparable results are to be expected. Clayey soils require much more lime than sandy soils of the same pH value.

American soils may range in pH value from about 3.2 for some peat soils to 9.5 for some "black alkali" soils. Few crop plants will grow below a pH of 3.5 or above a pH of 9.0. The fertile cultivated lands of the United

States commonly range from a pH value of 6.5 to 7.5. The reaction of the root sap of most agricultural plants ranges from about pH 4.0 to pH 6.0, so that plant tissues appear to be so constituted as to be able to tolerate the acidities ordinarily found in soils. Hoagland (1944), working with a number of species of plants, found that all plants were injured when grown in culture solutions having a pH as low as 3.0.

Reaction of American Soils

The rainfall and temperature of a region play a very important part in determining the chemical reaction of its soils. In general, the acidity of soils decreases with the annual rainfall to which the soil has been subjected. Also, it decreases from a humid cool to a humid tropical region and increases with an increase in altitude.

The soils of the United States may be divided into two soil orders. West of a line which may be represented roughly by the one hundredth meridian are found soils whose reaction usually lies between the pH values 7.0 and 8.0, although pH values of 8.0 and 9.0 are not uncommon. These soils are referred to as *pedocals*, and largely because of the low rainfall under which they originated—25 inches or less—they have retained much of their basic elements, which in turn accounts for their alkalinity. East of the one hundredth meridian are found soils whose pH values generally lie between 5.0 and 6.0, although pH values of 4.5 or 4.0 are not uncommon. These soils are called *pedalfers*. Pedalfers are formed in regions of high rainfall and are soils that have lost much of their basic elements through leaching. Because of the differences in the soil reactions of the pedocals

FIG. 120. Influence of fertilization on the growth of cotton (*Left*) One thousand pounds 4–8–4. (*Right*) No fertilizer. (Courtesy, South Carolina Pee Dee Experiment Station.)

and the pedalfers it naturally follows that the fertilizing practices to be followed in these two soil areas might be very different.

The pedalfers comprise a great part of the cultivated soils of the United States. They are now grouped by the Bureau of Plant Industry, Soils and Agricultural Engineering into five soil groups or provinces, prairie soils, podsols, gray-brown soils, red and yellow soils, and laterites. The prairie soils include an irregular strip of the country, a hundred or so miles wide, running north and south adjacent to the pedocals and roughly paralleling the one hundredth meridian. The reaction of the prairie soils approaches that of the pedocals. The other four soil provinces lie roughly in an east-west direction from the prairie country to the Atlantic Ocean. Roughly, the podsol province is found along the Canadian border, the gray-brown province in the mid-Atlantic and midwestern states, the red and yellow province in the southern states and the laterites in Florida and Cuba. The podsols may be extremely acid and pH values of 4.0 to 5.0 are very common. On the other hand, the laterites are almost neutral in reaction and usually have a pH value around 6.5. In general, it might be said that the acidity of the cultivated soils of the United States increases from the Gulf of Mexico to the Canadian border as does also their nitrogen and organic content. Bailey (1945) has recently made a study of the pH values of various horizons of soil solums of many representative soils of the United States. These natural differences in the soil reactions of the cultivated soils of America have played an important part in bringing about the variations often observed in the fertilizing practices followed in the various sections of the United States.

Soil Reaction as Influenced by Parent Material of Soil

Not only is soil reaction influenced by the climatic conditions under which the soil has originated, but also by the mineral or organic materials from which the soil has been formed. Pedalferic soils that have been formed from secondary minerals which contain a high percentage of silicon, or soils formed from organic matter, may be very acid, whereas pedalferic soils formed from limestone or diabase are usually only slightly, if at all, acid.

Soil Reaction as Influenced by Soil Management Practices

When virgin soils in a humid climate are put under cultivation they rapidly lose much of their organic matter through oxidation, and much of their nitrogen and bases through leaching and crop removal. Many American soils have passed and are now passing through this process. In general, it might be said that very ineffective efforts have been made by American farmers to retard the process.

As the basic elements are removed from soil complexes through leaching and crop removal they are replaced by hydrogen, so that a great increase in the acidity of the soil usually results. In some localities in America this trend of soil reaction toward high soil acidity has been further hastened through the use of acid-forming fertilizers. As a result there are in the United States today thousands of acres of land on which a profitable production of crops is no longer possible even with the help of fertilizers, unless the acidity of the soil is first reduced. This conclusion was borne out recently by the soil acidity determinations made on 2,250,000 soil samples taken from about 95 per cent of all farms in the state of South Carolina.

Adjusting Soil Reaction through the Use of Agricultural Lime

The use of agricultural lime and other basic materials for the purpose of reducing the acidity of the soil has been followed to a limited extent in the United States since Colonial times, but the liming practice in the past has not been extensive enough to retard greatly the drift of American cultivated soils toward greater acidity, although today the practice is becoming more extensive and effective. A discussion of the use of agricultural lime has been given in another chapter. It is well to emphasize here, however, that agricultural lime should be applied judiciously. Too much lime, or even the liming of the wrong crop, may sometimes be as ill advised as failure to use lime when it is advantageous to do so. It cannot be too strongly stressed that although almost innumerable liming experiments have been carried out by competent soil scientists there is much about the influence of lime on soils and crops that is not understood.

As pointed out elsewhere, agricultural lime is usually composed of compounds of calcium or magnesium. Since calcium and magnesium compounds are relatively cheap they should be more generally used for soil improvement. Calcium is the most important nutrient element in the soil in determining the reaction of the soil, and magnesium may be regarded as being second to calcium in importance. In addition, calcium and magnesium perform many other beneficial functions which may be explained in part by the fact that their compounds are readily soluble in an

FIG. 121. Fertilized corn. (*Left*) Limed. (*Right*) Unlimed. (Courtesy, Prof. A. W. Blair and the New Jersey Agricultural Experiment Station.)

acid solution; that they may combine with most plant nutrient elements; that they possess comparatively high displacement energy and may therefore be the principal cations in base-exchange compounds; that they are nutrient elements; and that they produce a good physical condition when added to the hydrogen clays. In addition these elements influence directly or indirectly root development and quality of crops, and may make possible the absorption and assimilation by plants of nitrogen and phosphorus.

Adjusting Soil Reaction Through the Use of Nonacid-forming Fertilizers

Up until a few years ago most of the complete fertilizers found on the American market were acid-forming, and their use led to an increase in the acidity of the soil. From 1900 to the early 1930's commercial mixed fertilizers became increasingly more acid-forming owing to a more extended use of ammonium salts. In 1900 the nitrogen in all mixed fertilizers was derived to the extent of 2.1 per cent from ammonium salts, 6.9 per cent from nitrates, and 91.0 per cent from organic compounds of plant or animal origin. In 1931, on the other hand, 61.2 per cent of the nitrogen of mixed fertilizers was derived from ammonia and its compounds, 11.6 per cent from nitrates, 18.8 per cent from organic ammoniates, and 8.4 per cent from calcium cyanamide and urea. In 1934 Mehring and Peterson (1934) reported that the acidic residue of the average mixed fertilizer was equivalent to 150 pounds of calcium carbonate per ton. As pointed out in a previous paragraph, many soils of America after being put under cultivation soon became too acid for the profitable production of crops. To meet this situation liming programs, together with the use of nonacid fertilizers, were advocated by agronomists.

Nonacid-forming complete fertilizers are manufactured by so compounding the fertilizer mixture that physiologically acidic and physiologically basic carriers of nutrient materials will balance each other. Frequently, in compounding mixtures, this is either not possible or it is not the most economical procedure, so that some basic material has to be added to the mixture, in amounts sufficient to counteract any excess of acidity. Pierre's (1931) values for estimating the equivalent acidity and basicity of fertilizer materials are now generally used by the fertilizer industry in formulating nonacid-forming fertilizers. The equivalent acidity or basicity of various inorganic or organic carriers of nitrogen as found by Pierre (1931) are given in Table 95. Equivalent acidity may be defined as the pounds of lime required to neutralize the acidity produced by a quantity of fertilizer containing 20 pounds, or a unit of nitrogen.

Lime of various forms and phosphatic rock are the most common materials substituted for sand or other inert filler. Hester and Shelton (1936) have shown that the neutralizing effect of phosphate rock filler is negligible under actual growing conditions, and Clevenger and Willis (1935) have called attention to the fact that after the application of a mixed fertilizer to the soil the presence of natural organic materials in the mixture may serve as neutralizing agents almost as extensively as a lime supplement. They also call attention to the fact that the immediate effect of an application of a nonacid-forming fertilizer to the soil is to produce a "salt effect" which results in a slight increase in soil acidity. This is, however, a temporary influence which disappears quickly.

Dolomitic limestone is the form of lime most commonly used for the neutralization of acid fertilizer mixtures. Calcitic limestone may be used, but because it will readily revert soluble phosphates it is not preferred. Experimental data show that for most conditions dolomite gives best results when ground to pass through a 60-mesh sieve. The addition of dolomite to acid fertilizer mixtures containing superphosphate results in very little or no reversion of water-soluble phosphoric acid to citrate-insoluble phosphoric acid, but if the mixture has been ammoniated the condition may be otherwise. The work of Beeson and Ross (1937) has shown that monoammonium phosphate is more reactive toward dolomite than is monocalcium phosphate. This explains why dolomite decomposition is greater in ammoniated mixtures. The magnesium compound

Table 95

EQUIVALENT ACIDITY OR BASICITY OF VARIOUS INORGANIC AND ORGANIC NITROGEN FERTILIZERS EXPRESSED AS POUNDS OF $CaCO_3$

Fertilizer Materials	*Net Equivalent Acidity or Basicity per Unit of Nitrogen*
Inorganic:	
Ammonium sulfate	107
Ammonium chloride	107
Mono-ammonium phosphate	107
Ammo-Phos	107
Sodium nitrate	36*
Cal-Nitro	22*
Calcium Cyanamid	57*
Urea	36
Organic:	
Animal tankage	3
High-grade tankage	15
Low-grade tankage	144*
Packing house tankage	30*
Garbage tankage	54*
Fish scrap	31
Acid fish	2
Cottonseed meal	29
Tòbacco stems	86*
Peruvian guano	19
Whale guano	9
Dried blood	35
Cocoa shell meal	12*
Castor pomace	18
Milorganite	34

* Basicity.

formed by reaction of dolomite in nonammoniated mixtures is dimagnesium phosphate and in ammoniated mixtures magnesium-ammonium phosphate. The water-soluble magnesia, in mixtures that contain dolomite as the only source of magnesium, is low in both types of mixtures and decreases with a decrease in the acidity of the mixture.

Adjusting Soil Reaction Through the Use of Acid Fertilizers, Sulfur, and Aluminum Sulfate

The use of mixed fertilizers that leave an acid residue in the soil is to be advocated on the alkaline and calcareous soils of the American southwest.

In fact, it appears that the acidity resulting from the use of the ordinary mixed fertilizers sold in that section is not sufficient, and in many cases the mixture should be fortified with an acidulated organic base or with 50 to 200 pounds of sulfur per ton of fertilizer. In addition, where it is found that the crops grown are subject to chlorosis it is advisable to add small amounts of zinc or iron salt. McGeorge (1936), of the California Experiment Station, has found that this practice is very beneficial to long-lived crops such as fruit trees.

As has been pointed out in other chapters, the use of ammoniacal compounds will result in an increase in soil acidity and that these compounds as well as sulfur are commonly used by Irish potato growers for increasing the acidity of the soil. Horticulturists also use aluminum sulfates (alums) for increasing the acidity of the soil for such plants as shortia, rhododendrons, blueberries, and hydrangeas. A high active aluminum content, 20 to 50 pounds per acre, is often desirable for such plants as rhododendrons, azaleas, and blueberries. The content of active aluminum may be detected by means of a quick chemical test.

Soil Reaction and Its Influence on Nutrient Availability

Many changes take place in the soil when the soil becomes acid or alkaline, or when the acidity of the soil is increased through the use of acid fertilizers or reduced by liming. Some of these changes are of first importance to the farmer, while others may be of only secondary importance, or of no importance at all.

The literature which concerns the influence of soil reaction on plant growth is voluminous, and a great deal of experimental data which bears on the subject is now available. Nevertheless, the subject, including all of the ramifications, is not well understood by the agriculturists. It must be confessed, also, that facts are limited concerning many related subjects.

The effect of soil reaction and the influence of liming and fertilizer practices on the availability of plant nutrients are among the most widely discussed subjects in agriculture. The reaction of the soil may greatly influence the availability of some plant nutrients but on the other hand the availability of other nutrients may be influenced mildly or not at all. The availability of some plant nutrients may be influenced to a greater extent by the conditions created by the reaction of the soil than by the reaction. It must be kept in mind that the term availability is used to refer to a result in the soil of usually many complex reactions and that these reactions may involve the processes of adsorption, assimilation, ionization, membrane permeability, ion-antagonism, solution and others.

Influence of Soil Reaction on the Availability of Nitrogen

Field crops appear to absorb most of their nitrogen in the nitrate form. Under the usual conditions that prevail in the soil, nitrates appear to be as available to crops at one pH value as at another, but most of the nitrogen applied to crops in mixed fertilizers is in forms other than that of nitrates, and in most cases these other forms of nitrogen have to be converted into nitrates by ammonifying and nitrifying organisms. Nitrifying organisms are affected adversely by an acidity greater than that represented by a pH value of 5.5 and for this reason crop response to many nitrogenous fertilizers is usually greater when the fertilizers are applied to soils that have been limed. On the other hand, nitrification can be hindered by too much alkalinity, but with humid soils this danger from overliming is slight.

Influence of Soil Reaction on the Availability of Phosphorus

Phosphates appear to be most available to plants in a majority of the soils of humid America when the soil reaction has a pH value between 5.5 and 7.5. When the soil reaction gets above or below this range the phosphates in the soil become less available. Soluble iron and aluminum from the insoluble soil iron and aluminum compounds begin to appear in the soil solution at about a pH value of 5.5, and great quantities will become soluble if the pH value of the soil is allowed to drop to 5.0 or 4.8. As soluble phosphates will unite readily with soluble iron and aluminum, especially aluminum, and become precipitated as insoluble phosphate compounds, crops growing on very acid soils often show phosphorus starvation. This is especially true when the iron and aluminum phosphates have passed the freshly precipitated state and have passed over into the crystalline form. In addition, there is good evidence to indicate that soluble aluminum itself is toxic to many plants, and Pierre (1931) found it an important factor in the relation of soil acidity to soil fertility.

Soil aluminum also comes into solution at pH values above 7.0, but under alkaline conditions it will not fix the soluble phosphates as it does when present in acid soils. Apparantly aluminum to some extent enters the plant at all pH levels. Soluble phosphates remain soluble in overlimed soils but for some reason plants appear to be prevented from absorbing or assimilating them. McGeorge and Breazeale (1931) found that plant seedlings did not absorb phosphates at pH values above 7.6. Excessive liming may precipitate water-soluble phosphorus compounds as citrate-soluble tricalcium phosphate, which if not utilized will pass over into a relatively unavailable fluorapatite crystalline form. In general, it may be said that acid clay soils give greater responses to phosphorus fertilizers than other soils of the same region.

In general, it can be said that when the soil has a pH of about 3.0 to 4.0 phosphates are held in the soil largely by iron. In soils having a pH range of 5.0 to 6.0 phosphates appear to be held to the clay complex by aluminum. Absorbed phosphate appears to be brought into solution by OH-ions when the soil has a pH of 6.0 to 7.0, but above a pH of 7.0 it is converted into calcium phosphate which becomes insoluble above pH 8.0. It thus seems advisable to maintain the pH of the soil by liming at such a point that neither insoluble iron, aluminum, nor calcium compounds are produced.

Influence of Soil Reaction on the Availability of Potassium

The experimental data available do not furnish evidence that the soil reaction within reasonable limits has any great influence on the availability of potassium. Apparently, potassium is uninfluenced, or at least little influenced, by the hydrogen ion that mobilizes other cations in the plant. A marked crop response is often secured from the application of potash to very acid soils but in most cases this is due to a potassium deficiency. In some cases plant injury due to soluble iron, aluminum, or manganese, may be such that a response to applied potash may not be secured even though potash is needed.

Marked crop response to a potash fertilizer is more often secured when applications of potash are made to overlimed soils or to soils that have been abundantly fertilized with calcium fertilizers. This has been shown by the work of Hester, Parker, and Zimmerly (1936), Cooper and Wallace (1936), and Gilligan (1938). MacIntire *et al.* (1930) concluded from the study of lysimeter experiments that liming depressed the solubility of soil potassium, and similar results have been secured by other investigators. Nevertheless, MacIntire (1941) points out that it has not yet been demonstrated that liming diminishes the uptake of potassium by clovers and other lime-loving plants of extensive root development. An excess of activ

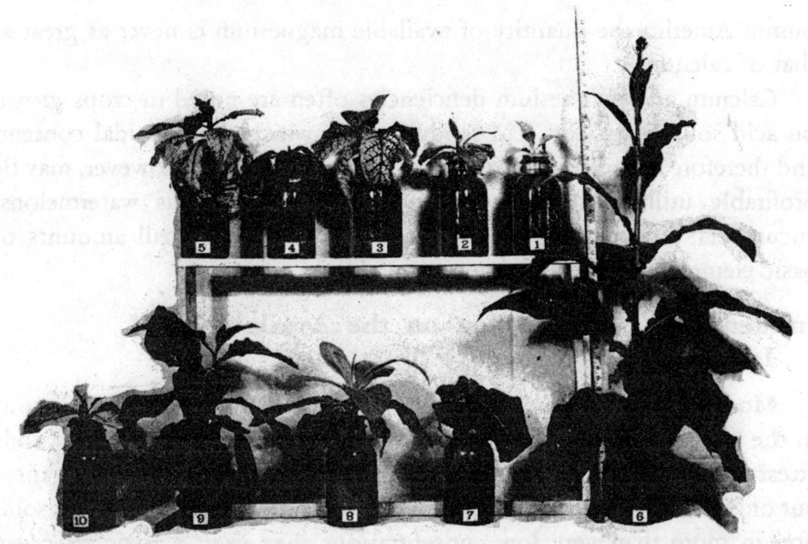

FIG. 122. Tobacco plants grown in cultural solutions. (1) No nitrogen added; (2) no phosphorus added; (3) no potassium added; (4) no calcium added; (5) no magnesium added; (7) no boron added; (8) no sulfur added; (9) no manganese added; (10) no iron added; (6) all of the above elements added. (Courtesy, Dr. J. E. McMurtrey, Jr. Reproduced from *U.S.D.A. Technical Bulletin 340.*)

calcium in the soil appears to interfere both with potassium and phosphorus absorption by the plant.

Influence of Soil Reaction on the Availability of Calcium and Magnesium

Acidity in the soil is correlated with a lack of bases in the colloidal complex. For this reason acid soils can yield but little available calcium or magnesium. The quantity of available calcium in soils usually increases rapidly with a rise in the pH value up to pH 8.0 and above, and the absorption of calcium by plants increases up to about a pH of 6.0. Hoagland and Arnon (1941) report that at acid reactions the plant seems to require a greater supply of calcium than at reactions approaching neutrality. Crane (1926), however, has shown that high acidity even in the presence of adequate soluble calcium interferes with the ability of plants to take up and retain soluble calcium. Hoagland (1944) found a loss of calcium to take place from the roots of plants to the cultural solution at very low pH levels. The quantity of available magnesium also will increase up to a pH value of 7.5 but above this value the quantity decreases. In the soils of humid America the quantity of available magnesium is never as great as that of calcium.

Calcium and magnesium deficiencies often are noted in crops grown on acid soils, especially acid sandy soils in which the colloidal content, and therefore absorbed basic content, is low. These soils, however, may be profitably utilized in the production of crops such as watermelons, cucumbers, strawberries, or other crops that tolerate small amounts of basic elements for their optimum growth.

Influence of Soil Reaction on the Availability of Iron and Manganese

Most soils contain an abundance of iron and manganese compounds in the insoluble form. In fact the red and chocolate colors of many soils attest to their presence in large amounts. Plants require iron and manganese but only in very small amounts. In fact, if they are present in the soil solution in more than very low concentrations they exert a toxic effect on plant growth. Apparently soils with the usual reactions between pH 5.5 and pH 7.0 supply plants with ample quantities of both iron and manganese. As already pointed out, excess quantities of these elements may become available at pH values below 5.5 and toxic influences on plant growth may result. It appears that manganese becomes toxic to plant growth before iron and aluminum come into solution in sufficient quantities to be toxic. In Table 96 are given, according to Hester (1942), data illustrating the influence of liming on the reduction of free aluminum in the soil.

Table 96

INFLUENCE OF LIMING ON THE REDUCTION OF SOLUBLE ALUMINUM IN THE SOIL

Soil pH	*Lime Applied (lbs.)*	*Aluminum in Soil (lbs. per acre)*
5.45	None	50 to 60
5.80	1500	5 to 10
6.00	3000	1 to 2

At pH values above 6.5, especially in sandy soils, iron and manganese may become unavailable to such an extent that deficiency symptoms in plants may appear. Rogers and Shive (1932) have suggested that plants growing in a neutral or near neutral soil seem to be unable to absorb and assimilate iron but are able to do so if the soil is slightly acid.

Influence of Soil Reaction on the Availability of Boron, Zinc, and Copper

In general it may be said that high pH values in the soil tend to revert soluble boron, zinc, and copper into insoluble forms. Camp (1945) states that it has been reasonably well established that the availability of zinc declines as the pH value of the soil rises, the critical point being somewhere between pH 5.5 and 6.5.

Problems Arising from Overliming

It is commonly thought that the problems arising from overliming a soil are more difficult to correct than those arising from an insufficient amount of lime, but this view should be corrected. Nevertheless, plants do appear to be unable to assimilate phosphorus, potassium, and even calcium itself in the presence of excess calcium carbonate. Following heavy liming, manganese deficiency often develops in sandy soils of low manganese content of the Coastal Plain south of Virginia but it seldom occurs north of Virginia. This may be due to the higher organic content of the soils north of Virginia. Decaying organic matter may furnish available manganese to a crop even though the pH value of the soil is high. Overliming of a soil is known also to hinder absorption by plants of zinc in Florida and of boron in Alabama. Also, Midgley and Dunklee (1940) found that overliming injury in Vermont appeared to be due largely to the fixation of boron.

Injury resulting from overliming may be corrected by the addition to the soil of those elements which have been made unavailable, and through the use of acid-forming fertilizers and amendments.

Many agronomists feel that the problems of overliming have been stressed too much. This is probably true in view of the great extent of acid soils and the high lime requirement of many subsoils. It seems that overliming would rarely become an important factor if reasonable attention were given to the proper use of the amendment.

Adjusting Soil Organic Content to Crop Requirement

The nitrogen content of the soil is found almost entirely in the organic matter of the soil. Nitrogen in organic matter must be converted to an available form by soil microörganisms before it can be utilized by plants. Those factors which control the well-being of ammonifying and nitrifying organisms such as soil moisture, presence of energy material, and proper soil reaction are also the principal factors controlling the availability of soil nitrogen. Organic matter is of maximum value to the growing crop only when it is in a process of decomposition, but this process leads to its elimination from the soil. A virgin soil may often lose 25 to 30 per cent of its organic matter during the first quarter of a century it is put under cultivation. Because the cultivation of the soil leads to the destruction of its organic matter the profitable production of crops can be maintained only through the frequent additions to the soil of suitable organic materials or by the application of available forms of commercial nitrogen. Nearly all soils are benefited by the additions of suitable organic matter, no matter how much they may contain. In the older fertilizer-consuming sections of the United States, the practice of applying available commercial nitrogen has been followed by most farmers, rather than the practice of turning under suitable organic matter. It must not be overlooked, however, that organic matter in the soil may have a beneficial effect other than that of supplying plant nutrients, and that maximum yields, especially during dry years, may not be secured without it.

One of the most important influences of organic matter on the soil is its effect on the water-holding capacity of the soil. Organic matter has about four times the water-holding capacity of clay and about eight times that of a sandy soil. For this reason the results secured from an application of commercial fertilizer during a dry year are usually more noticeable when applied to soils in which the organic matter has been maintained near the saturation point.

On many light-colored sandy pedalferic soils it is the organic matter content that largely determines the ionic exchange capacity, and therefore the soil's fertility. The reason for this may not always be apparent but its presence, in general, may be said to make the soil a better medium for plant growth. Optimum yields of crops may result from the influence of the organic matter on the water-holding capacity, or on other physical characteristics of the soil, or from its influence on phosphate availability, aluminum solubility, or ionic exchange capacity, or to all of these influences and others. Hester (1937) has shown that the addition of suitable organic matter to the sandy truck soils of the Coastal Plain may so reduce the fixation of phosphorus to an unavailable form that optimum yields of crops may be produced with much less superphosphate than is commonly used. Jensen (1917) found that the addition of organic matter to a soil increased the solubility of both lime and phosphoric acid 30 to 100 per cent.

16

Principles Underlying the Purchase of Fertilizers

The farmer who is in the market for commercial fertilizers often does not know what materials are best for him to purchase, and particularly is this true if he has had no personal experience with fertilizers. When this is the case, he should get in touch with his state agricultural experiment station or extension service for advice that will be unbiased and based upon the results secured by experimentation or upon the best practical information available.

What Is a Fertilizer?

Fertilizer may be defined broadly as any material which if added to the soil will result in a better growth of crops. Such a definition might imply a direct or indirect action on the crop.

Nutritive fertilizer is a term sometimes applied to those fertilizers that contain one or more of the commonly accepted essential elements for plant growth that are taken from the soil by plants. These fertilizers may be referred to also as *direct fertilizers*. On the other hand, the terms *stimulant* or *catalytic fertilizer* are sometimes used to refer to those fertilizers which do not carry one or more of the 12 elements that are taken from the soil by plants and that are generally accepted as essential, but that do carry other elements which may have some beneficial influence on plant growth. They are sometimes referred to as *indirect fertilizers*.

All fertilizer materials that may be present on the fertilizer market and that are sold within the same trade are called *commercial fertilizers*. Many terms such as *plant food*, *phosphate*, and *guano* are used locally in various sections of the United States as terms synonymous with he term fertilizer. This indiscriminate and loose usage of terms should be discouraged. *Plant food* is a term which has been employed in recent years to

FIG. 123. Do not waste fertilizer on land like this. (Courtesy, *American Hortigraphs and Agronomic Review.*)

refer to the combined content of nitrogen, phosphoric acid, and potash that is present in a fertilizer material.

Mixed, Chemical, and Special Fertilizers

In the past, the term *complete fertilizer* has been used to refer to those fertilizers which contain the 3 plant nutrients—nitrogen, phosphoric acid, and potash. A mixture of 2 or more fertilizer materials is referred to as *mixed fertilizer* or *mixed goods.* Two-thirds of the commercial fertilizers sold in the United States are in the form of mixed goods.

In recent years experimental work has indicated that for most soils and crops a complete fertilizer should carry the six elements, nitrogen, phosphorus, potassium, calcium, sulfur, and magnesium.

So-called *special fertilizers* often have been put on the market with the claim by the manufacturer that the fertilizer is particularly adapted to some specific crop. Unfortunately, in many cases the claims made for these fertilizers have been misleading. Only to a limited extent can a complete fertilizer be made which is primarily suited only for a given crop, and when one is so compounded it is suited to the crop in question only when the crop is grown under specific soil and climatic conditions.

Some commercial fertilizers are referred to as *chemical fertilizers.* As generally used, the term chemical fertilizer refers to those fertilizers which are composed of crude chemical compounds of standard composition which contain the plant nutrients in a concentrated and available form. Among such compounds are ammonium sulfate, superphosphate, and high analysis potash salts. These materials and others like them are called *standard* materials because they vary only slightly in their chemical composition. A ton of any one of these materials has about the same agricultural value as any ton of the same fertilizer. Standard materials are always

FIG. 124. Land like this responds to fertilizers. (Courtesy, *American Hortigraphs and Agronomic Review.*)

used in compounding high analysis complete fertilizers. On the other hand, such fertilizer materials as ground fish, tankage, and bone would not be classified as standard, for they may vary widely not only in their composition, but also in their availability.

Ordinary Grade, High-analysis, and Concentrated Fertilizers

The *grade* of a fertilizer is represented by the minimum guarantee of the analysis of its plant nutrienrt. The word *analysis*, as applied to fertilizer, is used to designate the percentage composition of the product expressed in terms of nitrogen, phosphoric acid, and potash in their various forms.

According to their plant food content fertilizers are commonly classified into three groups known as ordinary grade, high-analysis and concentrated.

If the total percentage of plant food in a fertilizer is below 25 the fertilizer is said to be of ordinary grade; if the percentage is from 25 to 30 inclusive the fertilizer is said to be of high analysis; if the percentage is above 30 the fertilizer is said to be a concentrated fertilizer. In the second group we find most of the so-called "double-strength" fertilizers. These are called double-strength fertilizers to distinguish them from the single-strength fertilizers of the ordinary analysis group. In 1919 goods analyzing a minimum of 14 per cent were accepted as high-grade goods, but these grades have largely disappeared from the American market owing in the main to the legislative action taken by various state legislatures. At the present time the sale of mixed fertilizers containing less than 16 per cent of total plant food is prohibited in Pennsylvania, Delaware, West Virginia, Alabama, Mississippi, Arkansas, Louisiana, Oklahoma, Ohio, Kentucky,

Wisconsin, Montana, and South Carolina, With a few exceptions the grades commonly recommended in the Middle West, Middle Atlantic, and New England states contain a minimum of 20 per cent of total plant food.

In some sections there is a tendency to classify all mixed fertilizers containing 20 per cent and less of plant food as ordinary or low-analysis mixtures and those having above 20 per cent of plant food as high-analysis mixtures. There has been some objection to the term concentrated, as applied to fertilizers, because the term frequently conveys to the farmer the idea that the fertilizer might be injurious to his crops. Such an assumption is not justified. It should be kept in mind that there is little direct relation between the total percentage of plant food in a fertilizer and plant injury due to plasmolysis. Most salts of the same molecular concentration, regardless of whether they are fertilizer salts or not, can injure a crop to about the same extent. For instance, sodium chloride is as likely to produce plant injury as is potassium chloride, although sodium chloride contains no plant food, whereas potassium chloride may contain from 48 to 62 per cent. Furthermore, injury resulting from the use of double-strength fertilizer is usually less than that resulting from the use of single-strength fertilizer as only half as much of a double-strength mixture per acre is required to give an equal application of plant food. After four years of investigation Lipman and Blair (1929) reported that there was no injury to corn from a broadcast application of 500 pounds or less of a complete fertilizer containing 50 units of plant nutrients even when used on a soil that was almost wholly sand. The fertilizer analyzed 16 per cent ammonia, 18 per cent phosphoric acid, and 18 per cent potash. The mixture was made from urea, triple superphosphate, and muriate of potash.

The content of plant food found in mixed goods usually ranges from about 14 per cent to 25 per cent. Formerly, mixed goods containing from 10 to 12 per cent of plant food were commonly found on the market. At the present time the general trend is toward the manufacture of complete fertilizers of much higher analysis. Since World War I the plant food content of the average complete fertilizer sold in the United States has increased 20 per cent but because of the use of high-analysis carriers the filler content has had to be increased 65 per cent. Table 97 illustrates the changes that have taken place in the composition and plant food content of fertilizer mixtures during the last half century according to Ross and Mehring (1938). As mentioned above, many states have fixed a total of 16 per cent nitrogen, available phosphoric acid, and potash as the minimum

Table 97

COMPOSITION OF REPRESENTATIVE MIXED FERTILIZERS IN 1880, 1910, AND 1937

Period and Formula	*Pounds per Ton*
1880 (2–9–2):	
Superphosphate, 12.5 per cent P_2O_5	1,053
Fish scrap, 6 per cent N; 8 per cent P_2O_5	600
Nitrate of soda, 15.5 per cent N	27
Kainite, 12.5 per cent K_2O	320
Total	2,000
1910 (3–9–3):	
Superphosphate, 16 per cent P_2O_5	1,125
Sulfate of ammonia, 20 per cent N	100
Nitrate of soda, 15.5 per cent N	130
Cottonseed meal, 7 per cent N	285
Manure salts, 20 per cent K_2O	300
Filler	60
Total	2,000
1937 (4–9–5):	
Superphosphate, 19 per cent P_2O_5	947
Ammonia, 2.3 per cent of superphosphate	21
Urea, 46.6 per cent N	18
Sulfate of ammonia, 20.5 per cent N	166
Nitrate of soda, 16 per cent N	63
Tankage, 7 per cent N	170
Potassium chloride, 60 per cent K_2O	166
Dolomite	224
Filler	225
Total	2,000

analysis of complete fertilizers that may be manufactured and sold. The South Carolina law now requires a minimum analysis of 20 per cent. According to Mehring *et al.* (1945), the average plant food content of the fertilizer materials used in the United States in 1944 was about 20.6 per cent. The average plant food content of complete mixtures sold in 1939 in Maine was 29.76 per cent and in South Carolina 16.03 per cent.

Extensive expansion in the use of the high-analysis and the concentrated fertilizers cannot be brought about suddenly. Farmers must first become educated to their use, and the natural inertia of such a large industry operating on the present standards, as well as the inertia of customers who are used to buying and applying fertilizers of the present grades, must be overcome. Furthermore, the drilling condition of many of the concentrated fertilizers is poor, but this defect is being remedied. The savings resulting from the use of high-analysis materials is a large item and may bring about the extensive use of concentrated fertilizers sooner than is now expected. It has been estimated that the present annual freight expenditures for moving finished fertilizers from the factories to the consumers is about $30,000,000 and the cost of bags and bagging is about $20,000,000. It would appear that at least one-half of the expense of these two items alone could be saved if concentrated fertilizers were employed universally.

Table 98

FERTILIZER AND PLANT FOOD CONSUMPTION IN THE UNITED STATES, 1900, 1910, 1920, AND 1930–1944*

Year	All Fertilizers, 1,000 Tons	Plant Food (tons)				Calculated Average Plant Food Content of All Fertilizer (%)			
		Nitrogen	Phosphoric Acid	Potash	Totals	N	P_2O_5	K_2O	Totals
1900	2,730	62,000	246,200	86,500	394,700	2.27	9.02	3.17	14.46
1910	5,547	145,900	499,200	211,000	856,100	2.63	9.00	3.80	15.43
1920	7,296	227,800	660,100	257,500	1,145,400	3.12	9.05	3.53	15.70
1930	8,425	376,600	792,800	353,800	1,523,200	4.47	9.41	4.20	18.08
1931	6,541	300,900	610,900	274,700	1,186,500	4.60	9.34	4.20	18.14
1932	4,545	213,600	412,900	191,600	818,100	4.70	9.08	4.22	18.00
1933	5,110	240,200	463,500	222,300	926,000	4.70	9.07	4.35	18.12
1934	5,794	275,286	529,529	262,699	1,067,514	4.75	9.14	4.53	18.42
1935	6,534	311,816	597,321	306,567	1,215,704	4.77	9.14	4.69	18.60
1936	7,222	350,390	672,849	350,075	1,373,314	4.85	9.32	4.85	19.02
1937	8,433	411,482	793,943	416,024	1,621,449	4.88	9.41	4.93	19.22
1938	7,758	384,056	743,657	393,450	1,521,163	4.95	9.59	5.07	19.61
1939	7,993	398,207	789,414	409,077	1,596,698	4.98	9.88	5.12	19.98
1940	8,656	419,093	912,255	435,016	1,766,364	4.85	10.57	5.04	20.46
1941	9,381	458,051	993,571	466,748	1,918,370	4.88	10.59	4.98	20.45
1942	10,131	398,563	1,130,576	547,022	2,076,161	3.93	11.16	5.40	20.49
1943	11,754	505,632	1,243,306	643,177	2,392,115	4.30	10.58	5.47	20.35
1944	12,468	626,208	1,305,410	641,722	2,573,340	5.02	10.47	5.15	20.64

* Including Hawaii, Puerto Rico, and government-distributed fertilizers.

In Table 98 is given, according to Mehring, Wallace and Drain (1945) the fertilizer and plant food consumption in the United States for some recent years.

Table 99

PRINCIPAL GRADES OF MIXED FERTILIZERS DISTRIBUTED IN THE CONTINENTAL UNITED STATES, 1942–1943*

Grades	Rank	States Using	Total Distribution	Percentage of Total: Actual	Percentage of Total: Cumulative
2–12–6	1	30	826,053	11.56	11.56
3–9–6	2	11	791,736	11.08	22.64
0–14–7	3	30	517,044	7.24	29.88
3–12–6	4	27	367,851	5.15	35.03
3–8–5	5	14	353,954	4.95	39.98
0–12–12	6	26	323,074	4.52	44.50
4–10–7	7	9	319,075	4.46	48.96
4–8–4	8	24	296,150	4.14	53.10
4–10–10	9	15	234,872	3.29	56.39
2–10–6	10	5	233,439	3.27	59.66
4–8–6	11	24	194,109	2.72	62.38
4–8–8	12	19	167,006	2.34	64.72
4–10–4	13	29	141,612	1.98	66.70
4–12–4	14	41	139,783	1.96	68.66
3–8–7	15	48	136,817	1.91	70.57
4–8–12	16	14	135,582	1.90	72.47
4–10–5	17	11	113,801	1.59	74.06
4–7–5	18	25	95,817	1.34	75.40
3–9–9	19	7	93,874	1.31	76.71
5–7–5	20	5	92,829	1.30	78.01
3–12–12	21	13	92,798	1.30	79.31
2–8–10	22	17	79,546	1.11	80.42
3–8–8	23	5	73,336	1.03	81.45
5–10–5	24	22	72,358	1.01	82.46
4–12–8	25	14	68,328	.96	83.42
4–9–7	26	10	56,282	.79	84.21
0–14–10	27	7	55,150	.77	84.98
4–9–3	28	5	53,785	.75	85.73
4–6–8	29	2	53,080	.74	86.47
4–10–6	30	19	46,506	.65	87.12
6–12–18	31	1	39,556	.55	87.67
0–9–27	32	11	35,304	.49	88.16
3–9–18	33	13	35,089	.49	88.65
0–10–10	34	15	33,565	.47	89.12
0–14–14	35	22	31,911†	.45	89.57
10–0–10	36	10	29,774	.42	89.99
0–20–20	37	20	29,690	.42	90.41
6–9–15	38	1	29,426	.41	90.82
4–16–4	39	19	22,052	.31	91.13
8–8–4	40	3	20,396	.29	91.42
6–15–15	41	8	19,616	.27	91.69

Table 99—(Continued)

Grades	Rank	States Using	Total Distribution	Percentage of Total	
				Actual	Cumulative
0–20–10	42	13	19,157	.27	91.96
10–10–5	43	2	18,212	.25	92.21
6–3–6	44	4	17,316	.24	92.45
3–8–12	45	6	15,829	22	92.67
8–0–8	46	2	15,540	.22	92.89
2–8–16	47	5	15,456	.22	93.11
2–16–8	48	8	14,816	.21	93.32
3–18–9	49	12	14,655	.21	93.53
10–6–4	50	28	13,705	.19	93.72
6–8–4	51	27	13,485	.19	93.91
4–5–7	52	1	11,854	.16	94.07
6–9–6	53	2	11,304	.16	94.23
5–10–10	54	17	10,806	.15	94.38
0–10–20	55	9	10,443	.15	94.53
3–6–10	56	1	10,038	.14	94.67
3–10–10	57	15	9,312	13	94.80
4–4–8	58	3	9,166	.13	94.93
0–0–5	59	2	8,596	.12	95.05
17–7–0	60	1	8,260	.12	95.17
0–0–6	61	3	7,908	.11	95.28
2–10–4	62	10	7,715	.11	95.39
5–8–8	63	2	7,703	.11	95.50
8–0–12	64	2	7,588	.11	95.61
6–10–4	65	4	7,228	.10	95.71
6–6–6	66	1	7,164	.10	95.81
5–6–10	67	1	7,125	.10	95.91
6–4–8	68	1	6,785	.09	96.00
5–5–8	69	1	6,264	.09	96.09
12–0–10	70	1	5,776	.08	96.17
6–12–8	71	10	5,237	.07	96.24
0–16–8	72	13	5,224	.07	96.31
3–15–0	73	2	5,162	.07	96.38
5–6–8	74	4	5,154	.07	96.45
0–14–5	75	2	5,120	.07	96.52
4–24–12	76	11	5,115	.07	96.59
518 other grades	..	..	187,225	2.63	99.22
Misc. grades unspecified	..	..	55,876	.78	100.00
Continental United States	..	..	7,145,345	..	..

* Tonnage actually reported. Estimated to be 95 per cent complete.
† In addition the Agricultural Agency distributed 150,262 tons of 0–14–14.

Fertilizer Grades Offered for Sale in the United States

About 600 different grades of mixed fertilizers were sold in the United States during the 1943 season. If minor variations, such as source of nutrients, be taken into account and counted as separate grades the number of grades sold would probably be over twice this number.

The state of Florida employs more different grades than any other state. Mehring, Deming, and Willett estimated in 1940 that Florida employs 400 to 425 grades. Of the states using relatively large tonnages of fertilizer, Alabama uses only nine grades and Mississippi only four grades. In general, it can be said that in most states the leading grade used by the state accounts for one-fifth to one-half of the state's total consumption, and that the 10 leading grades in most states account for two-thirds or more of the total consumption of mixed fertilizers. In Table 99 is given the rank of the principal grades of mixed fertilizers consumed in the United States in 1943 as compiled by Mehring, Bailey and Wallace (1944). During the period of World War II directives from the War Production Board limited the number of grades that could be sold in each state.

Carriers of Plant Nutrients

The term *carrier* is used to indicate the material in which the plant nutrient is found. For instance, sodium nitrate, superphosphate, and potassium sulfate are carriers of nitrogen, phosphoric acid, and potash respectively.

Physical Characteristics of Fertilizer Materials

The physical characteristics of fertilizer materials have been given in pervious chapters. It will be recalled that the physical characteristics of

Table 100

RELATIVE HUMIDITY AT WHICH SOME FERTILIZER MATERIALS BEGIN TO ABSORB WATER

Material	*Per Cent*
Ammonium nitrate	59.4
Ammonium sulfate	79.2
Calcium nitrate	46.7
Potassium chloride	84.0
Sodium nitrate	72.4
Urea	72.5

many salts are influenced greatly by their content of absorbed water. In Table 100 are given, according to Adams and Merz (1929), the relative humidities at which some fertilizer materials begin to absorb water at an air temperature of 86° Fahrenheit.

Granulated Materials

Many fertilizer materials are either too coarse or too fine for proper distribution and are improved by granulation. A granulated product is one that has been concentrated into comparatively small nonpulverized particles of approximately uniform size. The granulation of mixed goods prevents the segregation of compounds and greatly reduces their tendency to cake or to become sticky. As a rule granulated fertilizers can be distributed uniformly by most fertilizer distributors. This is not true of most pulverized fertilizers. The use of granulated materials for compounding mixed fertilizers is objectionable, however, for segregation may be promoted unless all the granulated materials used are of like size. Granulated materials will not "film" fertilizer distributing machinery and they may be introduced into the soil at the correct depth.

The processes of granulation now being used vary with the materials to be granulated, but in general the processes may be grouped into five methods, as follows: Compressing, graining, shredding, spraying, and rotary drying. Spraying gives the most uniform product but rotary drying seems to offer the most promise for the commercial granulation of fertilizer mixtures. Progress is being made in the granulation of complete fertilizers and the time may come when most mixed fertilizers will be granulated and bagged as soon as mixed. The expense of basing will thus be eliminated, and at present this amounts to at least $1.50 a ton.

The relative changes that take place in the composition of granular and powdered fertilizers in the soil have been studied by Sayre and Clark (1938).

Production of Neutral Fertilizers

At the beginning of World War II, for the first time since about 1909, the average complete fertilizer on the American market was physiologically neutral, that is, its use would not materially change the reaction of the soil.

In the illustration on p. 361 is shown, according to Mehring (1938), the average equivalent acidity or basicity of complete fertilizers sold in this country since 1880. From this figure it will be noted that the acidity of the average complete fertilizer increased from about 1900 to about 1933, when

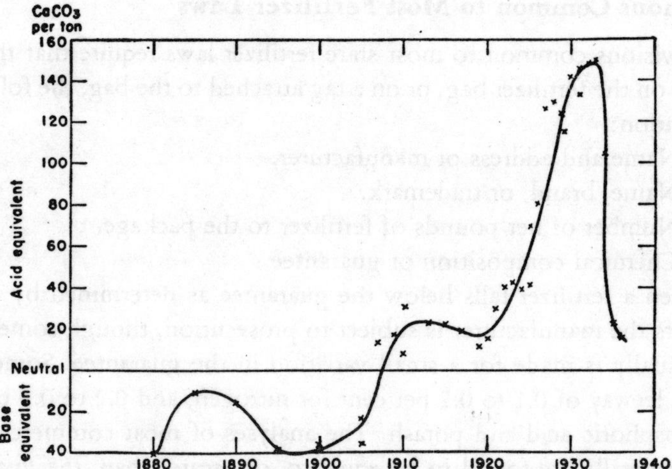

FIG. 125. Average equivalent acidity or basicity of complete fertilizers sold in the United States since 1880. (Courtesy, A. L. Mehring.)

it reached an equivalent acidity of 152 pounds of $CaCO_3$ per ton. Since 1933 it has been decreased so that today approximate neutrality has been reached. This reduction in acidity has been accomplished by the use of nonacid-forming materials to supply nitrogen and to the greatly increased use of dolomite as filler.

Lime has been used in the preparation of fertilizers since about 1890. It is true that at first this material was used only as a conditioner to improve drillability, and to neutralize any free acids that might otherwise destroy fertilizer bags. Later the use of Cyanamid and ammonia liquors made the use of lime for this purpose unnecessary.

Liming materials have been used also in the manufacture of calcium cyanamide and precipitated bone for diluting superphosphate, and as a source of magnesium.

Sale of Fertilizer Governed by Law

The introduction of mixed fertilizers on the fertilizer market resulted in the sale of low-grade mixtures in competition with those of higher analysis. In the United States, in an attempt to protect the farmers against fraudulent goods, every state except Nevada has passed laws governing the sale of fertilizers, and many states have provided for systems of inspection and analysis. State laws generally specify that all fertilizer analyses be made by the official methods of the Association of Official Agricultural Chemists.

Provisions Common to Most Fertilizer Laws

Provisions common to most state fertilizer laws require that there be printed on the fertilizer bag, or on a tag attached to the bag, the following information:

1. Name and address of manufacturer.
2. Name, brand, or trademark.
3. Number of net pounds of fertilizer to the package.
4. Chemical composition or guarantee.

When a fertilizer falls below the guarantee as determined by control chemists the manufacturer is subject to prosecution, though some allowance usually is made for a small variation in the guarantee. Some states allow a leeway of 0.1 to 0.2 per cent for nitrogen, and 0.2 to 0.3 per cent for phosphoric acid and potash. The analyses of most commercial fertilizers generally are found to be equal to, or greater than, the guarantee. Although the farmer is protected by his state laws, it is probable that the publicity derived from the annual publication of the findings of the agency of fertilizer control is more effective than the laws in checking fraudulent guarantees.

The chemical examination of fertilizers and the enforcement of the law are delegated usually to the state agricultural experiment station, to the state department of agriculture, to the state chemist, or to any two of these agencies. From time to time the various fertilizers offered for sale, whether mixtures or separate carriers, are sampled under official supervision on the market throughout the state, and the samples sent to the control laboratory and analyzed. Periodically the findings of the chemists are published in bulletin form and copies of this bulletin may be secured upon request by anyone interested. These publications usually contain lists of registered brands, guaranteed and found analyses, the forms of nitrogen in the samples analyzed and other useful information. Expenses incident to the inspection and control are usually defrayed by a tonnage tax or license fee. Partially as a result of these laws, it seldom happens that a fertilizer which has been compounded purposely to defraud, as far as the total quantity of plant nutrients is concerned, is put on the market. Nevertheless, some low-grade materials do get to the farm at a price per unit paid by the farmer for that of highest quality.

Value of State Fertilizer Laws

The state fertilizer laws that require a guarantee not only protect the farmer, but help also to protect those manufacturers who are attempting to conduct their businesses on a high ethical plane. These laws require the manufacturer or dealer in fertilizers to state the actual percentages of available nitrogen, available phosphoric acid, and water-soluble potash contained in the fertilizers which are offered for sale. Often they require that the carriers of the various nutrients be stated. The guarantee helps to prevent materials of low analysis from competing on an equal basis with those of high analysis.

Although the guarantee may be required by law, this does not protect the farmer entirely. Frequently, farmers pay without protest exhorbitant prices for their fertilizers, either because the required analysis does not give sufficient information or because the farmers do not understand, and make no effort to understand, the guarantee. More often than not the farmer is guided in his choice of a fertilizer by the cost per ton. Fertilizers having a low retail price are usually low grade from the standpoint of total plant nutrients present, or from the standpoint of availability, or both. The cost per unit of fertilizer nutrients in low-analysis complete fertilizers is always higher than the cost per unit of fertilizer nutrients in high-analysis complete fertilizers, even when the carriers of the various plant nutrients are the same. This is because the cost of labor in mixing, and the cost of freight, hauling, and handling is the same for the low-analysis as for the high-analysis mixtures. Before purchasing, farmers always should study the fertilizer guarantee in relation to the selling price of the fertilizer.

Methods Employed in Reporting Guarantee

The present method of reporting the chemical analyses of fertilizers is antiquated and, to the uninitiated, misleading. It must be admitted, however, that some progress is being made, for until recently in some states the element nitrogen, regardless of the chemical form in which it might be found in fertilizer materials, was reported as ammonia. The use of the term *phosphoric acid* is confusing and should be changed to *phosphorus*. The fertilizer chemist uses the term phosphoric acid to refer to the chemical compound phosphorus pentoxide. No actual phosphorus pentoxide is found in any fertilizer material. In fertilizer analytical work the chemist determines the amount of phosphorus that is present and then calculates its equivalent in phosphorus pentoxide. Likewise, no actual potash is found in any fertilizer material, and in analytical work the chemist determines first the potassium content and then calculates and reports an equivalent in terms of potassium oxide or potash.

Many students of fertilizers are looking forward to the day when the general concensus of opinion, as well as state laws, will demand that the *phosphorus* and *potassium* be reported on the elemental basis instead of as phosphorus pentoxide and potash. All state laws now require that the nitrogen be reported on the elemental basis.

State laws also govern the form in which the guarantee may be stated. Usually the nitrogen must be reported only in terms of *available nitrogen*. Phosphorus must be reported in terms of available, that is *citrate-soluble phosphoric acid* and potassium in terms of *water-soluble potash*. Thirty-six state laws now require that the percentages of nutrients in mixed fertilizers must be expressed in whole numbers only.

Unfortunately, in former years the fertilizer laws of some states allowed the fertilizer nutrients to be reported as certain equivalents. For instance, nitrogen could be reported as its equivalent of ammonia, phosphoric acid as its equivalent of bone phosphate of lime, and potash as its equivalent of muriate of potash. These equivalents often lead the consumer to believe that he was receiving more than when the actual nutrients were guaranteed.

In all states the required order to be followed in reporting a fertilizer guarantee is nitrogen, phosphoric acid and potash. According to this standard a 4–10–6 analysis means 4 per cent of nitrogen, 10 per cent of available phosphoric acid, and 6 per cent water soluble potash. Formerly, most of the Southern states used the order phosphoric acid, ammonia equivalent of nitrogen, and potash. In recent years their state laws were changed to allow the order to correspond with that followed internationally and in the United States generally. This standardization was brought about largely through the efforts of the National Fertilizer Association.

State laws vary as to the information the guarantee must include. Some do not require all the information that may be desired, while others require much information that is of little value and often much that is confusing to the farmer. For all practical purposes an analysis should give the following information:

1. Percentage of water-soluble nitrogen.
2. Percentage of water-insoluble nitrogen.
3. Percentage of available insoluble nitrogen.
4. Percentage of available phosphoric acid.
5. Percentage of water-soluble potash.
6. A statement as to whether the fertilizer is acid-forming or nonacid-forming.

In some cases the percentages of chlorine (maximum per cent), manganese, zinc, copper, and magnesium may be desired.

The guarantee of fertilizers, as usually given, shows simply the percentages, in whole numbers, of the various nutrient elements present and gives little information as to their sources. The analysis of a complete fertilizer as represented by the guarantee does not, and cannot, give all that may be desired in regard to the source of materials. Fertilizer mixtures in particular can be easily adulterated and detection may be very difficult or impossible even when the mixture is subjected to a chemical analysis. Now that very high-analysis organic nitrogen compounds, such as calcium cyanamide and urea, are available, the unscrupulous manufacturer may mix these with low-grade water-insoluble organics and produce a mixture that will pass inspection in some states.

Fertilizer Formula

The Association of Official Agricultural Chemists defines the term *formula* as expressing the quantity and grade of the crude stock materials used in making a fertilizer mixture; for example, 800 pounds of 20 per cent superphosphate, 800 pounds of 9–20 tankage, and 400 pounds of sulfate of potash-magnesia. When the analysis as well as the formula is given on the tag or bag, such a guarantee is called an *open formula* guarantee. The South Carolina state laws require that the open formula guarantee be given.

Fertilizer Ratios

A fertilizer ratio is the ratio existing between the percentages of nitrogen, phosphoric acid, and potash found in a fertilizer. For instance, a 3–9–3 or a 4–12–4 grade would have a 1–3–1 ratio. In recent years much discussion has revolved around the subject of fertilizer ratios and it is hoped that these discussions will do much toward bringing about a reduction in the number of fertilizer grades now found on the market. It is of interest to note that a study of the fertilizer recommendations for field crops made in the Southeast reveals a tendency toward a 1–1–1 ratio, provided the nutrients are expressed as *nitrogen*, *phosphorus*, and *potassium*. The most common ratios now employed for the older cropped soils of the East and South are a 1–2–2 and a 1–1–1, whereas formerly it was much higher. On the newer soils of the West the most common ratio was 1–8–1 and 1–6–1, but now, according to Romaine (1940) the ratios are changing to a 1–2–1, 1–4–4, and 1–2–2.

Hibbard (1927) concluded that as yields of wheat, oats, and corn were the same when grown under somewhat wide variations of nutrient ratios, nothing could be gained by attempting to find a definite ratio or physiological balance through the commonly employed triangular system of fertilizer tests. This view however, is not generally held by fertilizer specialists.

FIG. 126. Extrusion machine for making fertilizer drillable. (Courtesy, Swann Chemical Company.)

Varied Sources of Factory-mixed Fertilizers

The materials that go into factory-mixed fertilizers are many and varied. The fertilizer manufacturer uses satisfactorily many materials that cannot be used by the farmer in the manufacture of home-mixed fertilizers. Because of this fact alone the industry may be said to perform an important economic service and to justify its existence.

Processes Involved in the Manufacture of Complete Fertilizers

The processes involved in the manufacture of complete fertilizers are: assembling, pulverizing, acidulating, and blending. Assembling consists simply of getting together the ingredients, such as superphosphate, ammonium sulfate, fish scrap, and muriate and sulfate of potash, or other materials from which the complete fertilizers are to be compounded. Many of the raw materials need pulverizing, and in such cases are usually ground. Often the raw materials contain their plant nutrients in a form relatively unavailable for plant use and must be acidulated or processed. They are then so compounded that proper recognition is given to chemical reactions as well as to the demands of the farmer.

If a manufacturer of complete fertilizers buys his superphosphate instead of making it, he manufactures his fertilizers according to what is known as the *dry-mix* process, which consists of mixing the various necessary fertilizer materials in their dry state. If the manufacturer makes his superphosphate, he usually employs what is known as the *wet-mix* process, which consists of adding his sulfuric acid directly to the mixture of materials that must be acidulated, and then allowing the mass to *cure*. After the materials have been treated and allowed to dry and cure they are ground thoroughly, mixed with other fertilizer materials if necessary, bagged, and stored or shipped. A fertilizer factory which employs the wet-mix process generally consists of the following units: First, a sulfuric acid manufacturing plant; second, storage, grinding, mixing, and shipping plant: third, wharves, sidings, repair shops, offices, and chemical laboratories.

There are four types of commercial fertilizer factories: First, the complete plant that manufactures its own sulfuric acid; second, the plant that buys its sulfuric acid, but grinds its own phosphate rock and acidulates it; third, the plant that makes only its ammoniates and buys its other materials; fourth, the dry-mixing plant which buys its materials and mixes them much as the farmer does when he home-mixes. In Table 101 is given, according to the American Fertilizer Handbook for 1938, the number of fertilizer factories, by classes, in the United States during that year and in Table 102 is given, according to the Fertilizer Review (March-April, 1938), the location of the fertilizer plants and phosphate deposits in the United States in 1938. These tables represent the latest available information.

These tables are representative of a prewar year. Many new plants and many enlargements to old plants were built during the period of World War II but this information has not been compiled since the close of the war.

Table 101

Location of Fertilizer Plants and Phosphate Deposits in the United States in 1938

State	Sulfuric Acid, Superphosphate, and Mixed Fertilizers*	Superphosphate and Mixed Fertilizers*	Mixed Fertilizers*	Concentrated Superphosphate†	Synthetic Nitrogen‡	By-product Nitrogen (Coke Plants)	Potash	Phosphate Rock Fields
Alabama	5	11	37	1	..	8	..	..
Arizona	..	..	1	..	..	..	..	..
Arkansas	1	1	12	..	..	..	..	X
California	3	..	52	..	2	..	1	..
Colorado	..	..	2	..	..	1	..	..
Connecticut	..	1	6	..	..	1	..	..
Delaware	..	..	8	..	..	..	..	..
Dist. of Columbia	..	..	1	..	..	..	..	..
Florida	4	3	46	2	..	..	..	X
Georgia	20	16	148	..	..	..	..	..
Idaho	..	..	3	..	..	..	..	X
Illinois	1	4	4	..	..	8	..	..
Indiana	..	8	10	..	..	6	..	..
Iowa	..	..	3	..	..	..	..	..
Kentucky	..	1	4	..	..	1	..	X
Louisiana	5	1	8	..	..	..	..	..
Maine	..	..	14	..	..	..	..	..
Maryland	6	3	31	..	..	1	1	..
Massachusetts	3	1	6	..	..	3	..	..
Michigan	1	1	7	..	2	10	..	..
Minnesota	..	..	4	..	..	3	..	..
Mississippi	3	5	12	..	..	..	..	..
Missouri	..	..	2	..	..	1	..	..
Montana	1	..	..	1	..	..	..	X
New Jersey	2	2	22	..	..	2	..	..
New Mexico	..	..	1	..	..	..	..	..
New York	1	1	19	..	3	8	..	..
North Carolina	8	7	70	..	..	..	..	..
Ohio	6	12	12	..	..	15	..	..
Oregon	..	..	4	..	..	..	..	..
Pennsylvania	2	1	27	..	..	13	..	..
Rhode Island	..	..	1	..	..	1	..	..
South Carolina	10	5	76	1	..	..	..	X
Tennessee	5	6	3	3	..	1	..	X
Texas	3	..	19	..	..	..	..	..
Utah	..	..	1	..	..	1	..	X
Vermont	..	..	1	..	..	..	..	..
Virginia	8	3	29	..	1	..	..	X
Washington	..	..	12	..	..	1	..	..
West Virginia	..	..	..	..	1	4	..	..
Wisconsin	..	..	4	..	..	2	..	..
Wyoming	..	..	..	..	..	..	..	X
Canada §	..	..	..	1	2	..	..	..
Total	98	93	722	9	11	91	2	10

* The plants listed in column 1 are known as complete fertilizer plants; those in column 2 as acidulating plants; and those in column 3 as dry-mixing plants.

† Column 4 includes eight privately-owned concentrated superphosphate plants, and the government plant at Muscle Shoals. Certain other plants that do not serve the fertilizer industry are omitted.

‡ Three of the plants included in column 5 are among the seven or eight largest nitrogen fixation plants in the world. They produce synthetic nitrate of soda, ammonia liquors, urea, cyanamide, ammonium phosphate, ammonium nitrate, and other products.

§ These particular plants sell most of their products in the United States.

Table 102

NUMBER OF FERTILIZER FACTORIES, BY CLASSES, IN THE UNITED STATES, 1938

State	Number of Cities with Plants'	Class*					Number of Plants
		A	B	C	D	E	
NORTHEASTERN:							
Maine	10	..	..	1	13	..	14
Vermont	1	..	..	..	1	..	1
Massachusetts	10	6	1	..	6	2	15
Rhode Island	1	..	..	..	1	..	1
Connecticut	6	..	1	..	8	..	9
New York	15	1	1	1	14	..	17
New Jersey	24	2	1	..	23	..	26
Pennsylvania	28	1	2	8	24	..	35
Delaware	6	..	..	..	8	..	8
Maryland	20	8	4	2	36	..	50
SOUTHEASTERN:							
Virginia	24	7	4	3	36	..	50
West Virginia	1	..	..	..	1	..	1
Kentucky	2	1	2	..	2	2	7
Tennessee	10	5	5	..	2	..	12
North Carolina	54	9	7	2	68	2	88
South Carolina	54	9	5	..	76	..	90
Georgia	105	20	19	..	145	1	185
Florida	32	5	3	..	47	..	55
Alabama	34	6	10	..	41	..	57
Mississippi	14	3	5	..	10	..	18
LAKE:							
Ohio	15	7	13	3	8	..	31
Indiana	15	..	7	1	10	..	18
Illinois	5	4	5	5	2	1	17
Michigan	6	1	1	1	4	..	7
Wisconsin	5	..	..	2	3	..	5
SOUTH CENTRAL:							
Arkansas	7	1	2	..	9	..	12
Louisiana	7	5	1	..	8	..	14
Texas	17	3	..	2	19	..	24
California	26	3	2	5	51	1	62
All other States	33	2	..	17	27	..	46
Total, 1938	587	109	101	53	703	9	975
Total in 1932	539	118	106	6	591	14	889
Total in 1928	482	106	99	54	545	28	832

* *A*, complete fertilizer plants, with acid chambers, superphosphate, and mixing equipment.

B, superphosphate and mixing plants, with rock grinding, acidulating, and mixing equipment (without acid chambers).

C, ammoniate and mixing plants, manufacture ammoniates only but have mixing equipment (without acid chambers, rock grinding or acidulating equipment).

D, dry mixing plants, with mixing equipment but buying all ingredients.

E, unclassified.

How Fertilizers Are Shipped

Generally, mixed fertilizers are shipped in burlap bags, the standard sizes of bags used being 100, 112, 125, 167, and 200 pounds, according to the demands of the trade. Burlap is imported from India in bales containing 2,000 yards each. It is manufactured into bags at bag factories in this country. Cotton bags are replacing burlap bags to a limited extent. Many separate fertilizer materials are bagged in paper bags. The five-ply double laminated paper bags are excellent for shipping and storing ammonium nitrate, and are cheaper than burlap and cotton bags. In the Southeast mixed fertilizer is generally shipped in 200-pound bags, but in the North and West it is frequently loaded in bags of smaller or irregular sizes, that is, 100, 125, and 167 pounds.

A recent survey conducted by the Division of Cotton Marketing of the United States Department of Agriculture, showed that approximately two-thirds of the bagged fertilizer handled in this country, that is, mixed and non-mixed fertilizers, was handled in 100-pound bags, and one-fourth was handled in 200-pound bags.

Unit Basis of Purchase

The purchase of fertilizers may be made on the *unit* or on the *ton* basis. A unit is a term used to designate 1 per cent of a ton, or 20 pounds. In the trade, sales of fertilizer materials are nearly always made on the unit basis. This is an excellent basis of trading, for it cannot be otherwise than fair to both consumer and dealer.

Sometimes the ton basis of purchase of fertilizer is employed. With the standard high-analysis products this is a satisfactory method of trading as such materials do not vary greatly in their chemical composition. Sales made according to this method usually are based upon the minimum guarantee of the plant nutrients contained in the material.

Agricultural Value of a Fertilizer

The term *agricultural value*, as applied to fertilizers, represents the value of the increase in crop yield produced by the use of a ton of the fertilizer in question. It is a variable factor depending not only upon the availability of the plant food constituents of the fertilizer, but also upon the soil and climatic conditions under which the crop is grown, the rate of fertilizer application, and the market value of the crop tested. Experimental evidence secured by Conner (1931) at the Indiana Experiment Station indicates, as might be expected, that the agricultural value of some

FIG. 127. A drillable, concentrated fertilizer. This fertilizer is in the form of cylindrical granules resembling rifle powder, about one-eighth of an inch in length. (Courtesy, Swann Chemical Company.)

fertilizer materials may be influenced by the fineness of division of the material. Slowly acting materials can seldom be applied profitably to quickly growing crops. Likewise, expensive materials cannot always be applied profitably to crops of relatively low market value. As weather conditions are constantly changing, the agricultural value of a fertilizer must necessarily vary somewhat from year to year, even when the same variety of crop is grown on the same field, and the same rate of fertilizer application is used.

At one time color played a part in the farmer's selection of fertilizers. Dark-colored fertilizers held the highest place in popular favor, but without reason. Occasionally, even today, farmers are known to call for red kainite. It is said that in some cases in order to meet this popular fancy some fertilizer manufacturers formerly dyed their product. Today there are found on the market a few brands of top dressings that have been dyed for the purpose of giving to the product an identification mark that might readily be recognized by the farmer. Odor, too, influences some farmers in making their fertilizer selections, but neither color nor odor should be used as a criterion for determining the agricultural value of a fertilizer.

Commercial Value of a Fertilizer

The term *commercial value* of a fertilizer should not be confused with the term agricultural value. The commercial value of a fertilizer is determined entirely by the market or trade conditions. Fertilizers should be bought on a basis of their guaranteed analysis.

The wholesale price which the manufacturers paid for the ingredients that went into their mixed fertilizers in a recent year was about $4.15 a unit for water-insoluble organic nitrogen, $1.76 a unit for water-soluble nitrogen, $.70 a unit for available phosphoric acid, and about $.75 a unit for water-soluble potash. From these, or similar figures, the approximate wholesale commercial value of the ingredients that go into a mixed fertilizer can be determined as follows if the guarantee of the fertilizer is known.

We arrive at the approximate wholesale commercial value of a ton of mixed fertilizer by multiplying the value per unit of each nutrient by its corresponding per cent—that is, by the number of units of the nutrients given—and then add the products in order to find the commercial value per ton of 2,000 pounds. For example, take a 4–8–4 fertilizer containing 2 units of its nitrogen as water-soluble inorganic and 2 units as available water-insoluble organic, and use the above prices as a basis.

Available water-insoluble nitrogen......	2 by $4.15 gives $8.30	
Water-soluble nitrogen................	2 by $1.76 gives $3.52	
Phosphoric acid.......................	8 by $ .70 gives $5.60	
Potash................................	4 by $ .75 gives $3.00	
Wholesale commercial value per ton of 2,000 pounds..............................		$20.42

To ascertain the approximate retail value an amount sufficient to cove such charges as cost of mixing, bagging, shipping, selling of goods, and profit must be added. This amount is usually about 20 per cent of the wholesale cost.

Many states publish each year a schedule of trade values for the various plant nutrients. These prices represent the average wholesale values of the various fertilizer ingredients, usually for a few months previous to the annual date of publication. This date may be as early as October 1 in some states and as late as March 1 in others. In Table 103 is given, as an example, the schedule of trade values published on October 1, 1937 and 1941, for the State of South Carolina.

When a schedule similar to those given above is used to approximate the commercial value of mixed fertilizers, it should be kept in mind that such an evaluation is entirely commercial and is not intended to indicate even a possible agricultural value. It is well also to keep in mind that trade values of fertilizer ingredients are continually changing and that any schedule of trade values may soon be out of date. Furthermore, many farmers do not understand the meaning of such calculated evaluations. They are likely to place more emphasis on the evaluation itself rather than upon the kind, form, and proportion of constituents in the fertilizer upon which the agricultural value must depend to a great extent.

Table 103

AVERAGE CASH PRICES PER TON, AT THE PORTS, OF UNMIXED INGREDIENTS USED IN THE MANUFACTURE OF COMPLETE FERTILIZERS IN SOUTH CAROLINA, BASED ON THE MOST RELIABLE QUOTATION AVAILABLE FOR SEASONS 1937 AND 1941*

	Per Unit		*Per Pound*	
	1937	*1941*	*1937*	*1941*
Phosphoric acid (available)	$ .75	$ .70	$ .03¾	$.03½
Ammonia in ammonium sulfate	1.25	1.50	.06¼	.07½
Ammonia in sodium nitrate	1.70	1.82	08½	.09¼
Ammonia in cyanamide	1.25	1.50	.06¼	.07½
Ammonia in dried blood	4.15	3.79	.20¾	.19
Ammonia in nitrogenous tankage	4.10	3.17	.20½	.15¾
Ammonia in domestic fish	4.00	4.23	.20	.21¼
Ammonia in cottonseed meal	4.90	5.18	.24½	.26
Potash in kainite	.80	.74	.04	.03¾
Potash in manure salts	.75	.64	.03¾	.03
Potash in muriate	.65	.57	.03¼	.03
Potash in sulfate	.85	.86	.04¼	.04½
In mixed fertilizers the ingredients were reckoned as follows:				
Phosphoric acid (available)	.80	.70	.04	.03½
Available water-insoluble ammonia	4.00	4.19	.20	.20¼
Water-soluble ammonia	1.40	1.76	.07	.08¾
Potash soluble in water	.75	.75	.03¾	.03¾

* In 1937 South Carolina laws required that nitrogen be expressed as ammonia, in 1941 as nitrogen.

The approximate commercial value of a mixed fertilizer may be calculated also by the buyer when he has the guarantee, if he will use the cost of standard materials as the basis for his calculations. As an illustration, let us suppose that the wholesale price of sodium nitrate is $32.00 a ton and that an average ton of sodium nitrate contains 320 pounds of nitrate nitrogen. The wholesale cost per pound of nitrogen in this case would be 10 cents. If we add to this price 20 per cent, the retail price should be about 12 cents a pound. By figuring the number of pounds of mineral nitrogen in a ton of the fertilizer in question, and multiplying this figure

Table 104

PRICES PAID BY FARMERS OF THE UNITED STATES FOR DIFFERENT GRADES OF FERTILIZER IN A REPRESENTATIVE PRE-WORLD WAR II YEAR.
(PER NET TON OF 2,000 POUNDS)

Grade	*Low Price*	*High Price*	*Grade*	*Low Price*	*High Price*
5-7-5	$30.00	$45.00	0-10-10	$28.50	$36.70
1-8-4	22.00	29.55	2-10-2	23.70	37.50
2-8-2	21.18	32.00	3-10-3	26.00	42.65
2-8-5	26.10	38.00	4-10-4	26.00	41.75
2-8-10	30.79	48.00	0-12-5	22.80	30.10
3-8-3	21.80	36.35	2-12-2	27.00	40.00
3-8-4	30.00	44.00	2-12-6	31.00	43.00
3-8-5	26.00	42.50	3-12-3	28.00	37.00
3-8-6	34.50	41.60	4-12-4	28.00	48.00
4-8-7	34.00	48.00	0-14-4	28.00	34.00
5-8-5	30.50	41.00	5-15-5	39.50	51.00
5-8-7	37.00	50.85	2-16-2	33.40	50.00
1-9-4	23.50	35.50	16 % superphosphate	13.00	29.15
2-9-3	22.84	37.85	Nitrate of soda	40.00	70.65
3-9-3	22.58	29.50	Muriate of potash	40.00	60.00
0-10-4	21.30	27.50	Kainite	13.00	26.00

by 12 the commercial value of the mineral nitrogen in the fertilizer may be calculated. In a similar manner the commercial value of the phosphoric acid and potash may be calculated if the wholesale prices of some standard carriers of these nutrients are known. The total of the commercial values of the individual plant nutrients found in the mixed fertilizer represents the commercial value of the fertilizer.

FIG. 128. Partial view of complete fertilizer plant of Armour Fertilizer Works, Chicago Heights, Illinois. (*Left to right*) Sulfuric acid plant, office, acidulating, mixing, and storage plants. (Courtesy, *The Fertilizer Review.*)

There are four groups of factors that enter into the retail price of mixed commercial fertilizers: (1) the cost of the raw materials at the nearest port; (2) the cost of mixing, bags, freight, bagging, labor and hauling; (3) the cost of selling the product; and (4) the profit. The last three factors may be grouped as cost of distribution. The farmer should keep in mind that the cost of distribution remains about the same regardless of the percentage plant food content of the fertilizer.

Table 105

WHOLESALE PRICES PER TON OF TYPICAL GRADES OF MIXED FERTILIZERS IN DIFFERENT GROUPS OF STATES, 1919–1940

Year	*New England States*	*Middle Atlantic States*	*South Atlantic States*		*South Central and Southwest States*	*Mid-West and Northwest States*
	5–8–7	3–8–4	3–8–3	2–10–2	2–10–2	2–12–2
1919	$94.25	$61.57	$55.68	$44.45	$48.16	$49.93
1920	66.99	45.15	43.99	37.58	40.78	41.15
1921	61.79	44.46	39.52	34.61	39.25	34.79
1922	47.32	29.07	28.59	26.22	31.28	28.19
1923	37.08	25.04	26.26	23.66	28.53	28.65
1924	40.00	23.91	21.38	19.19	27.11	28.65
1925	41.34	26.44	22.80	20.44	26.01	28.10
1926	38.77	26.44	22.80	20.44	25.96	28.10
1927	28.09	21.32	19.00	16.70	20.46	25.41
1928	34.50	28.87	22.90	20.44	23.84	26.20
1929	34.15	28.79	22.40	19.96	24.78	26.19
1930	34.10	..	21.59	..	..	25.65
1931	31.36	..	19.13	..	..	21.90
1932	26.91	..	16.00	..	..	18.17
1933	23.89	..	14.77	..	..	19.10
1934	25.86	..	17.35	..	..	20.36
1935	24.54	..	17.44	..	..	20.17
1936	25.32	..	15.89	..	..	20.20
1937	27.53	..	17.87	..	..	21.32
1938	28.26	..	17.34	..	..	20.48
1939	28.58	..	18.05	..	..	20.09
1940*	29.21	..	18.39	..	..	..

* Average 11 months.

The lowest and highest prices per ton of some standard grades of fertilizer actually paid by farmers of the United States in a representative recent year, as reported by Howard (1931), are given in Table 104. It is very unfortunate that such variations in prices as are shown in Table 104 should occur during one season. Certainly it is an index of unhealthy conditions in the fertilizer industry. During the period of World War II O.P.A. ceilings prevented such fluctuations in prices.

The wholesale prices for some commonly used grades of mixed fertilizers for the years 1919 to 1940 inclusive, as reported by The Fertilizer Review (1940), are given in Table 105.

In Table 106 are given the average wholesale prices of some fertilizer materials for the last 25 years.

Table 106

TREND OF WHOLESALE PRICES OF FERTILIZER MATERIALS

	Ammoniates			*Phosphates and Potash*	
	Nitrate of Soda per Unit N Bulk	*Sulfate of Ammonia per Unit N Bulk*	*Cottonseed Meal S.E. Mills per Unit N*	*Superphosphate Baltimore per Unit*	*Muriate of Potash Bulk, per Unit, c.i.f. Atlantic and Gulf Ports*
1910–14	$2.68	$2.85	$3.50	$0.536	$0.714
1922	3.04	2.58	6.07	.566	.632
1923	3.02	2.90	6.19	.550	.588
1924	2.99	2.44	5.87	.502	.582
1925	3.11	2.47	5.41	.600	.584
1926	3.06	2.41	4.40	.598	.596
1927	3.01	2.26	5.07	.535	.646
1928	2.67	2.30	7.06	.580	.669
1929	2.57	2.04	5.64	.609	.672
1930	2.47	1.81	4.78	.542	.681
1931	2.34	1.46	3.10	.485	.681
1932	1.87	1.04	2.18	.458	.681
1933	1.52	1.12	2.95	.434	.662
1934	1.52	1.20	4.46	.487	.486
1935	1.47	1.15	4.59	.492	.415
1936	1.53	1.23	4.18	.476	.464
1937	1.63	1.32	4.91	.510	.508
1938	1.69	1.38	3.69	.492	.523
1939	1.69	1.35	4.02	.478	.521
1940	1.69	1.36	4.64	.516	.517
1941	1.69	1.41	5.50	.547	.522
1942	1.74	1.41	6.11	.600	.522

Standardization of Grades

In recent years there has been a marked tendency not only to reduce the number of brands, but also the number of analyses as well. A program for the elimination of unnecessary grades has been carried on by the National Fertilizer Association and its predecessor organizations since 1918. In many sections the manufacturers, representatives of the agricultural colleges, and others interested, have agreed upon a standardization of the analyses and ratios of complete fertilizers offered for sale within their territory. Where this has been done usually about a dozen analyses have been selected to represent a so-called standard group for the state or section in question. This is a commendable move which can result only favorably for all concerned. In addition, because of wartime stringencies, during the period of World War II the War Food Administration passed many directives of a very constructive nature that greatly reduced the number of grades. It is to be hoped that in the years following World War II these gains will be maintained.

Fertilizer Brands

A fertilizer *brand* is a term, design, or trademark used in connection with one or several grades of fertilizer. A brand name is a specific designation applied to an individual fertilizer. In former years manufacturers found that a well-known brand name would aid in the sale of a fertilizer more than would the analysis of the fertilizer as given on the bag or tag. In time this brought about a multiplicity of brands on the market. Often a given analysis would be put out by a company under two or more brand names. Brand names are frequently of such a nature that they indicate usefulness for some special crop. In the main, brand names should be ignored by the purchaser and only the guaranteed analysis considered.

The term *standard brands* is used to refer to those brands that are commonly found on the market. As a rule, they are of low to medium grade from the standpoint of the total amount of plant nutrients present, but the present tendency appears to be toward the manufacture of goods of higher analysis. In fact, it is not unlikely that the future may see a greater use of individual fertilizer carriers and of diminution in the use of mixtures.

In former years it was sometimes the custom of certain manufacturers, in order to sell their brands of fertilizer at approximately the same price each year, to change the analysis annually to conform to the changes in the commercial value of plant nutrients. Such changes often reacted detrimentally to the farmer, who based his opinion of the agricultural value of a fertilizer upon his past experience with the brand in question.

Fertilizers that have been made from the same materials, regardless of who the manufacturer may be or whether the fertilizer has been home mixed or factory mixed, should give similar results when used under like conditions.

Advantages and Disadvantages Derived from the Purchase and Use of Complete Fertilizers

The advantages derived from the purchase and use of complete fertilizers, rather than single fertilizer materials, are: (1) the mechanical characteristics of complete fertilizers are usually of such a nature that they are in excellent condition for immediate use; (2) complete fertilizers can be secured from local dealers at such times and in such amounts as the farmer may desire; (3) complete fertilizer of a grade best suited to the particular crops grown in the locality can usually be obtained. For continuous use a complete fertilizer is generally to be preferred to the use of single nutrient materials. The one-sided fertilization of pastures with single nutrient materials which has at times been advocated, is likely, if followed, to have the same unfortunate ultimate result as that which has been widely noted in orchards where such a system of fertilization has been practiced.

The disadvantages incident to the purchase and use of complete fertilizers are: (1) the farmer may be led to purchase his fertilizers purely on the basis of cost per ton without sufficiently considering the real value of its constituents; (2) it is impossible to judge by a superficial examination the composition of the mixture.

The Purchase of Fertilizer

Farmers should follow the practice of purchasing their fertilizers so as to receive the greatest value for every dollar expended. If a farmer can use carload lots of fertilizer, and can pay cash, he may effect a considerable saving by purchasing his fertilizers direct from the broker, importer, or manufacturer. The average farmer buys his fertilizer *on time*. That is, he secures his fertilizer from the dealer before the crop is planted, but does not pay for it until after his crop has been sold. The price he pays for his fertilizer is such that the dealer is fully repaid for this extension of credit. This system of purchasing fertilizer should be discouraged as it operates to the disadvantage of both seller and buyer. The seller assumes a financial risk and gambles on crop yields and price. The buyer is forced to take at the seller's price the fertilizer that the seller prefers to sell, regardless of the buyer's crop and soil requirements. There are sufficient established credit agencies and banks to finance the cash purchase of fertilizer on agricultural security at legal interest rates. It would be well if the sale of fertilizer could be put on a cash basis, and on an f.o.b. factory basis instead of on a delivered price basis, as is now the custom of the trade.

In many sections coöperative buying of fertilizers by farmers has been employed successfully for the purchasing of mixed goods and separate materials. In this way the small farmer can secure his fertilizer at as low a price as the large farmer. Such transactions should be based upon the actual quantity or quality of plant food purchased. In most cases of coöperative buying, brand names, which are often misleading and confusing, are given little consideration.

Fertilizer Filler

Few fertilizer materials which are extensively used contain more than 25 per cent of plant nutrients, so that at least 75 per cent of most carriers may be practically useless to the plant. Farmers generally overlook this fact when they use such materials as ammonium sulfate and sodium nitrate. Furthermore, it is seldom that a fertilizer mixture of a definite formula made from given materials equals 2000 pounds or one ton. Many

FIG. 129. A promising potato field, in Aroostook County, Maine, resulting from good soil, good seed, good fertilizer, and good cultural care. (Courtesy, National Fertilizer Association.)

of the standard grades contain a total of only from 1400 to 1800 pounds of fertilizer materials. The other 200 or 600 pounds is composed of some make-weight material which in the fertilizer trade is called *filler*. Sand was formerly used for this purpose to a greater extent than any other material, but dolomitic limestone is extensively employed today and gypsum, floats, fuller's earth, rock salt, coal ashes, grit removed in the preparation of refined chalk, spent foundry sand, peanut hulls, sawdust, and various other organic materials may be used. Muck is sometimes used in complete fertilizers, not only as a filler, but also for the purpose of preventing the mixture from "setting up," that is, from becoming hard on standing. If filler is not desired, kainite or other low analysis potash salts may be substituted for a portion of the chloride or sulfate of potash which is generally employed. But even when a complete fertilizer is so compounded that the addition of filler is not required, there remains a large amount of make-weight substances which may be as useless as filler. Formulas that demand that the mixture contain a very high percentage of plant nutrients usually cannot be made from the ordinary materials but they may be made from high-analysis or concentrated materials. The average complete fertilizer contains only a total of about 300 to 450 pounds of nitrogen, phosphoric acid, and potash.

The quantities of filler consumed in certain years, as estimated by Mehring (1942), are given in Table 107.

Table 107

QUANTITIES OF FILLER CONSUMED IN THE UNITED STATES DURING CERTAIN YEARS

Year	*Short Tons*	*Per Cent of Total Fertilizer Consumed*
1880	5,000	0.7
1890	50,000	2.9
1900	112,000	3.5
1910	258,000	4.6
1920	635,000	8.8
1930	730,000	8.7
1937	510,000	6.1

During 1945 about 12 million tons of complete fertilizers were sold in this country. It is apparent that enormous amounts of make-weight material must have been handled, bagged, and shipped during that year.

This make-weight material would have been greatly reduced, and a very great financial saving made, had farmers bought only high-analysis or concentrated fertilizers.

Conversion Factors

The consumer may desire to convert the guarantee from nitrogen and phosphoric acid to ammonia and bone phosphate of lime, respectively, and vice versa. This can be accomplished by multiplying the reported percentages of nitrogen and phosphoric acid by certain figures known as *conversion factors*. For example, the percentage of nitrogen can be converted into its equivalent percentage of ammonia by multiplying the percentage of nitrogen by the factor 1.214. On the other hand, the percentage of ammonia can be converted into percentage of nitrogen by multiplying the percentage of ammonia by the factor .823. The percentage of bone phosphate of lime may be converted into the equivalent

Table 108

CONVERSION FACTORS

Per Cent	*Per Cent*	*Factor*
KCl to K_2O	multiply KCl	by 0.6317
KCl to Cl	multiply KCl	by .4756
$NaCl$ to Cl	multiply $NaCl$	by .6065
K_2O to KCl	multiply K_2O	by 1.5830
K_2SO_4 to K_2O	multiply K_2SO_4	by .5405
K_2O to K_2SO_4	multiply K_2O	by 1.8490
$MgSO_4$ to MgO	multiply $MgSO_4$	by .3349
MgO to $MgSO_4$	multiply MgO	by 2.9806
$MgCO_3$ to $CaCO_3$	multiply $MgCO_3$	by 1.1868
$CaCO_3$ to $MgCO_3$	multiply $CaCO_3$	by .8424
$CaCO_3$ to CaO	multiply $CaCO_3$	by .5603
$MgCO_3$ to MgO	multiply $MgCO_3$	by .4762
MgO to $MgCO_3$	multiply MgO	by 2.0915
N to $NaNO_3$	multiply N	by 6.0673
N to KNO_3	multiply N	by 7.2176
$NaNO_3$ to N	multiply $NaNO_3$	by .1647
N to $(NH_4)_2SO_4$	multiply N	by 4.7167
$(NH_4)_2SO_4$ to N	multiply $(NH_4)_2SO_4$	by .2120
P_2O_5 to P	multiply P_2O_5	by .4366
P to P_2O_5	multiply P	by 2.3217
K_2O to K	multiply K_2O	by .8297
K to K_2O	multiply K	by 1.2051
SO_3 to S	multiply SO_3	by .4004
S to SO_3	multiply S	by 2.4975

percentage of phosphoric acid by multiplying the percentage of B.P.L. by the conversion factor .46. Similarly, phosphoric acid can be converted into bone phosphate of lime by multiplying the percentage of phosphoric acid by the conversion factor 2.18. The percentage of actual potash in potassium sulfate and muriate of potash may be obtained by multiplying the percentage purity by the conversion factors 54 and 63 respectively. Sodium nitrate and some of the potash salts are crude salts, and their guarantee is usually given on the basis of the purity of the substance. For example, muriate of potash is usually guaranteed as about 80 to 85 per cent pure muriate, that is, 80 to 85 per cent potassium chloride and 15 to 20 per cent of impurities. Other useful conversion factors are given in Table 108.

Advantages and Disadvantages of Purchasing the Single Fertilizer Materials, and Using Them as Such

Often the standard fertilizer materials are bought and applied directly to crops without first mixing with other materials. In fact, in 1945 over 2,500,000 tons of various fertilizer materials were sold directly to the consumer. This tonnage was made up largely of superphosphate, sodium nitrate, ammonium sulfate, ammonia liquors, floats, Cyanamid and muriate of potash in the order named. The advantages accruing to the farmer from the purchase of the single fertilizer materials, and the use of them as such, are three-fold. (1) He secures his plant nutrients at the lowest possible cost. These materials do not have added to their cost such extra overhead expenses as handling, mixing, and rebagging that must be included in the cost of complete fertilizers. (2) The farmer is in a position to use one or more of the plant nutrients in such quantities as he finds his soils and crops demand. (3) The farmer secures a better understanding of the characteristics of the materials he is using and the response his crops exhibit following their use.

The disadvantages of purchasing the single fertilizer materials and using them as such are: (1) The materials may not readily be obtained from local dealers. When obtainable from the wholesale dealer the small farmer ordinarily cannot afford to buy small lots because of the high cost of freight charges. (2) The mechanical condition may be inferior to that of mixed goods. Again, the single fertilizer materials may often be of a higher salt content than mixed goods, and therefore greater precautions must be exercised in applying them to crops. (3) There is a tendency among farmers to use only one carrier of nitrogen, whereas experimental data have shown the value of using several sources of nitrogen for certain soils and crops.

Composition of Complete Fertilizers Sold in the United States

During 1940 the National Fertilizer Association, in coöperation with the Bureau of Plant Industry, Soils and Agricultural Engineering of the United States Department of Agriculture, conducted a survey in which they attempted to find out, among other things, the grades of the complete fertilizers most commonly used during 1939. Similar surveys were made in 1917, 1925, and 1934. The information secured indicated that more complete fertilizer of 3–8–5 analysis was sold than of any other analysis. This was followed by a 2–12–6, 4–8–4, and a 3–8–3, respectively. A survey today would certainly show a gain in the use of the higher analysis fertilizers during the period of World War II. The average plant food content of all complete fertilizers sold in the United States is now above 20 per cent, whereas in 1914 it was probably less than 13 per cent. The average complete fertilizer sold in 1939 would have averaged 3.76–9.08–5.78.

At the present time it is possible to make from certain synthetic nitrogen compounds, treble superphosphate, and high-analysis potash salts, a fertilizer mixture which will carry a total of 50 units and more of nitrogen, phosphoric acid, and potash combined. It is to be hoped that the future will see a greater use of fertilizers carrying the higher percentages of plant food. When a farmer buys a 4–12–4 instead of a 3–8–3 he gets approximately 10 per cent more plant food for his money. If he buys a 5–15–5 he gets approximately 15 per cent more. Knight (1937) has estimated that if the unnecessary filler commonly used in a 3–8–3 was eliminated the resulting product would be a 4–10–4 and this change would result in the saving to the farmers of one state alone of more than $2,000,000 annually. Fifteen years ago, Blair, of the New Jersey Experiment Station, expressed the opinion that no complete fertilizer should be used that contained less than 18 to 20 per cent of plant food. He concluded that the lower grades would too often become the dumping ground for low-grade and inert products. This opinion is now generally held by agronomists.

17

Principles Underlying the Use of Fertilizers

The practice of applying fertilizers to crops should be a profitable one. Not only is the farmer concerned in making applications of fertilizers a paying practice, but the fertilizer manufacturers also desire that the farmer should make a profit through the use of their products. They realize that the profitable use of fertilizers by farmers determines largely the volume of their sales.

Primary Function of Fertilizers

The primary function of fertilizers is to supply plant nutrients to growing crops. Besides supplying nutrients fertilizers may often influence plants in other ways, and may also bring about marked changes in the characteristics of the soil. These other effects may be either beneficial or detrimental, depending upon the fertilizer, the soil, and the crop. The farmer should not overlook such secondary influences as they often play an important part in determining the character of the soil and crop management practices that should be followed.

Object in Using Fertilizers

Fertilizers are used primarily with the object of producing a profit. Profit is represented by the difference between the value of the increase of crop resulting from the use of fertilizer, and the cost of the fertilizer plus the cost of applying the fertilizer and harvesting the increase.

During 1940 the National Fertilizer Association conducted a nationwide survey in an attempt to secure from the farmers of the nation an expression of their opinion of the agricultural value of the fertilizers they used in 1939. Table 109 was computed by Smalley, Engle and Willett (1939) from the answers of more than 48,000 farmers. This table shows that in this country in 1939, for every dollar spent for fertilizers farmers estimated a return of $3.60. A survey conducted today would probably show similar results. The return for such crops as tobacco, lettuce, and strawberries is probably now between $10.00 and $15.00.

Table 109

CROP RETURNS FROM THE USE OF COMMERCIAL FERTILIZER, BY CROPS

Crop	*Value of Crop Produced by Fertilizer*	*Tons of Fertilizer Used on Crop*	*Estimated Cost of Fertilizer*	*Increase in Value of Crop for Each Dollar Spent for Fertilizer*
Corn	$ 84,387,890	1,632,600	$ 40,672,300	$2.07
Cotton	171,709,646	1,501,800	38,424,600	4.47
Wheat	33,355,404	739,300	18,915,600	1.76
Potatoes	55,004,062	555,500	17,671,050	3.11
Fruits and vegetables	62,501,271	556,360	16,342,400	3.82
Tobacco	135,937,201	505,400	15,033,000	9.04
Citrus fruits	31,848,310	256,000	8,819,000	3.61
Hay and alfalfa	11,426,347	239,300	5,161,000	2.21
Oats	8,108,377	247,000	5,635,500	1.44
Sweet potatoes	19,613,563	133,000	3,545,150	5.53
Tomatoes	12,108,895	85,400	2,641,200	4.58
Peanuts	6,210,370	57,000	1,469,000	4.23
Miscellaneous	77,598,839	807,740	22,617,870	3.43
Total	$709,810,175	7,316,400	$196,947,670	$3.60

Basing their estimates on the replies of the above farmer survev Smalley and Engle (1943) made an estimate of the quantity of crop produced by a ton of average fertilizer. Tl.eir estimates are given in Table 110.

Warner (1931) has estimated from data secured from fertilizer experiments and reliable field tests and demonstrations of 30 fertilizer-using states, that $1.00 spent for fertilizer will show a return on the average of $2.00 to $5.00. Brand (1944) estimates it now to be about $5.00, for prices of farm commodities are high and fertilizer prices are low.

In most localities the greater the profit obtained from the use of the first 100 pounds of fertilizer, the greater is the number of pounds of fertilizer commonly used per acre. This is shown in the fact that the average amount of fertilizer used per acre on truck crops is two or more times as great as that applied to field crops like cotton.

Table 110

AVERAGE YIELD INCREASES PRODUCED BY ONE TON OF FERTILIZER

Crop	*Yield Increase*
Corn	125 bu.
Cotton	2 bales
Wheat	85 bu.
Oats	140 bu.
Tobacco	1,370 lbs.
Potatoes	185 bu.
Sweet potatoes	285 bu.
Soybeans	50 bu.
Peanuts	200 bu.
Tomatoes	215 bu.
Snap beans	130 bu.
Apples	700 bu.
Milk	8,000 lbs.
Beef	1,000 lbs.

Chemical Composition of the Soil as Related to the Use of Fertilizers

The chemical composition of the soil has a direct bearing on the proper fertilization of crops. It is well known that soils differ widely in their chemical composition, because of the varied minerals from which they were formed as well as the weathering to which they have been subjected. Because of this fact some knowledge of the chemical com-

FIG. 130. Erosion plats of the Missouri agricultural experiment station. These plats have been called the "birthplace of soil conservation in the United States." (Courtesy, Better Crops and the Missouri Agricultural Experiment Station.)

position of the soil to which fertilizers are to be applied is desirable if the fertilizers are to be used judiciously. Sandy soils are usually derived from those minerals that are deficient in potash. As most crops require potash in relatively large amounts, potash fertilization is nearly always essential in order to procure the highest yields on these classes of soils. Clay soils, on the other hand, are usually derived from those minerals that are rich in the feldspars and therefore in potash, and they are more often deficient in phosphoric acid and lime than in potash. Soils of limestone origin generally contain enough lime, and usually sufficient amounts of phosphoric acid and potash to produce fair crop yields. On these soils nitrogen is often the first limiting element to plant growth.

In the humid sections of the United States the total nitrogen content of the soil decreases from north to south, and this change is a negative exponential function, so that for every 18° F. fall in mean annual temperature the average nitrogen content of the soil increases two or three times. This, together with the availability of soil nitrogen, must be taken into consideration when planning a fertilizer program.

All states do not have the same percentage of good land. This is illustrated in Table 111.

Table 111

NATURAL PRODUCTIVITY OF TILLABLE LAND IN REPRESENTATIVE STATES

	Per Cent of Tillable Land in Each Productivity Class			
	Excellent	*Good*	*Fair*	*Poor*
Iowa	74	19	4	3
Illinois	43	20	18	19
Ohio	17	25	42	16
New Jersey	0	29	29	42
South Carolina	0	2	49	49
Florida	0	5	22	73

Corn, oats and wheat generally respond more to fertilization when planted on poor soils, while cotton, Irish potatoes and hay usually give the highest returns from fertilization when grown on the more fertile soils. If soils contain adequate amounts of calcium and magnesium it may be found that acid-forming fertilizers will give good results, but if soils are strongly acid the basic fertilizers may prove to be the best.

Physical Composition of the Soil as Related to the Use of Fertilizers

With reference to the size of their predominating particles, soils may be classified roughly into sands, silts, clays, and loams. Seldom do we find a large area of pure sand, silt, or clay. Most soils are mixtures of soil particles of varying sizes, but in most cases one soil texture predominates.

Soil texture has a direct influence on crop fertilization. Sandy soils, because they are open and porous, and because they do not contain much fine or colloidal material, cannot absorb large quantities of plant nutrients. Such soils are subject to excessive loss of plant nutrients by leaching. Nitrates should never be applied to these soils in large quantities at any one application, for during periods of heavy rainfall they may be leached from the soil within a short time.

Clay soils, although generally of a reasonably good structure, may in some cases be so impervious to air and water that although receiving liberal fertilization, profitable crops cannot be grown. This is especially true of clay with a high silicon content. But crops may readily absorb and hold limited quantities of soluble nitrogen in the ammoniacal form, and potash, and fix the soluble phosphoric acid of fertilizer applications. The phosphate requirement of acid well-drained soils tends to be proportional to the clay content for clay acids are the principal soil acids.

Silts and loams offer the best physical conditions for the growth of most field crops and when these soils are fairly well supplied with organic matter they give promise of the most profitable returns from the judicious use of commercial fertilizer.

Greater attention should be paid to creating a much more favorable structure in the B and C soil horizons. Ten to forty per cent of the root systems of many crops is below the plowed soil and it is the physical structure of the lower soil horizons that often limit plant growth.

Many soils have been seriously eroded and these offer serious problems to farmers. In general, row crops should not be planted on soils of more than 12 to 18 per cent slope, although some laterites do not erode badly even at 70 per cent slope and some Rendizna soils do erode badly at 1.5 per cent slope.

The greatest return from the use of commercial fertilizer is obtained from those soils that are in the best physical condition for plant growth. The best results cannot be expected from fertilizers when they are applied to soils that are too fine or too coarse, too compact or too porous, too dry

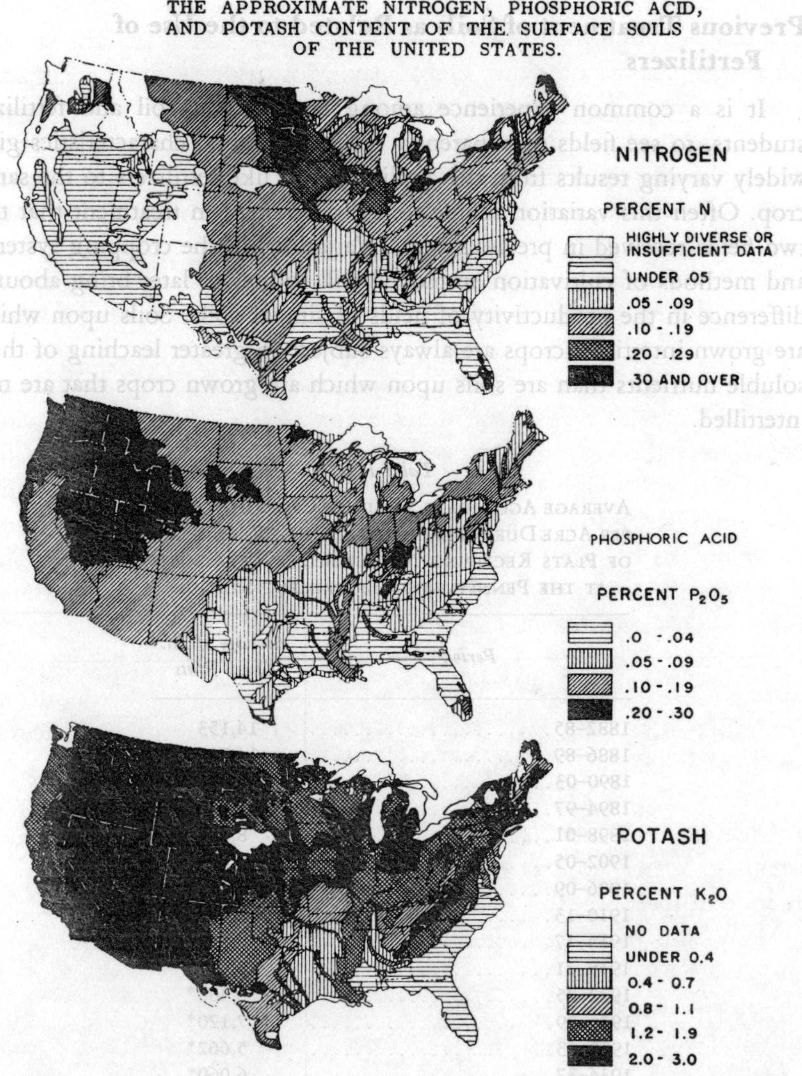

FIG. 131. The approximate N, P_2O_5, and K_2O content of the surface soils of the United States. (Courtesy, National Planning Association and *The Fertilizer Review.*)

or too wet. The farmer should always consider carefully the physical characteristics of his soil before he makes large expenditures for fertilizers.

Previous Treatment of Soils as Related to the Use of Fertilizers

It is a common experience among farmers and soil and fertilizer students, to see fields of apparently the same natural characteristics give widely varying results from the application of like fertilizers to the same crop. Often this variation is due to the differences in treatment that the two fields received in previous years. Variations in the cropping systems and methods of cultivation employed will sooner or later bring about a difference in the productivity of fields of similar soils. Soils upon which are grown intertilled crops are always subject to greater leaching of their soluble nutrients than are soils upon which are grown crops that are not intertilled.

Table 112

AVERAGE AGGREGATE YIELDS OF CROPS IN POUNDS PER ACRE DURING EACH ROTATION OF FOUR YEARS OF PLATS RECEIVING NO FERTILIZER TREATMENT AT THE PENNSYLVANIA EXPERIMENT STATION

Period	*Average of All Check Plats*
1882–85	14,153
1886–89	14,784
1890–03	14,474
1894–97	11,567
1898–01	8,685
1902–05	10,119
1906–09	8,969
1910–13	9,073
1914–17	8,559
1918–21	7,909
1922–25	6,400*
1926–29	7,120*
1930–33	5,662*
1934–37	6,069*

* Average tiers 1 and 3 only.

Fertilizing practices also bring about a variation in the productivity of soils. It is well to keep in mind that because soils are physical-chemical-biological complexes profound changes may be brought about in them as a result of the use of commercial fertilizers. The addition of farmyard or green manures often produces crop yields much out of proportion to the quantity of nutrients added. The same can be said concerning the application of lime. As a result, the productivity of the soil may be reduced at a more rapid rate than if no manure or lime had been added. An application of lime often results in the freeing of fixed phosphoric acid to the extent that crop response to applications of phosphoric acid cannot be secured for some years following application of the lime. On some soils of the southeastern states superphosphate has been applied for years in quantities greater than that demanded by the crops grown, and as a result phosphate has accumulated in the soil. It is not surprising that such soils often do not show immediate results from the application of phosphatic fertilizers.

The decrease in crop yields resulting from the continuous growing of crops without the application of fertilizer is well known and is clearly illustrated in Table 112. According to information furnished the author by Gardner and Noll, this table represents a summary of the results secured over a period of 55 years at the Pennsylvania Experiment Station.

Character of Crop and Use of Fertilizers

The returns secured from an application of commercial fertilizer are always influenced by the character of the crop. Most of our field crops are rich in plant nutrients and are of low commercial value. On the other hand, fruit and vegetable crops are relatively low in their plant nutrient content, and have a high commercial value. These factors are nearly always the principal ones which determine the agricultural value of a fertilizer.

The Theory of Mineral Up-take by Plants

Plant nutrients are released from soil-forming minerals by weathering. Among the more important processes are the hydrolytic action of water and the influence of hydrogen clay on soil minerals. The action of hydrogen sorbed in the exchange atmosphere of colloidal clay on some soil minerals may easily transfer calcium, as has been shown by Graham (1940, 1941), to the exchange atmosphere of colloidal clay and thus store it for plant use.

Recent studies seem to show that the continued intake of water by plants necessitates the continuous extension of their absorbing roots to provide fresh contacts with unexhausted moisture films, and that root soil contact may be equally important to the intake of many plant nutrients.

Most investigators, though not all, now agree that the release of the mineral elements in the soil by weathering from soil minerals is too slow a process to take care of the requirements of rapidly growing vegetation. The requirements of a rapidly growing crop must be taken from the reservoir of available nutrients that are held by electrostatic forces on the surfaces of the soil colloids. In soils that are heavily cropped this reservoir must be refilled from time to time through the application of soluble fertilizers. The mechanism of transfer of absorbed ions from the soil colloids is not thoroughly understood, but it is assumed that the nutrient cations are absorbed by plant roots in exchange for H ions, and in like manner the nutrient anions are exchanged for HCO_3 ions.

Both the kinds of colloid and the nature of the absorbed ions influence the replacement of the cations with hydrogen. For instance, calcium is more tightly held by clay than is sodium, and hydrogen ions displace potassium more readily from colloids containing calcium as a complementary ion than from those containing sodium. In addition, soils that are known to give up potassium from nonreplaceable forms are usually rich in the micacious minerals.

Influences of Nitrogen on Plant Growth

Nitrogen is a constituent of protein, and therefore a constituent of every living cell. When nitrogen is not present in the soil in quantities sufficient for optimum crop growth the kernels of cereals and the seeds of other crops do not attain their normal size, but become shriveled and light in weight. Nitrogen starvation in fruit trees is characterized by the early shedding of leaves, the death of lateral buds, and the development of unusually colored fruit. The most apparent influence of nitrogen on plant growth is its influence on the development of the vegetative portion of the plant. A nitrogen deficiency appears to reduce in some way the water content of the plant. Following an application of nitrogenous fertilizer to plants suffering from nitrogen starvation the green color of the leaves may be deepened. In many cases the green color of the plant indicates that the plant is in a healthy condition. It is well to emphasize the fact, however, that the deepening of the green color of plants together with a thickening of the leaves appears to be correlated to a greater or less extent with an excess of nitrogen, sulfur, iron or manganese in the soil. These are the nutrient elements that may change their valence. Bright tobacco growers find that an excess of any of these four elements in the soil will produce tobacco of poor quality.

In general, plant characteristics showing nitrogen deficiency are about the same for most field crops. Plant leaves first assume a light green color, yellowing and drying up or shedding as they grow older. The chlorosis of a leaf indicates a disturbance of nitrogen metabolism, as well as of iron metabolism, as has been pointed out by Bennett (1943).

Normal applications of nitrogen appear to hasten maturity slightly. On the other hand, excessive applications of nitrogen may delay maturity and often may cause the plant to produce excessive vegetative growth. When excessive quantities of nitrogen are supplied, the flavor of peaches is known to deteriorate, as do also the malting qualities of barley. Likewise, excessive nitrogen has been observed to bring about sap impurity in sugar plants, lodging of small grains, and an increase in the nicotine content of tobacco; it has been observed to affect the keeping quality of many fruits and vegetables, as well as to injure the color and odor of hops. An excessive quantity of nitrogen is known to decrease the resistance of plants to disease. The small grains, for instance, succumb readily to rust when excessive quantities of nitrogen are present.

The percentage nitrogen content of plants is increased when the nitrogen is available in abundant quantities. This has been noted particularly with the grass family, but also with many vegetables. The protein content of wheat may be increased by the application of nitrogen at the heading stage. Thatcher and Arny (1917) have called attention to the data secured at the Ohio Experiment station, which showed that oats receiving nitrogenous fertilizer usually produced grain having a higher percentage of nitrogen than oats from plats receiving no nitrogen.

The influences of nitrogen on plant growth are in many respects similar to the influences of water. Large applications of nitrogen may cause the plant to become succulent, a condition which may or may not be desirable. With such crops as lettuce, radishes, and the grasses, where succulence may be desirable, or at least not objectionable, larger quantities of nitrogen may be applied than is normally the custom. With most field crops the application of nitrogen should be sufficient only to keep the plant growing in a normal manner. Plants receiving sufficient nitrogen appear to be better able to utilize phosphoric acid and potash.

Influences of Phosphoric Acid on Plant Growth

Low crop yields are more often due to a lack of phosphoric acid than to a lack of any other nutrient. Phosphoric acid has often been called the master key to agriculture. Phosphoric acid appears to be concerned in the production of nucleoproteids and it appears that phosphoric acid influences the production of seed or grain more particularly than does nitrogen or potash. Russell (1932) found, however, that in the case of cereals it does not decrease the ratio of stem to grain. Phosphoric acid has a marked influence on the hastening of the maturity of some plants, and thus serves a useful purpose in cold climates where growing seasons are short. The more soluble forms of phosphoric acid appear to hasten maturity to the greatest degree. At the Rothamsted Experiment Station phosphoric acid has been shown to be of particular value in hastening the maturity of crops during wet years. The hastening of maturity during dry seasons may be objectionable in that maturity may take place before there has been sufficient rainfall to meet the demands for a satisfactory yield of crop. This is often apparent with corn and tobacco. The influence of phosphoric acid on hastening maturation of cabbage has been noted by Noll (1923), and a similar influence on cotton has been noted by Blackwell and Buie (1921), and Williams, Buie, and Stacy (1937).

The presence of phosphoric acid in the plant has been found to influence cell division as well as the formation of fat and albumin. In its absence starch will not change to sugar. The most noticeable plant symptom of phosphoric acid deficiency in most field crops appears to be slow growth with small and very dark green leaves but there are no symptoms that can be accepted for diagnostic purposes with assurance.

Phosphoric acid also stimulates the development of roots, particularly of the root crops. It has been found to have a beneficial influence on the development of a good lateral root system of fall-sown grains, and to increase tillering. Both Russell (1915) and Noll (1923) have shown that phosphoric acid promotes the formation of lateral and fibrous roots. In many respects phosphoric acid counteracts the influences of nitrogen. Bear and Salter (1916) have presented data from the West Virginia Experiment Station showing that the phosphorus content of the soil is related to nitrogen fixation, or to nitrogen conservation, or to both.

In Great Britain a disease of apples known as Baldwin spot, or bitter pit, has been found to be due to a deficiency of phosphoric acid. Bronze-purple colored leaves are often noted at the top of new shoots of phosphate-starved apple trees.

Influences of Potash on Plant Growth

As far as is known, plants do not build potassium into the structure of any of their parts. Nevertheless, absorbed potassium cannot be leached out of the plant as long as it is living, but it is readily washed out after the plants are killed.

Among the more marked influences of potash on plant growth is its influence upon the development of the woody parts of stems and the pulp of fruits. It seems to slow down the effects resulting from excessive nitrogen and to prevent the too rapid maturation resulting from too much available phosphoric acid. It is certainly an important element affecting the oxidizing and reducing processes that go on in the plant. Potash is essential both for the photosynthesis and for the translocation of starch. Nightingale (1943) has secured evidence showing that potash may be essential for nitrate reduction and protein synthesis. It probably also plays an important part in oil and albuminoid synthesis. It appears to be essential in the development of chlorophyll and in regulating the water content of the cytoplasm. It may also influence cell wall formation and stiffness of stems. Although potash may increase the efficiency of photosynthesis it does not increase the size or the number of leaves. When potash is deficient the older leaves of plants become mottled or chlorotic with more or less necrosis along the margins. Potash imparts firmness and coherency to the ash of tobacco, and is necessary for continued burning of tobacco without flame. McMurtrey (1941) states that potash shortage in tobacco results in leaves that are off color and that lack body, elasticity, and aroma, and that have poor fire-holding capacity.

Janssen and Bartholomew (1929) have shown that the potash present in the plant is for the most part readily soluble in water. Potash, like sodium, calcium, and chlorine, is found in largest amounts in the stalks of plants, but it is concentrated also in the growing tips.

Potash appears to aid the plant in resisting certain diseases, and in this way may counteract the effect of too much nitrogen, which appears to decrease disease resistance. The common legumes require relatively large amounts of potassium in spite of the fact that potassium does not enter into organic combination in plant tissue. The influence of potash on the control of cotton wilt has been shown by Young, Janssen, and Ware (1932), of the Arkansas Experiment Station, and by Young, at the Texas Experiment Station. Potash also furnishes the necessary principle for alleviating rust in cotton and small grains and bacterial leaf-spot in tobacco. Potash is known to delay maturation and to retard transpiration. During dry years potash salts appear to induce a longer period of growth so that in some cases this may be an advantage in that the crop may not mature until sufficient rainfall has occurred. Potassium not only appears to protect plants from excessive losses of water during periods of drouth, but it appears to lessen the injuries due to low temperature. Pastures well supplied with potash stay green longer during dry periods and in some parts of the country the quantity of soluble potash determines the stand of clovers. Kraybill (1930) has called attention to the work done with sweet potatoes at the New Jersey Experiment Station by Schermerhorn. He has shown that the application of potash resulted in the production of a stocky shaped potato high in protein. This observation has also been made by Nightingale (1943). When insufficient potash was available the plants were unable to synthesize sufficient protein.

Apparently an increase in the absorption of nitrogen by plants is followed by an increase in leaf area, and this in turn must be followed by an increase in the absorption of potash if carbohydrates are to be made in sufficient quantities for normal protein production comparable to the amount of nitrogen absorbed. For this reason neither potassic nor nitrogenous fertilizers give their full effect without adequate supplies of the other. Wall (1940) has reported that if ammoniacal nitrogen enters a potash-deficient tomato plant rapidly the synthesis of some forms of organic matter appears to stop, and that ammonia accumulation will produce injury to the tissues of the plant.

The absorption of large quantities of soluble potash by the plant does not appear to interfere with the normal development of most crops. An optimum supply of potash appears necessary for the production of the best quality of grain and fruit. In fact, potash is sometimes referred to as the quality element in fertilizers. An abundance of potash appears to improve the shape, size, and cooking qualities of Irish and sweet potatoes, the sugar content of sugar beets, the shape of cucumbers, the color and firmness of tomatoes, the crispness of celery, the color and flavor of strawberries, the color and flavor of apples and peaches, the thickness of rind and the percentage of juice of oranges, the quality of tobacco leaf, the length and strength of cotton fiber, the lodging of corn, and the tenderness of peas. Potash has a marked influence on the flavor of grapes and wines and, like phosphorus, may improve the keeping quality of certain fruits and vegetables.

Potash appears to prevent the accumulation of iron in the joint tissues of corn, accumulations that interfere with the transportation of foods which may increase the plant's susceptibility to root rot. This may probably also hold true for the legumes so that concern over potash deficiency may arise in those areas where heavy applications of lime and phosphate have been made.

Fertilizing Value of Vitamins and Hormones and Growth Regulators

Much publicity has been given to the use of Vitamin B for fertilizer purposes, and the use of growth regulators for treating seeds or incorporating with fertilizers. Tests conducted at the Boyce Thompson Institute for Plant Research show, according to Zimmerman (1941) that these growth substances have no fertilizing value when used as generally recommended for horticultural practice. Carpenter (1943), however, has shown that vitamin B_2, riboflavin, and other vitamins are present in soils and may be absorbed by plants. This would suggest that plants might supplement synthesized vitamins from the soil. As pointed out by Hoagland (1945), it might be assumed that all the vitamins used by animals may have a role in plant growth and metabolism.

Hamner (1945) came to the conclusion that soil type and nutrient supply have much less influence on the vitamin content of plants than do variety and climate. The principal use of growth substances to date has been to propagate plants from cuttings, to inhibit buds of potato tubers,

FIG. 132. A big crop of celery removes more plant food from the soil than almost any other crop. (Courtesy, Better Crops with Plant Food.)

to induce seedless tomatoes, and to prevent preharvest drop of apples. Many of these hormones are halogen, methyl, and nitro derivatives of phenoxy and benzoic acids.

Zimmerman (1943) came to the conclusion that the practice of adding vitamins and plant hormones to fertilizers has more sales than practical value.

Response of Crops to Plant Nutrients

Ever since fertilizers have been commonly used it has been recognized that crops respond in varying degrees to the individual nutrient elements. Furthermore, there are varietal differences in response to mineral deficiencies. This has been shown to be the case with garden beets by Walker *et al.* (1945). Nevertheless, a classification of crops on the basis of their response to nutrients has been attempted by many investigators. The classification suggested by Hartwell (1922) and based on the experimental response of crops at the Rhode Island Experiment Station is probably the best that has been suggested so far. This classification is given in Table 113. In this

Table 113

RESPONSE OF CROPS TO FERTILIZER NUTRIENTS

Group	*Nitrogen*	*Phosphoric Acid*	*Potash*
I	Rye	Carrot	Corn
	Bean	Buckwheat	Rye
	Corn	Millet	Cabbage
	Cucumber	Oat	Turnip
	Cabbage	Pea	Bean
	Pea	Bean	Oat
	Potato	Tomato	Pea
II	Wheat	Corn	Millet
	Sunflower	Potato	Wheat
	Turnip	Rye	Buckwheat
	Tomato	Wheat	Carrot
	Beet	Sunflower	Potato
	Carrot	Barley	Tomato
	Oat	Lettuce	Barley
III	Millet	Cabbage	Squash
	Parsnip	Beet	Sunflower
	Buckwheat	Cucumber	Beet
	Lettuce	Onion	Onion
	Barley	Parsnip	Parsnip
	Squash	Squash	Lettuce
	Onion	Turnip	Cucumber

FIG. 133. Response of evergreen cuttings to application of root hormone. (Courtesy, Henry H. Kirkpatrick and the Boyce Thompson Institute for Plant Research.)

table group I shows the least response, and group III the greatest. According to this classification corn falls in group I as to nitrogen and potash, and in group II as to phosphoric acid. Similar variations will be noted with the other crops. It should be kept in mind that the response of these crops to the three nutrients may not be the same when grown under other soil and environmental conditions.

As might be expected, the response of crops to the various nutrients appears to be extremely varied. This is well illustrated in the general summary of all published information on the influence of fertilizers on crop quality which was presented by Hartwell (1932) at the 1932 convention of the National Fertilizer Association. A careful survey of this paper should be made by anyone who is interested in the influences exerted on plant growth by the various nutrient elements.

In Table 114 is given a key, by McMurtrey (1938) showing the plant characteristics in tobacco which may indicate a deficiency of some of the essential elements.

Laurie and Wagner (1940), after studying the deficiency symptoms of greenhouse flowering plants, suggested the key to nutrient deficiency symptoms given in Table 115, and Davidson (1941), after making an intensive study of trees, prepared the key given in Table 116, showing plant nutrient deficiency symptoms in deciduous fruit.

Fruit trees appear to have a high nitrogen requirement but seem to be able to tolerate a rather low supply of phosphoric acid and potash.

Table 114

KEY TO DEFICIENCY EFFECTS ON TOBACCO PLANTS GROWN IN SOLUTION CULTURES*

			Element	Symptoms
Decreased growth. More or less localized effects. Causal parasites absent	*Group 1.* General on whole plant, or localized effects on older or lower leaves	General on whole plant. Also yellowing and drying or "Firing" of lower leaves	Nitrogen	Plant light-green. Lower leaves yellow and dry to a light-brown color. Stalk slender and short. Roots long with few lateral branches, white in color
			Phosphorus	Plant dark green. Lower leaves may yellow and dry to a greenish-brown to black color. Stalk slender and short. Roots long with few lateral branches, reddish-brown color
		Local. Occurring as mottling or chlorosis with or without necrotic spotting of lower leaves. Little or no drying of lower leaves	Potassium	Lower leaves mottled with necrotic spots at tips and margins, with tips and margins tucked or cupped under. Stalk slender, necrotic areas in extreme cases. Roots long with few lateral branches and of a yellowish slimy appearance
			Magnesium	Lower leaves chlorotic and typically show no spots. Tips and margins turned or cupped upward. Stalk slender, necrotic areas in extreme cases. Roots long with few lateral branches and slimy in appearance
	Group 2. Localized effects on newer or bud leaves	Terminal bud remains alive. Chlorosis of newer or bud leaves with or without necrotic spots. Veins light- or dark-green	Iron	Young leaves chlorotic, typically show no spots; veins typically green, that is, principal ones. Stalk slender and short. Roots short with abundant short laterals brown in color
			Manganese	Young leaves chlorotic with necrotic spots scattered over leaf. Smallest veins tend to remain green producing a checkered effect on leaf. Stalk slender. Roots not so abundant and brownish in color
			Sulfur	Young leaves light green, no necrotic spots. Veins lighter green than inter-vein tissue. Stalk short and slender. Roots white, abundant and much branched
		Terminal bud dies, preceded by peculiar distortions at the tip or base of young leaves making up bud	Calcium	Young leaves making up terminal bud first typically hooked, then die back at tips and margins so that later growth of such leaves shows a cut-out appearance at tips and margins. Stalk finally dies back at terminal bud. Roots short, much branched, dark brown in color with more or less decomposition
			Boron	Young leaves making up terminal bud first light green at base, then more or less break-down takes place at base of young leaf, and if later growth follows, leaf shows twisted growth. Stalk finally dies back at terminal bud. Roots show many short laterals, brown color with some decomposition

* Symptoms under these conditions are similar to those observed on the tobacco plant in the field, except for iron and manganese deficiency which have not been seen on tobacco in the field.

Table 115

KEY TO NUTRIENT DEFICIENCY SYMPTOMS FOR GREENHOUSE FLOWERING PLANTS

A. Effects general on whole plant or localized on older, lower leaves.
   I. Effects usually general on whole plant, although often manifested by yellowing and dying of older leaves.
      a. Foliage light green. Growth stunted, stalks slender, and few new breaks. Leaves small, lower ones lighter yellow than upper. Yellowing followed by a drying to a light brown color, usually little dropping. **Minus nitrogen.**
      b. Foliage dark green. Retarded growth. Lower leaves sometimes yellow between veins but more often purplish, particularly on petiole. Leaves dropping early. **Minus phosphorus.**
   II. Effects usually local on older, lower leaves.
      a. Lower leaves mottled, usually with necrotic areas near tip and margins. Yellowing beginning at margin and continuing toward center. Margins later becoming brown and curving under and older leaves dropping. **Minus potassium.**
      b. Lower leaves chlorotic and usually necrotic in late stages. Chlorosis between the veins, veins normal green. Leaf margins curling upward or downward or developing a puckering effect. Necrosis developing between the veins very suddenly, usually within 24 hours. **Minus magnesium.**

B. Effects localized on new leaves.
   I. Terminal bud remaining alive.
      a. Leaves chlorotic between the veins; veins remaining green.
         1. Necrotic spots usually absent. In extreme cases necrosis of margins and tip of leaf, sometimes extending inward, developing large areas. Larger veins only remaining green. **Minus iron.** Note: Certain cultural factors, such as high pH, overwatering, low temperature, and nematodes on roots, may cause identical symptoms. Nevertheless, the symptoms are still probably due to iron deficiency in the plant owing to unavailability of iron caused by these factors.
         2. Necrotic spots usually present and scattered over the leaf surface. Checkered or finely netted effect produced by even the smallest veins remaining green. Poor bloom, both size and color. **Minus manganese.**
      b. Leaves light green, veins lighter than adjoining interveinal areas. Some necrotic spots. Little or no drying of older leaves. **Minus sulfur.**
   II. Terminal bud usually dead.
      a. Necrosis at tip and margin of young leaves. Young leaves often definitely hooked at tip. Death of roots actually preceding all the above symptoms. **Minus calcium.**
      b. Breakdown at base of young leaves. Stems and petioles brittle. Death of roots, particularly the meristematic tips. **Minus boron.**

In general the legumes contain a higher percentage of nitrogen than nonleguminous plants. This appears to be due not only to the fact that they secure the greater part of their nitrogen indirectly from the unlimited supply in the soil atmosphere, but also because their roots seem to have a favorable influence on nitrification and can probably secure from the organic matter of the soil larger amounts of nitrogen than can the non-

Table 116

KEY TO PLANT-NUTRIENT DEFICIENCY SYMPTOMS IN DECIDUOUS FRUIT TREES

	Element Deficient
A. Symptoms which, in the early stages of the deficiency, are general on the whole tree or tend to be localized in the older leaves of the current season's growth.	
B. Symptoms rather generally distributed but usually most prominent on lower leaves of current season's growth. No areas of dead tissue developing on foliage except in advanced and severe cases.	
C. Leaves yellowish green; discoloration begins on old, mature leaves and progresses toward the tip. Accumulations of reddish and purplish-red pigments usually obvious. Under prolonged deficiency, twigs hard and slender and foliage small.	**Nitrogen**
C. Young and nearly matured leaves dark green, while mature leaves are bronzed or ochre-dark green. Old leaves usually mottled with light-green areas developing between dark-green veins. Stems and leaf-stalks developing abnormal amounts of purplish pigment, especially during cool summer weather. As the deficiency continues, new twigs are slender and leaves small (apple) or strapshaped (peach).	**Phosphorus**
B. Symptoms appearing first on mature or lower portion of current season's growth and occurring as mottling or chlorosis, with or without spots, blotches, marginal scorching, or other necrotic (dead) areas in foliage.	
C. Foliage necrosis (death of tissues) varying in size from very small spots or dots to patches or extensive marginal areas. Foliage, especially of peach, usually crinkled. Necrotic areas developing first on mature leaves near the middle or lower half of the current season's growth. Twigs usually slender	**Potassium**
C. Foliage necrosis occurring as fawn-colored patches on most mature, large leaves. Affected leaves dropping progressively toward the tips of the current season's twigs. Severe defoliation common, leaving a tuft or rosette of thin, dark-green leaves at the terminals.	**Magnesium**
C. Foliage small, narrow, more or less crinkled, and chlorotic at tips of new growth; twigs slender, with very short internodes near tips, producing rosettes of leaves. Defoliation progressive from bases to tips of twigs.	**Zinc**
A. Symptoms appearing first on young tissues and tending to be localized at the terminals of twigs.	
B. Twigs dying back from terminals. Newly developing or nearly mature leaves showing severe necrosis.	
C. Immature leaves, especially those at terminals, dying back from tips and margins or along the midribs. Following severe injury to foliage at terminals, twigs die back for an indefinite distance. These symptoms are always associated with injury to root tips.	**Calcium**
C. Leaves more or less chlorotic and wrinkled; sometimes abnormally thick and brittle. In severe cases, there is dieback of twigs and spurs. With bearing trees, deficiency is associated with necrotic areas in the flesh or on the surface of fruits even when no abnormal growth is apparent in vegetative parts. . . .	**Boron**

FIG. 134. Potash irons out the tobacco leaf, gives it a smooth surface and improved texture. (*Left*) No potash; (*center*) 4 per cent potash; (*right*) 16 per cent potash. (Courtesy, Better Crops and American Potash Institute.)

leguminous plants. There are, however, nonlegumes that may contain as high a percentage of nitrogen as the legumes. This is true of rape and some of the common weeds, such as the ox-eye daisy and purslane.

Cooper (1930) called attention to the fact that when certain legumes are grown in soils where liberal amounts of available phosphorus and nitrogen are present a negative correlation may exist between the amounts of these materials present in the plant. As both nitrogen and phosphorus are in the same family of elements, they would be expected to have somewhat similar functions to perform in the nutrition of organisms. An application of nitrogen may often reduce the absorption of phosphorus. This has been shown by Archibald and Nelson (1929). They have reported the phosphorus content of pasture plants to be 10 per cent higher on unfertilized check plats than on plats liberally fertilized with a complete fertilizer and calcium nitrate.

The permanent grass plats of the Rothamsted Experiment Station show that leguminous pasture plants respond much more readily to phosphoric acid and potash than do the pasture grasses. In fact, in these tests the application of nitrogenous fertilizers over a long period of time has resulted in a curtailment in the growth of legumes. Other Rothamsted results show that swedes are very responsive to phosphates and mangels to potash. The yields of these crops may be used as a measure of the quantities of available phosphate and potash in the soil.

It appears that the feeding power of plants for potash is related somewhat to their rate of growth. Plants that grow slowly can probably utilize the less soluble forms of potash.

Compound Deficiency Symptoms

Fortunately there appears to be little or no blending of the plant nutrient deficiency symptoms. Where there is more than one nutrient deficient the plant will usually show the deficiency symptoms of the first limiting nutrient to dominate the rest. When the plant's need for the first limiting nutrient is satisfied through fertilization then the deficiency symptoms of the second limiting nutrient become dominant and so on. This characteristic of plants simplifies the problem of diagnosis and treatment.

Estimating Fertilizer Needs]

Often the appearance of the growing crop will in a general way indicate its nutrient needs. For instance, if it is noted that there is an early and full development of seeds, there is little doubt but that the soil contains a satisfactory amount of available phosphoric acid. Likewise, a late and imperfect development of seed and fruit, especially of plants that have had a dark green color with a tendency to develop purplish coloration, may indicate the lack of sufficient soluble phosphoric acid.

When the leaves of crops present a normal green color and the stems and leaves are well developed, it may be assumed that the soil has sufficient nitrogen. If there is a rank and excessive growth of leaf and stem it may again be assumed that the available nitrogen is excessive as compared with the available phosphoric acid and potash. This excessive vegetative growth, due to an oversupply of nitrogen, is often accompanied in some plants by the imperfect development of flower buds.

A deficiency of potash often manifests itself on sandy soils and sometimes on clay soils. In the plant a deficiency in potash is shown by a characteristic chlorosis of the leaves. The edges of the leaf turn yellow, roll, and die, and with many crops the dead tissue falls away from the leaf, leaving a ragged edge.

The need for magnesium also shows itself in crops by a characteristic coloring of the leaf. The veins remain green but the rest of the lamina turns yellow, red, or purple, depending upon the crop.

In general it might be said that when a farmer finds that his crop yields are not satisfactory, and that water has not been a limiting factor to crop growth, he should make an effort to determine whether or not some nutrient element is lacking in his soil. For this purpose plant symptoms of nutrient deficiency in conjunction with soil characteristics should be studied. A consideration of the following soil and plant characteristics may help in determining what nutrients are deficient.

Where nitrogen may be deficient:

1. If the soil is grayish in color.
2. If the soil is sandy.
3. If the soil is deficient in organic matter.
4. If the soil is extremely acid.
5. If the soil has been supplied with organic material which has a wide nitrogen-carbon ratio.
6. If the top-soil has been removed by erosion.
7. If leguminous green manure has not been turned under regulariy.
8. If the land has been burned over at frequent intervals.
9. If plants appear stunted and have pale green or yellowish leaves.
10. If the leaves of corn "fire."
11. If plants mature prematurely.
12. If leaves of trees are shed early.

Where nitrogen is usually abundant:

1. If soils contain plenty of organic matter.
2. If plants mature slowly.
3. If plants have a healthy green color and are making a good growth.
4. If small grain lodges badly.

Where phosphoric acid may be deficient:

1. If the soil has a grayish color.
2. If the soil has a fine texture.
3. If the soil is poorly drained.
4. If the soil is extremely acid.
5. If the soil is low in organic matter.
6. If the growth of small grain has been good but the yield has been poor.
7. If plants appear stunted where nitrogen and potash have been applied.
8. If tomato or corn leaves have a purplish color.
9. If pastures are devoid of clovers.

Where phosphoric acid is usually abundant:

1. If plants develop good root systems.

2. If plants mature at the proper time.
3. If the ratio of straw to grain is normal.
4. If plants are making normal growth.

Where potash may be deficient:

1. If the soil has a grayish color.
2. If the soil is sandy.
3. If the soil is peat or muck.
4. If the soil is of limestone origin.
5. If corn plants fall badly.
6. If plant leaves show symptoms of potash deficiency.

Where potash is usually abundant:

1. If plants have good root systems.
2. If plant growth has been normal.
3. If plants have vigorous appearance with vigorous stems and branches.
4. If plant leaves do not show symptoms of potash deficiency.

It might be thought that the above considerations would be a rather crude way to estimate plant nutrient deficiencies. Nevertheless, it should be kept in mind that fertilizer recommendations and opinions as to soil adaptability for a given crop are guesses at best, and that any information that will increase the validity of the guess is worthy of consideration.

Choice of Nitrogenous Fertilizers

Nitrogen is the most expensive plant nutrient which the farmer has to buy, although there is a considerable variation in the price the farmer has to pay for nitrogen in the various sources available to him. The wholesale prices of these different forms vary at the present time from about 10 cents to 30 cents a pound, so that it is essential that a farmer give careful consideration to the efficiency of the various forms of nitrogen in producing crops. Fortunately for the farmer this problem is not a new one and much experimental evidence is already available to aid him in his selection.

The choice of a nitrogenous fertilizer, like the selected method and time of application, must necessarily depend upon many factors. Four major forms of nitrogenous fertilizers must be recognized; the water-insoluble proteid organics, the nitrates, the ammoniacal forms, and the soluble organic nonproteid forms. The water-insoluble organic sources of nitrogen do not readily leach from the soil, and do not increase soil acidity. They are low in total availability and are expensive. The nitrates are completely available and in the form of calcium and sodium nitrate tend to reduce soil acidity. Following application they are subject to leaching and are usually best applied in split applications. The ammoniacals are completely available to crops and are not easily leached from the soil. They are notorious for increasing soil acidity, but this acidity may be corrected with limestone. When properly used they are excellent forms of nitrogen to apply at planting. The nonproteid organic, such as urea, is completely available, is resistant to leaching, and only slightly increases soil acidity. It is an excellent form of nitrogen to use in complete fertilizer, and its efficiency is high, especially when used in conjunction with a little limestone. The use of the nonproteid organic calcium cyanamide has already been discussed.

As stated above, nitrogen in the nitrate form readily leaches from the soil. The reason for this lies in the fact that the nitrate ion is negatively charged and therefore may not be held by the soil complex. When the nitrate ion is leached from the soil it carries with it valuable soil bases such as calcium, magnesium, and potassium. Nevertheless, nitrates are ideal carriers of nitrogen where immediate availability is desired. In order to overcome in part their objectionable leaching qualities nitrates are most commonly used, as side-dressing and top-dressing and for immediate consumption of winter growing crops. Nitrates appear to give best results when applied to vegetables, or to short-season crops, or when applied during cold backward seasons. Ammoniacal nitrogen usually is cheaper than nitrate nitrogen.

All of the nitrogenous fertilizers not of plant or animal origin, with the exception of the nitrates and calcium cyanamide, are physiologically acid and should be used with lime on acid soils for acid intolerant crops. It appears that plants that are acid-tolerant are better able to utilize nitrogen in the form of ammonia than when it is in the nitrate form. Evidence is at hand which indicates that the presence in the soil of both nitrate and ammoniacal nitrogen may satisfy the requirements of many plants better than does the presence of only one form of nitrogen.

In general the application of nitrogen to pastures appears to discourage, for a while at least, the growth of legumes. Field observations show that plants started with a minimum of nitrogen usually develop an extensive root system.

Only under exceptional conditions will plants recover more than 50 per cent of the nitrogen applied the first year after application. As the soil micelle readily absorb ammonia, ammoniacal forms of nitrogen may be put under the crop without fear of loss by leaching, provided extremely heavy applications are not made. It must be remembered, however, that the ammoniacal nitrogen compounds usually nitrify to nitrates before they are absorbed by many field crops and that if the resulting nitrates are not absorbed by plants or microörganisms they may be lost by leaching.

Choice of Phosphatic Fertilizers

In general, it appears that the efficiency of phosphatic fertilizers is reduced when they are applied to very acid or very alkaline soils. Salter and Barnes (1935) found that the more acid phosphates, monoammonium and monocalcium, tended to increase in efficiency with increasing pH value of the soil. They found also that the availability of low-ammoniated superphosphate changes very little with change in soil pH value, whereas materials containing phosphorus in the form of tricalcium phosphate—either as such (highly ammoniated superphosphates) or in carbon or fluoride combinations (bone meal and phosphate rock, respectively)—showed decreasing efficiency with increasing pH and were not well adapted for use on neutral or alkaline soils. The efficiency of slag appeared to occupy an intermediate position between that of ammoniated superphosphates and phosphate rock.

The relative availability of eight phosphates for tomatoes, potatoes, and lima beans when grown in sand, as reported by Hester (1940), is given in the following table.

Table 117

AVAILABILITY OF PHOSPHATE COMPOUNDS TO PLANTS GROWN IN SAND*

Compounds	*Tomatoes*	*Potatoes*	*Lima Beans*
Iron phosphate	1	14	52
Aluminum phosphate	24	90	98
Manganese phosphate	83	122	0
Tri-calcium phosphate	145	110	87
Tri-magnesium phosphate	301	320	319
Di-calcium phosphate	296	302	307
Mono-calcium phosphate	320	307	371
Rock phosphate	5	31	33

* Milligrams P_2O_5 absorbed by plants from equal amounts of P_2O_5 applied as the above compounds.

DeTurk (1941) called attention to the fact that most legumes have the power to get phosphorus from soil sources which are too slowly available to supply grasses.

Choice of Potash Fertilizers

In view of the fact that most potash fertilizers appear on the market only in the chloride and sulfate forms the farmer has little choice. Muriate of potash, manure salts, and kainite contain potash in the chloride form, while potassium sulfate, of course, is in the sulfate form. The relative value of sulfates and chlorides has been discussed in a previous chapter. The real basis of choice of potash fertilizers is dependent upon the relative cost per unit of the potash found in the fertilizer.

Influence of Season on Nutrient Availability

Because soil potassium and sulfur compounds are only slightly soluble in the soil solution plant deficiencies of these elements are more likely to be noted during dry seasons. The same may be said concerning boron. On the other hand, because magnesium and manganese compounds are highly soluble their deficiency symptoms in plants are more likely to appear during wet seasons. Many plant nutrient deficiency symptoms may not appear during the winter months when the plants are growing slowly, whereas they would appear readily during the summer months when the plants are growing rapidly.

Choice of High- or Ordinary-analysis Fertilizers

As has been explained elsewhere, following World War I many new high-analysis and concentrated fertilizers came on the market. These materials made it possible for the manufacturer of fertilizer to make mixed goods of a very high analysis. Complete fertilizers of high analysis were recommended largely on the basis of the financial saving that would accrue from their use. When the high-analysis mixed goods failed to sell as readily as mixed goods of ordinary analysis, the manufacturers diluted them with filler and made ordinary-analysis goods that would sell. This has resulted during the last 25 years in about a 65 per cent increase in the use of filler in the average mixed fertilizer as compared with about a 20 per cent increase in total plant food. Mehring (1936) estimated in 1934 that 17.8 per cent of the total fertilizer tonnage in seven southeastern states was made up of filler, and that this filler in an average ton of mixed fertilizer cost the consumer $2.21 a ton, or a total cost to the farmers of these states of $5,482,000. Also, Bledsoe (1937) has estimated that the fertilizer manufacturer pays only about $2.00 a ton for fertilizer filler but that owing to certain fixed overhead charges such as cost of bags, labor, tax, and freight the farmer must pay $11.00 to $14.00 a ton for the filler.

Early results with the high-analysis complete fertilizers showed them to be somewhat less efficient than the standard strength fertilizers. This difference in favor of the standard strength fertilizers often has been found to be due to their source of nitrogen, or to their content of secondary elements such as calcium, magnesium, and sulfur, or to their content of the rarer elements manganese, zinc, copper, boron, and molybdenum.

More recently very extensive tests have been conducted by Skinner, Mann, Collins, Batten, and Bledsoe (1937) and others. They concluded that the differences due to the presence of the rarer elements were unimportant and that the sources of nitrogen did not vary widely in their effect on cotton yields where the fertilizers were properly supplemented with limestone. This was in line with the conclusions reached by Schreiner and Skinner in 1934 to the effect that inorganic sources of nitrogen, properly formulated, may be as effective for cotton production as higher priced organic nitrogen.

Ross and Mehring (1934, 1935, 1936) and others have called attention

FIG. 135. Influence of potash on the darkening of potatoes. (*Lower left*) Potatoes grown in potash-deficient soil. (*Upper and right groups*) Potatoes grown in soil having sufficient potash. (Courtesy, Wisconsin Agricultural Experiment Station.)

to the fact that double-strength fertilizers may be purchased at a saving of about 20 per cent per unit of plant food over those of single-strength, and Skinner (1938) estimated that potash in 60 per cent muriate could be bought at a 40 per cent saving over that of 20 per cent kainite. In 1940, the plant food in a 5–10–5 could be bought at a saving of about 8 per cent over that of a 4–8–4. It thus seems clear that an extended use of the high-analysis fertilizers that are properly supplemented should give excellent results and at the same time greatly reduce the cost of crop fertilization. All plant food elements in complete fertilizers should be completely available to the crop, and at the same time, as far as possible, resistant to loss by leaching.

It should not be overlooked, however, as has been pointed out by Volk *et al.* (1945), that the use of high-analysis fertilizers will hasten the development of nutritional problems involving calcium, magnesium, sulfur, acidity, and the rarer elements.

Effect of Fertilizer Grade on Salt Content of Soil Solution

It has long been recognized that the soluble fertilizer materials had to be applied properly if injury to plants, arising from too great a concentration around the roots, was to be avoided. Furthermore, it has long been known that among the commonly used fertilizer materials the least injury usually results from the phosphates, and the greatest injury from sodium nitrate and the potash salts.

The introduction on the market of higher grades than were formerly used was objected to by some because it was felt that they would have to be used with greater care if injury were to be avoided. This assumption, however, is not justified, for White and Ross (1939) have shown that damage of salt injury from typical present-day mixtures is less than that from mixtures formerly used, even when the nitrogen in the mixtures is applied to the crop in split applications. Furthermore, salt injury is not necessarily correlated with the plant food content of the fertilizer.

Necessity for Rarer Elements in Mixed Fertilizers

The rarer elements such as manganese, copper, zinc, boron, and molybdenum are sometimes incorporated in fertilizer mixtures by the manufacturers, either because they are convinced that these elements are a valuable addition to the fertilizer, or because they feel that the addition of these elements will aid in the sale of their brands, whether or not the elements added are needed in the territory they serve. Manganese, copper,

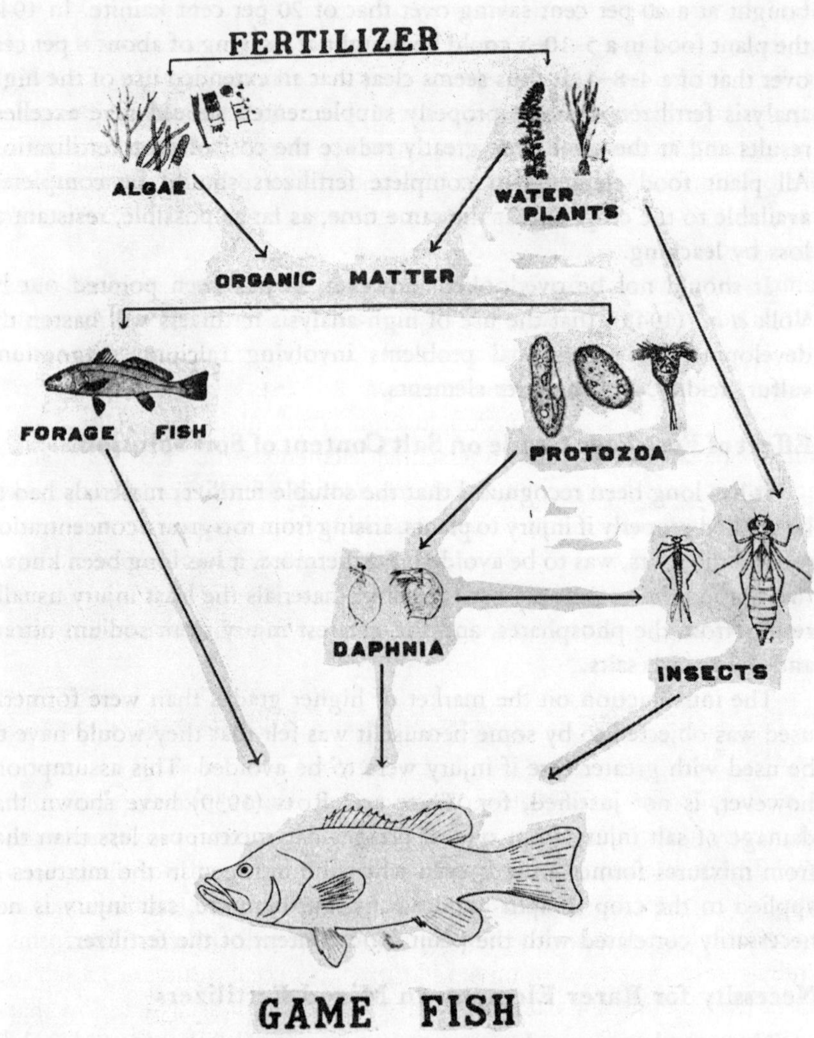

FIG. 136. Showing how fertilizer may increase the yield of fish in fertilized fish ponds. (Courtesy, Mississippi Agricultural Experiment Station.)

and zinc are usually added as sulfates and boron as a sodium salt. Objection cannot be made to the proper use of the rarer elements, for the advantage of adding manganese sulfate to the Florida everglade peats for tomatoes and beans has been demonstrated by Skinner and Ruprecht (1930), and the advantage of adding copper sulfate to peat for the purpose of correcting "rabbit's ear" of lettuce and poor coloration of onions has been amply shown by Felix (1927) and Knot (1933), respectively. Also, Barnette *et al.* (1936) have shown that zinc sulfate must often be added to the sandy soils of Florida for the control of "white bud" of corn and, according to Schreiner (1931), boron is now often successfully included in the fertilizer mixtures used under tobacco in Sumatra. Nevertheless, farmers are confronted with the fact that many fertilizer companies are advertising that their mixed goods carry the rarer elements. This is probably profitable sales advertising for the manufacturers and costs little, but it is very misleading to farmers for, with the exception of the soils of Florida and perhaps other areas in the Altantic and Gulf costal plain, few soils of the United States are deficient in essential elements other than nitrogen, phosphorus, potassium, calcium, sulfur, and magnesium. Furthermore, the indiscriminate use of the rarer elements may result in serious crop injury. However, investigational work with boron has now progressed to such a point, and boron deficiency symptoms in plants have been shown to be so widespread that it is very likely boron may soon be a common constituent of complete fertilizers.

Reducing Fertilizer Cost Through the Elimination of Organics

Years ago the wholesale commercial prices of nitrate, ammonia, and organic nitrogen of plant and animal sources were approximately the same. Today, however, there is a wide variation in the cost of the various forms of nitrogen, and the amounts of each used in mixed fertilizers has an important influence on the cost of commercial mixed fertilizers. The forms of nitrogen incorporated in mixed fertilizers in recent years are shown in Table 118. Experimental data which have already been discussed show that these three forms of nitrogen will produce similar crop yields provided the acid-forming nitrogen carriers are supplemented with sufficient lime to correct the acidity resulting from their use.

In view of the above facts a considerable saving may be made through the purchase of mixed fertilizers which contain the cheaper forms of soluble nitrogen. This may necessitate, however, an adjustment by the farmer in the rate or method of application to conform to the characteristics of the materials used to prevent loss of nitrogen from the soil before it is utilized by his crops.

Table 118

FORMS OF NITROGEN CONTAINED IN MIXED FERTILIZERS IN SPECIFIED YEARS

Year	*Ammoniacal, Per Cent*	*Nitrate, Per Cent*	*Organic, Per Cent*	*Amide, Per Cent*
1880......	7.83	3.91	88.26	..
1885......	17.50	5.00	77.50	..
1890......	14.41	7.63	77.96	..
1895......	7.50	17.92	74.58	..
1900......	5.74	23.44	70.82	..
1905......	10.19	25.00	64.81	..
1910......	26.32	18.42	55.26	..
1915......	24.54	23.15	49.07	3.24
1920......	24.55	29.55	37.26	8.64
1925......	38.13	21.94	30.94	8.99
1930......	54.71	12.77	24.62	7.90
1931......	57.44	11.61	23.51	7.44
1932......	56.33	14.02	18.33	11.32

Reduction of Fertilizer Costs Through the Elimination of Side Applications

In recent years it has been advocated that advantage be taken of the slow leaching and low price of urea and ammoniacal compounds and that they be substituted entirely for the commonly used higher priced nitrates. Such a change in practice, in the light of present experimental evidence, has much to commend it.

Chemical Tests for Availability

The availability to plants of nutrients in commercial fertilizers varies with the plant to which the fertilizer is applied and to an even greater extent with the soil in which the plant is grown. Because of this it approaches the impossible for chemists to devise a standard chemical test which will measure exactly the availability of a fertilizer for a given crop. Nevertheless, it is necessary that agricultural chemists use similar or standard methods in analyzing soils and fertilizer materials that are offered for sale on the market. Such methods have been developed and adopted as standard by the Association of Official Agricultural Chemists and are in common use.

The availability of nitrogen in commercial fertilizers is expressed usually in terms of that portion soluble in water and that portion insoluble in water. The water-soluble portion is largely nitrates and ammoniacal compounds. The water-insoluble portion is usually organic compounds of plant or animal origin. As explained in a previous chapter, the proteid-organic portions are subjected to tests of quality through the use of neutral or alkaline permanganate solutions.

Tests for available phosphoric acid are, in the case of the superphosphates, the determination of that portion soluble in water and in a neutral solution of ammonium citrate. For basic slag the test is for that portion soluble in a 2 per cent citric acid solution. For bone, the total phosphoric acid content is determined.

The test for the availability of the potash salts is their solubility in water.

Systems of Fertilization

Many investigators have attempted so to standardize fertilizer practices that definite and specific systems of fertilization can be advocated for any and all conditions. Most attempts of this nature have been only partially successful largely because of the great variation in conditions presented, but systems such as those proposed by Wagner, Ville, Hopkins, and others are worthy of consideration by the fertilizer student.

The Wagner System

The Wagner system is named after the German scientist, Wagner, who suggested it. He urged that as phosphoric acid and potash are readily fixed in the soil, and therefore not subject to easy leaching, these nutrients be added in sufficient amounts to meet the maximum needs of any crop grown. Also, that nitrogen be applied in an active form, but that care be exercised to apply nitrogen in such amounts and at such times as to insure a minimum loss of the element through leaching and a maximum growth of the crop.

The Ville System

The Ville system has received more popular recognition than the Wagner system. It was first advocated by the French scientist, George Ville (1909) about 1876. He maintained that artificial manures were more remunerative than barnyard manure, and that their application was the only way to maintain the productivity of the soil. His system is based upon the assumption that field crops may be divided into four groups according to their response to the four elements, nitrogen, phosphorus, potassium and calcium. Reference has already been made to a similar but more recent classification by Hartwell. According to Ville's theory of "dominant constituents" one group of plants is supposed to be benefited particularly by nitrogen, another by phosphoric acid, another by potash, and still another by lime. In other words, each group is supposed to demand one element in excess of the other three. According to this system phosphoric acid is considered the dominant element for corn, sorghum, sugar cane, cotton, and turnips; nitrogen is considered dominant for wheat, oats, rye, barley, beets, and most of the pasture grasses, as well as fruit trees; potash is considered dominant for potatoes, clover, and tobacco; and lime is considered dominant for the legumes. According to the Ville system, any complete fertilizer to be applied to a crop should contain in largest amount the nutrient to which the crop primarily responds.

The Hopkins System

In the system that was advocated by the late Dr. Cyril G. Hopkins, of the Illinois Experiment Station, and which is sometimes called "The Illinois System," it was proposed that the organic matter and nitrogen content of the soil be maintained by turning under sufficient quantities of leguminous crops and plant residues where the grain system of farming is followed, and that this practice be supplemented with animal manures where the livestock system of farming is followed; that phosphoric acid be applied in such cheap carriers as ground phosphate rock, and that to soils deficient in potash sufficient amounts of soluble potash fertilizer be applied to satisfy the crop needs. The system presupposes that the green-manuring crops that are incorporated in the soil will be the means of making available the insoluble phosphate rock that is added as well as the native insoluble phosphates and potashes that are present in most soils. Attention is given also to the liming of the soil. This system has been the means of securing excellent results in the building up and the maintaining of the fertility of the soil, both in experiment station tests and in general farm practice in the state of Illinois.

Fertilizing System Based on the Composition of Plant Ash

For many years younger chemists particularly have leaned to the view that the ideal system of fertilization would be based upon crop analysis.

Table 119

NUTRIENT CONTENT OF SOME FIELD AND HORTICULTURAL CROPS

Crop	*Yield*	*Part of Crop*	*N (lbs.)*	*P_2O_5 (lbs.)*	*K_2O (lbs.)*	*Totals*
Cotton	500 lbs.	lint }				
	1,000 lbs.	seed }	38	18	14	70
	1,500 lbs.	burrs, leaves, stalks	27	7	36	70
		Total	**65**	**25**	**50**	**140**
Tobacco	1,500 lbs.	leaves	55	10	80	145
		stalks	25	10	35	70
		Total	**80**	**20**	**115**	**215**
Corn	60 bu.	grain	57	23	15	95
	2 tons	stover	38	12	55	105
		Total	**95**	**35**	**70**	**200**
Wheat	30 bu.	grain	35	16	9	60
	1.25 tons	straw	15	4	21	40
		Total	**50**	**20**	**30**	**100**
Oats	50 bu.	grain	35	15	10	60
	1.25 tons	straw	15	5	35	55
		Total	**50**	**20**	**45**	**115**
Barley	40 bu.	grain	35	15	10	60
	1 ton	straw	15	5	30	50
		Total	**50**	**20**	**40**	**110**
Potatoes	300 bu.	tubers	65	25	115	205
		tops	60	10	55	125
		Total	**125**	**35**	**170**	**330**
Sweet potatoes	300 bu.	roots	45	15	75	135
		vines	30	5	40	75
		Total	**75**	**20**	**115**	**210**
Sugar beets	15 tons	roots	55	22	53	130
		tops	60	23	92	175
		Total	**115**	**45**	**145**	**305**
Tomatoes	10 tons	fruit	60	20	80	160
		vines	40	15	95	150
		Total	**100**	**35**	**175**	**310**
Cabbage	15 tons	all	100	25	100	225
Celery	350 crates	all	80	65	235	380
Spinach	9 tons or 1,000 bu.	all	90	30	45	165

Table 119—(Continued)

Crop	Yield	Part of Crop	N (lbs.)	P_2O_5 (lbs.)	K_2O (lbs.)	Totals
Apples	400 bu.	fruit	20	7	30	57
		leaves, wood	10	3	5	18
		Total	**30**	**10**	**35**	**75**
Peaches	500 bu.	fruit	30	15	55	100
		leaves, wood	55	10	45	110
		Total	**85**	**25**	**100**	**210**
Grapes	4 tons	fruit	10	6	20	36
		leaves, canes	15	4	15	34
		Total	**25**	**10**	**35**	**70**
Oranges	600 boxes	fruit	65	23	105	193
		leaves, wood	25	7	25	57
		Total	**90**	**30**	**130**	**250**
Soy beans	25 bu.	grain	110	35	40	185
	1.25 tons	straw	15	5	20	40
		Total	**125**	**40**	**60**	**225**
Peanuts	1 ton	nuts	60	10	10	80
	3 tons	vines	25	5	40	70
		Total	**85**	**15**	**50**	**150**
Pea beans	30 bu.	grain	73	23	24	120
		straw	22	7	31	60
		Total	**95**	**30**	**55**	**180**
Alfalfa	3 tons	all	140	35	135	310
Sweet clover	5 tons	all	185	45	165	395
Red clover	2 tons	all	80	20	70	170
Lespedeza	3 tons	all	130	30	70	230
Cow peas	2 tons	all	125	25	90	240
Timothy	1.5 tons	all	40	15	45	100

Systems of this type have been advocated from time to time. Probably the idea has arisen from a belief that the quantity of plant nutrients to be applied to the soil in the form of fertilizers should be as great as that removed from the soil by crops. Thus, according to such a system, the nitrogen, phosphoric acid, and potash contained in the complete fertilizer which is to be applied to a crop should be present in the same amount and in the same ratio as that found in the crop. Such a system disregards the quantity of nutrients lost from the soil by leaching, as well as the residual effects of previous fertilization. Furthermore, plants vary widely in their composition, and plant composition is not a true index of nutrient absorption. The ash constituents of plants are influenced greatly by the amounts of the available constituents in the soil. Pember (1917) found that, if phosphorus was supplied in excess, barley plants would absorb about three times the quantity of phosphorus required for optimum growth. Also, an increase in the water supply, within given limits, nearly always brings about an increased absorption of the ash constituents, although this is not true for nitrogen. Grain grown on irrigated soils has a higher mineral content and a lower protein content than grain grown without irrigation. The percentage of nitrogen in a crop usually decreases as the yield increases owing both to a natural development whereby a greater relative accumulation of carbohydrate material takes place in the plant as well as to the fact that the supply of available nitrogen in the soil is smaller than could be assimilated by the crops. Vandecaveye (1940) made a very exhaustive study of all reported work on the influence of fertilization on the mineral content of crops. He found that with one exception 29 studies showed that application of phosphate fertilizers affected and increased phosphorus content of herbage.

In Table 119 is given, according to Romaine (1940), the number of pounds of nitrogen, phosphorus, and potassium found in common yields of some field and horticultural crops.

Table 120

VARIATION IN MINERAL CONTENT OF WHEAT GRAIN

Ash Element	*Pounds per Ton of Dried Material*		*Per Cent Variation*
	Minimum	*Maximum*	
Magnesium	2.88	3.06	15
Phosphorus	6.74	8.06	16
Sulfur	4.26	5.16	17
Potassium	8.84	11.22	21
Calcium	0.86	1.14	24
Sodium	2.26	3.14	28
Chlorine	1.46	2.24	34

Leclerc and Breazeale (1908) showed that plants exude salts from the surface of their leaves and rain and dew may wash them back to the soil.

Forbes, Beegle and Mensching (1913) analyzed a number of wheat plants which were selected from fertilizer test plats of the Ohio Experiment Station. They found a variation of 15 to 34 per cent in the content of some ash constituents. Their results are given in Table 120. It is well to recall that probably not over 10 per cent of the ash elements in a plant may be considered a part of the permanent plant tissue. About 90 per cent is found in the cell vacuoles.

Cash Crop System of Fertilization

The cash crop system of fertilization is used to a greater extent than any other system, particularly in the older agricultural sections where fertilizers have been used for a long time. This system proposes that the fertilizer be applied in such liberal amounts to the money crop of the rotation as to assure maximum yield, and that the residue be utilized by the succeeding crops which are usually grown to be fed on the farm. This is a common method employed in the South. The extent to which this system is employed is indicated by the estimation that about 20 per cent of all the fertilizer used in the United States is applied to cotton. The chief objection to this system is that succeeding crops may not receive sufficient nutrients for their needs, and that large amounts of soluble nutrients may be leached from the soil before a succeeding crop has a chance to utilize it. If the crops following the cash crops do not, or cannot, utilize the fertilizer residue, such a system becomes uneconomical.

Availability of Commercial Fertilizer Salts

The availability of fertilizer salts may be influenced by many factors, but it is probably determined largely by the solubility of the salts in question, and the strength of the ions formed in the soil from the ionization of the salts. In Table 121 is given the solubility of some of the commonly used fertilizer salts.

It has long been recognized that soluble salts do not normally move into a plant merely in relation to the water absorbed. On the other hand, absorption of nutrients by plants should not be thought of simply as the passing of a substance through a membrane according to the chemical laws of mass action and equilibrium but rather as a complex phenomenon in which living cells are involved. As a matter of fact, a greater proportion of ions is absorbed by plants from a dilute solution than from a more concentrated one. Breazeale (1923) considers absorption an electrical phenomenon in that plant roots absorb nutrients in the form of mobile ions carrying charges which satisfy opposite charged colloids within the plant.

Table 121

SOLUBILITY IN WATER OF SOME COMMONLY USED FERTILIZER SALTS

Substance	*Formula*	*Solubility in 100 Parts at 0° C.*
Ammonium chloride	NH_4Cl	29.4
Ammonium nitrate	NH_4NO_3	118.0
Ammonium sulfate	$(NH_4)_2SO_4$	71.0
Calcium carbonate	$CaCO_3$	0.0013
Calcium nitrate	$Ca(NO_3)_2$	93.1
Calcium sulfate	$CaSO_4.2H_2O$	0.223
Diammonium phosphate	$(NH_4)_2HPO_4$	62.8*
Dicalcium phosphate	$Ca_2H_2(PO_4)_2.2H_2O$	0.028
Manganese sulfate	$MnSO_4.7H_2O$	76.9
Magnesium carbonate	$MgCO_3$	0.0106
Monoammonium phosphate	$NH_4H_2PO_4$	19.9
Monocalcium phosphate	$CaH_4(PO_4)_2.H_2O$	4.0†
Potassium chloride	KCl	34.0‡
Potassium nitrate	KNO_3	13.3
Potassium sulfate	K_2SO_4	11.11‡
Sodium nitrate	$NaNO_3$	87.7‡
Tricalcium phosphate	$Ca_3(PO_4)_2$	0.0013

* 10° C.
† 15° C.
‡ 20° C.

The order of absorption of ash constituents for certain plants appears to be the same as that found in the electromotive series. Apparently many plants selectively absorb the stronger ions. Thus, the ash of nearly all grains which are grown on fertile soils show a higher percentage of potash than any other nutrient. Cooper (1930) has shown that samples of pasture grasses grown on productive soils contained relatively large amounts of the elements which form strong ions, while the grasses taken from unproductive soils contained a relatively low ash content and were very high in elements which form weak ions. The electromotive series for the principal cations and anions with which plants are largely concerned, as shown by Cooper (1930), is given in Table 122. The ions are arranged in order of strength from top to bottom of the table.

It appears that plants that are capable of tolerating a low intensity and quality of light are generally capable of utilization of large amounts of

FIG. 137. Diagnosing the nutrient status of corn by testing the tissues. The box on the ground is the Purdue test kit, the box on the tripod is a convenient outdoor work table. The compartments are handy for carrying the clean and used vials. (Courtesy, Dr. George D. Scarseth, Dr. Mack Drake, and Better Crops with Plant Food.)

weak electrolytes such as compounds of silicon and iron, and may contain a relatively high percentage of these elements.

Hoagland (1917) concluded that alkalinity is decidedly more harmful to plants than acidity. Apparently hydrogen is necessary in the colloidal complex of the soil, or at least helpful in maintaining the mobility of other cations. Gregory, (1928) studied mathematically the general effect of

Table 122

STANDARD ELECTRODE POTENTIALS OF MATERIALS COMMONLY FOUND IN SOILS

Cations (+) Ion	*Charge*	*E.M.F. in Volts*	*Anions (−) Ion*	*Charge*	*E.M.F. in Volts*
Potassium	K^+	−2.922	Fluorine	F^-	+1.960
Sodium	Na^+	−2.712	Sulfate	SO_4^{--}	+1.900
Calcium	Ca^{++}	−2.500	Phosphate	$H_2PO_4^-$	+1.700*
Magnesium	Mg^{++}	−1.500	Nitrate	NO_3^-	+1.690*
Aluminum	Al^{+++}	−1.300	Chlorine	Cl^-	+1.350
Manganese	Mn^{++}	−1.100	Bromine	B^-	+1.080
Zinc	Zn^{++}	−0.760	Iodine	I^-	+0.550
Iron	Fe^{++}	−0.430	Hydroxyl	OH^-	+0.410
Hydrogen	H^+	0.000			
Copper	Cu^{++}	+0.170			

* Discharge potential of acid.

cations and anions on plant growth and concluded that the cations have a far greater effect than anions on growth processes. As absorption of plant nutrients from the soil takes place in the molecular form, it seems unreasonable to believe that cations are superior to anions or vice versa.

Biological Tests for Determining the Fertilizer Needs of Soils and Crops

Among the methods of biological examination of the soil and of chemical examination of the plant that have been developed recently in an attempt to determine the fertilizer needs of soils and crops four have been outstanding: the Neubauer, the Azotobacter, the Hoffer, and the Thornton tests.

The Neubauer test, developed by Neubauer and Schneider (1923), has been applied most intensively in European countries. It consists in growing rye seedlings for a period of 18 days on the sample of soil to be tested, and under standard greenhouse conditions. The plants are then analyzed, the nutrients removed from the soil during the test period are

Table 123

EVALUATION OF AZOTOBACTER PLAQUE TEST BY VARIOUS WORKERS

Worker	*Year*	*Source of Soils*	*Investigators' Appraisal*
Winogradsky	1925–27	France	Devised soil plaque technique
Niklas, Scharrer, and Strobel	1926	Germany	Recommended test; application to phosphate testing not emphasized
Ziemiecka	1929–32	Poland and Rothamsted	Reported test successful
Guittonneau	1929	France	Test confirmed by field tests and chemical methods
Walker and Sullivan	1929	Iowa	Test gave encouraging results
Itano and Arakama	1930	Japan	Test not applicable to rice fields
Kryuchkova	1930	Russia	Reported method successful
Sackett and Stewart	1931	Colorado	Recommended test highly
Stewart, Sackett, Robinson and Kozer	1932	Colorado	Reported Azotobacter test superior to Neubauer and Hoffer cornstalk methods
Dahlberg and Brown	1932	Colorado, Wyoming, Montana, Nebraska	Test gave satisfactory results
Jones	1932	Ontario	Test gave reasonably good correlation with crop returns; minimum phosphate requirements for Azotobacter higher than for growing plants
Pittman and Burnham	1932	Utah	Found test not reliable for soils tested; Azotobacter more responsive to phosphate than most crops
Keller	1932	Germany	Satisfactory results obtained
Greene	1932	Arizona	Correlation with field conditions in only 12% of cases
Young	1933	Utah, Colorado, Iowa	Test of little value for determining phosphate deficiencies in Iowa soils
Joshi and Ayyar	1934	India	Considered minimum phosphate requirements of Azotobacter higher than that of crops
Fuller	1934 1935	Massachusetts	Test indicated all soils tested were phosphate-deficient in spite of fact that all produced good crops

determined, and the results used as an indicator of the fertilizer needs of the soil. In general the Neubauer test gives results comparable to pot and field tests, especially with clay soils, and has proved of some practical value. Also, a reasonable agreement between the Neubauer test and rapid chemical methods of determining availability of phosphorus has been found by Mooers (1938). McGeorge (1944) came to the conclusion that while some modifications of the Neubauer tests should be made, the availability measured with rye seedlings is applicable to many other crops.

Sackett (1930), of the Colorado Experiment Station, developed a bacteriological method for determining soil deficiencies that appears to some extent to be comparable with the Neubauer method. The method, first suggested by Christensen in Denmark, consists of the addition of various fertilizers to a weighed amount of soil to be tested and, after inoculation with *Azotobacter*, the determination of the relative development of colonies as compared with the checks. The cultures showing the greatest bacterial development indicate the best fertilization. Results can be secured by this method in 72 hours. Halversen and Hoge (1942) summarized the appraisals of 17 investigators who have tested the Azotobacter plaque test. Their compilation is given in Table 123. Similar tests in which *Aspergillus niger*, *Cunninghamella* and other molds are used instead of Azotobacter yield similar results. Halversen and Hoge (1942) found also that in Idaho the Azotobacter growth indicated much greater response to phosphate than did field crops. Mooers (1938) found a general agreement between the Neubauer and the fungus methods when a large number of soils were tested by both methods, yet there were frequent and wide disagreements in individual samples. Mehlich, Fred, and Truog (1938) prefer the use of the Cunninghamella species because of its quick response to the supply of available phosphates.

Working with six Ohio soils, Olsen and Shaw (1943) found on comparison that the Mitscherlich, Neubauer, and chemical methods for determining available potassium placed all soils in the same order.

Plant Tissue Tests for Determining the Fertilizer Needs of Plants

The employment of chemical analysis of plant tissue as a guide in fertilizer practice has now reached a stage that would have been unpredictable a few years ago.

In the earlier days many agronomists came to the conclusion that the composition of plants is not a good index of a given soil's need for a particular fertilizer as indicated by field trials. Early investigations on the analysis of plants conducted for the purpose of determining the requirements of a given soil for specific nutrient elements were based on the storage of salts in the mature plant or plant part. A number of reasons are now known why such procedures are unsatisfactory.

Today emphasis is placed upon the internal concentration of nutrients in the leaf and especially upon the changes in levels during different stages of the growth cycle. As a result much headway has been and is being made.

Boynton and Compton (1945), working at Cornell with fruit trees, concluded that the chemical analysis of leaves cannot as yet take the place of careful observation of plant behavior and appearance, of the development of visible leaf or fruit symptons, and of past climatic and management conditions, but that when coupled with such observations the chemical analysis of leaves makes diagnosis possible which neither alone would have permitted.

The iron content does not vary greatly in the green parts of plants and the same can be said of the major essential elements. A tenfold variation of the major elements is rare but in the case of manganese and silicon or, better, aluminum and selenium and other nonessential elements, there may be variations of a thousandfold or even a hundred-thousandfold..

Whereas the quick soil tests tend to indicate something concerning the storage of readily available nutrients, most of the quick plant tissue chemical tests now commonly employed appear to be useful in determining the nutrient deficiencies of plants. These methods do not indicate the quantity of elements that have been built into tissue parts, but only the quantities of nutrients in the cell sap. The quick plant tissue chemical tests have many limitations that should be considered in evaluating their results but they are helpful in obtaining information as to the state of soil fertility.

The use of simple quantitative chemical tests for determining the presence of nitrates, inorganic phosphates, and potassium salts in the tissues of plants will aid greatly in the field diagnosis of symptoms induced by deficiencies of these nutrients. Well-nourished plants should carry reserves of these nutrients in their tissues.

Some rather promising chemical methods of testing plant tissue for soil nutrient deficiencies have been developed at the Indiana Experiment Station by Hoffer, Thornton, and Scarseth (1943); in more recent years these have became known as the Purdue tests. According to the Hoffer (1926) method a test is made for nitrates in the internodal tissue of growing cornstalks. A few drops of diphenylamine is applied to a cut surface of this tissue. The development of a dark blue color indicates that the plant is securing a sufficient supply of nitrogen, whereas a yellowish-green color is associated with a deficient supply. The nodal tissue is tested for the presence of iron by applying a few drops of a 10 per cent aqueous solution of potassium thiocyanate and then a drop or two of dilute hydrochloric acid. A red color indicates the presence of iron, and the presence of iron is supposed to be indicative of a deficiency of potassium for in the absence of potassium, iron is made insoluble by excessive oxidation and accumulates in the plant. This may result from the plant's demand for a stronger cation. If these tests reveal that neither nitrogen nor potassium is deficient, and yet the crop has made but little growth, it is assumed that an available supply of phosphoric acid is lacking. Willis and Piland (1934) have reported that the accumulation of iron in nodal tissue may be prevented by either the application of potash salts or copper sulfate. Hence it seems that excessive iron absorption is not always specific for potash deficiency.

The Thornton test consists of placing a small quantity of finely chopped green plant tissue in a test tube and then adding a small amount of a hydrochloric acid solution of ammonium molybdate and a crystal of stannous chloride, and noting the resulting color. The intensity of the color indicates the amount of inorganic phosphorus extracted from the plant tissue. Plants that are grown in soils deficient in phosphorus show little accumulation of inorganic phosphorus in their tissues. Yellow corn leaves, deficient in phosphorus, often show an abundance of nitrates. Apparently in this case nitrogen cannot be assimilated without an adequate supply of phosphorus. Bray (1945) also has developed a plant tissue test for nitrates.

In the Purdue test now used for potassium the finely divided tissue is shaken with a solution of cobaltnitrate, and after shaking ethyl alcohol is added. Two minutes after adding the alcohol this test is read. The amount of potassium present is indicated by the degree of turbidity. Hoffer (1941) described in detail some satisfactory quick chemical tests for nitrates, phosphates and potash in plant tissue.

Other similar chemical plant tests have been proposed by various investigators and some are now being developed at several experiment stations. The development of chemical plant tests for the determination of nutrient deficiencies offers a promising field for investigation. Ulrich (1943) pointed out that the sensitivity of plant tissue depends upon the part of the plant analyzed, the particular fraction of the nutrient deter-

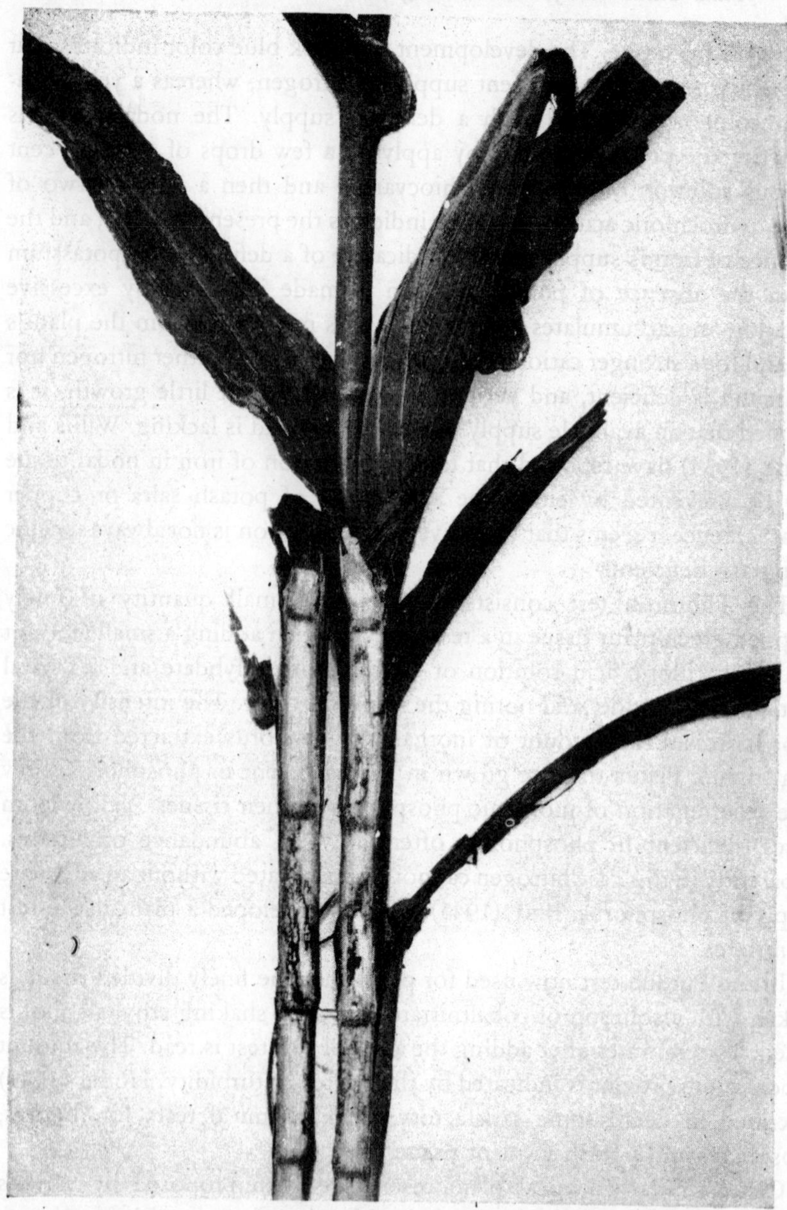

FIG. 138. When chemical tests are used to confirm nutrient deficiency symptoms in the field, they are made on the corn plant tissues as indicated. Tests for potash are made in the tissues in the base of the leaves. Iron accumulations in the joint tissues also indicate potash starvation. Tests for nitrates are made on the inter-nodal tissues throughout the plant, and for phosphates on stalk tissues just below the tassel. (Courtesy, Dr. George D. Scarseth, and Better Crops with Plant Food.)

mined (*e.g.*, nitrate or organic nitrogen, inorganic or organic phosphorus), and the position on the plant from which the sample is selected.

Rapid Chemical Tests Used in Determining Soil Deficiencies

The most popular methods now used for determining the fertilizer needs of soils are the recently developed rapid chemical tests. When these tests are standardized against field results and employed intelligently they do have a diagnostic value but reliability for all crops on all types of soils cannot be claimed for any single test. As Hoagland and Arnon (1941) pointed out, no one chemical criterion can be invoked in appraising the supplying power of all soils for a specific nutrient element. As no laboratory procedure can determine accurately the probable need for nitrogen fertilization, tests for ammonia or nitrates are not generally made. The methods employed for determining nutrient deficiencies in the main are colorimetric and turbidimetric, soil extracts being used. The rapid chemical tests are usually used in conjunction with tests for determining the pH values of the soils analysed. Constable and Miles (1941) have published an excellent paper on laboratory methods and equipment necessary to make he rapid chemical tests.

Much can be said in favor of the rapid chemical tests. In the hands of well-trained and experienced workers these tests have proved a valuable aid to better fertilizer recommendations. A testing service free to farmers is now maintained by many government agricultural organizations and by commercial agencies. As a general rule our state departments of agriculture, experiment stations, and extension services are better equipped and qualified to render this service than are commercial agencies. In fact, this service as performed by certain commercial agencies is probably resulting in more harm than good. Soil testing should be done in a well-equipped laboratory under good guidance, and not in poorly equipped laboratories by novices. Neither should they be made by persons having good chemical training but who have had little experience in making the tests. The individual who interprets the rapid chemical tests should not only know the chemistry involved, and the characteristics of the soil under consideration, but he should understand plant physiology and plant disease symptoms, and should have the results of hundreds of tests at his disposal.

Thomas (1936) found that the agricultural experiment stations in 40 states in this country make determinations of pH values on soil samples sent in by farmers. For these determinations 10 states used some form of

FIG. 139. In the author's laboratory at Clemson College hundreds of samples of South Carolina soils are tested every month for their available phosphoric acid content. (Courtesy, U.S. Soil Conservation Service.)

electric potentiometer while others used one or more of the commercial pH tests, such as the Truog-Hellige, La Motte, La Motte-Morgan, and Soiltex, which are modifications of the methods developed by Truog and Morgan. Thirty-six states used one or more methods for determining the available phosphorus content in the soil. The Bray, Truog, and Morgan methods were the tests most commonly used for determining the quantity of available phosphorus, although the Spurway and Thornton tests were used by some states. Thomas (1936) also found that soil samples sent in by farmers were tested for available potassium by 24 states. The Morgan and Spurway methods were employed most frequently in determining the quantity of available potassium, although the Truog, Bray, La Motte, and Thornton methods were used by some state experiment stations. When estimating the replaceable potash in a given soil it should not be overlooked that as the replaceable potash diminishes the more difficult becomes its availability to the plant, and that plants can secure more potassium from the soil than is contained in the soil complex.

Table 124

RAPID METHODS OF SOIL TESTING DEVELOPED IN DIFFERENT PARTS OF THE WORLD

Author	*Year and Place Originated*	*Extracting Reagent*	*Remarks*
Daubeny	1845, England	Carbonic acid	
Von Liebig	1872, Germany	Dilute hydrochloric or acetic acid	Potassium and phosphorus both are considered
Lechartier	1884, France	2% ammonium oxalate	For phosphorus only
Dyer	1894, England	1% citric acid	Soils with less than 0.01% soluble phosphoric acid (P_2O_5) probably need phosphatic fertilization
Assn. Official Agricultural Chemists	1907, United States	0.2 normal hydrochloric acid	For phosphorus only
Neubauer and Schneider	1923, Germany	Analysis of seedlings grown in soil	A single test may be for either phosphorus or potassium
Mitscherlich	1924, Germany		Test plants are grown in pots
Das	1926, India	Alkaline carbonate solutions	For calcareous soils potassium and sodium carbonates better than ammonium carbonate
Hoffer	1926, Indiana		Various reagents used on cornstalks indicate deficiencies of nitrogen, phosphorus, and potassium
Oden	1927, Sweden	Electrodialysis	
Murphy	1934, Oklahoma	Conventional base exchange procedure	

Anderson and Noble (1937) and Moser (1939) have shown considerable variation in the results secured from different methods when used on the same soil. An excellent discussion of the value of commonly used methods is that by Peech and English (1944). In Tables 124 and 125, prepared by

Table 125

RAPID CHEMICAL TESTS FOR PHOSPHORUS AND POTASSIUM IN SOILS, COMMONLY USED IN THE UNITED STATES

Author	*Agricultural Experiment Station Where Originated*	*Extracting Reagent*	*Method of Estimating Element*
		Phosphorus Tests	
Morgan....	Connecticut	Sodium acetate + acetic acid; 5 normal, pH 4.8	Sodium molybdate + stannous oxalate
Spurway...	Michigan	Dilute acetic acid; about pH 3.2	Ammonium molybdate + tin
Truog.....	Wisconsin	0.002 normal sulfuric acid buffered at pH 3.0 with ammonium sulfate	Ammonium molybdate + stannous chloride with precautions to eliminate effects of arsenic and silicon
Bray.......	Illinois	Ammonium molybdate in 0.1 normal hydrochloric acid solution	Add tin
Thornton..	Indiana	Ammonium molybdate in 0.1 normal hydrochloric acid solution	Stannous chloride or stannous oxalate
Fraps......	Texas	0.2 normal nitric acid	Gravimetric or volumetric molybdate method
Fraps......	Texas	0.2 normal nitric acid	Colorimetric molybdate method
Hester.....	Virginia Truck	Sodium acetate + acetic acid; 0.167 normal (acetate) pH 5.0	Colorimetric molybdate method
Harper.....	Oklahoma	0.2 normal sulfuric acid	
Hance.....	Hawaiian Sugar Planters Assn.	0.5 normal hydrochloric acid	
		Potassium Tests	
Morgan....	Connecticut	Sodium acetate + acetic acid; 5 normal	Sodium cobaltinitrite + isopropyl alcohol
Spurway...	Michigan	Dilute acetic acid; 1:3	Sodium cobaltinitrite + alcohol
Thornton..	Indiana	Sodium cobaltinitrite in acetic acid	Add alcohol
Bray.......	Illinois	Sodium acetate + nitric acid	

Schreiner and Anderson (1938), are shown the extracting reagents used by some of the most commonly employed rapid chemical tests.

Spectroscopic Analysis of Plant Parts

In recent years, especially in Florida, spectroscopic analysis of plant parts to determine nutrient deficiencies have proved more successful than the commonly used rapid chemical tests.

The analysis is made by placing in a crater of a carbon electrode a suitable quantity of the ash of the plant part to be analyzed. It is then arced for a definite time for an adequate exposure on a spectrographic plate. The exposed plate is then developed and the characteristic lines of the elements identified as to wave length from a comparison with a standard spectrum. The percentage of the elements present is determined from a comparison of the density of the lines with those of a suitable standard. The measurement of the line density is determined with the aid of a densitometer, which is a part of the spectrographic equipment.

Limiting Element to Plant Growth

In nearly every soil there are one or more essential elements of plant growth that are lacking in sufficient available quantities to produce the maximum yields possible under optimum environmental conditions. The element that is lacking to the greatest extent is spoken of as being in the minimum, and this element is often the controlling factor of plant growth.

Table 126

AMOUNTS OF PLANT FOOD ELEMENTS WHICH ARE CONSIDERED LOW, MEDIUM, AND HIGH AS COMPARED WITH THE GENERAL AVERAGE OF PENNSYLVANIA SOILS; EXPRESSED AS POUNDS PER ACRE (2 MILLION)

	pH	*Ca*	*Mg*	*K*	*PO_4*	*Al*	*NO_3*	*O.M.*
Very low	4.5	200	10	25	Tr.*	Tr.*	Tr.*	.5
Low	4.5 to 5.5	200 to 1000	10 to 50	25 to 50	0 to 5	0 to 5	0 to 25	.5 to 1.5
Medium	5.5 to 6.5	1000 to 1500	50 to 100	50 to 150	5 to 15	5 to 10	25 to 75	1.5 to 2.5
High	6.5 to 7.5	1500 to 2000	100 to 200	150 to 250	15 to 50	10 to 25	75 to 200	2.5 to 3.5
Very high	7.5 to 8.2	2000 to 4000	200 to 400	250 to 400	50 to 150	25 to 50	200 to 600	3.5 to 6.0
Abnormal	8.2+	6000+	400+	400+	150+	50+	600+	6.0+

* Trace.

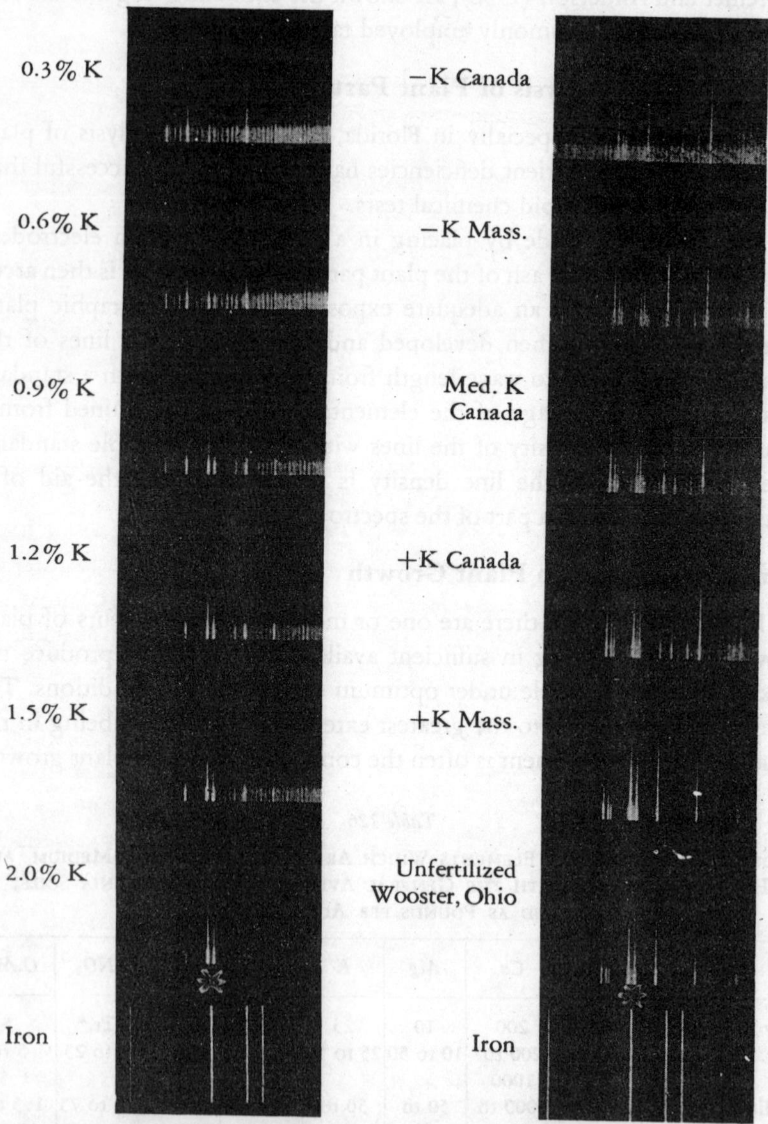

FIG. 140. (*Left*) Spectrogram of standard potassium mixtures. (*Right*) Spectrogram of apple leaf tissue from Canada, Massachusetts, and Ohio. (Courtesy, I. W. Wander and the Ohio Agricultural Experiment Station.)

* Potassium doublet at 4044.16 and 4047.22 Å.

Within certain limits an increase or decrease in the element in the minimum results in an increase or decrease in crop yields.

The amounts of available plant food which would be considered low, medium, or high might vary from locality to locality. Those accepted as general averages for Pennsylvania soils, according to Merkle (1940), are given in Table 126.

A deficiency of an element may be brought about by an excess of other nutrients which may interfere with the solubility, absorption, or utilization of the deficient element. It should be kept in mind that the complete absence of an essential element from the soil rarely occurs, and that seeds may contain sufficient quantities of some elements to supply for a time the needs of a growing seedling.

Scope of Fertilizer Studies

In the past there has been a tendency at some experiment stations to narrow the problem of crop fertilization to the fertilization of an individual crop. As all fertilizers have a direct or indirect influence upon the soil which often lasts for years, the fertilization of crops should be studied also from the standpoint of the rotation in which they are grown. The object of most fertilizer experiments is to determine the quantities of nutrients that should be applied to our crops under our varied soil and climatic conditions and in connection with our cultural practices. These factors vary so widely that it is probable that the ideal application is seldom approached closely. The kinds of fertilizer a farmer should apply, and the most profitable amounts for him to use, are always somewhat of a guess although his practice may be based on experimental evidence. Experiments must be continued, however, and as many factors as possible determined. Only in this way can a rational system for the use of fertilizers be worked out.

Use of Stable, Green and Brown Manures with Commercial Fertilizers

Experiments conducted by Thorne (1919) over long periods of time at the Ohio Experiment Station have shown that barnyard manure supplemented with commercial fertilizers gives better results than manure alone. Similar results have been secured over shorter periods of time by other experiment stations. Truck growers and florists use large quantities of both barnyard and green or brown manures in addition to large applications of commercial fertilizers. Truck farmers usually produce their crops on sandy soils, and they desire the organic matter of manure for the pur-

FIG. 141. Showing the value of using complete fertilizer in addition to barnyard manure for corn, at Wooster, Ohio. (*Left*) 8 tons of manure per acre. Final yield 34 bushels. (*Right*) 8 tons of manure and 200 pounds of 4–12–4 per acre. Final yield 54.5 bushels. (Courtesy, Ohio Agricultural Experiment Station.)

pose of increasing the water-holding capacity of their soils and to protect their crops against too high a salt concentration during periods of drouth. A rapid method for the determination of the biologically active organic matter of the soil is much needed.

The practice of green-manuring is very old, having been employed by the early Greeks and Romans. Many agricultural experiment stations have shown that the incorporation of green manures in the soil, in addition to a complete fertilizer, results in an increase in the yields of the succeeding crops. Most soils are benefited by the addition of suitable organic matter regardless of the quantity they contain. Johnson (1924), at the Virginia Truck Experiment Station, found that when 1 ton of a 6–4–8 fertilizer was applied, the average yield of potatoes was 155 bushels, but when a green manure crop was turned under previous to the addition of the fertilizer, the average yield was raised to 232 bushels.

Do We Use the Correct Fertilizer Ratio in America?

In European countries the general ratio of nitrogen, phosphoric acid, and potash is about 1.00 part of nitrogen, 0.75 part of phosphoric acid, and 1.00 part of potash, whereas in this country the general ratio is about 1.00 part of nitrogen, 2.00 parts of phosphoric acid, and 1.00 part of potash. There are soil scientists who feel that we use more phosphoric acid than is necessary. We have probably developed the habit of using excessive quantities of phosphoric acid because phosphoric acid can be produced so cheaply in this country. It is thought by some soil scientists that the future will see a decrease in the relative percentage of phosphoric acid used in American mixed goods.

Value of Legumes and Crop Residues in Maintaining Soil Fertility

It has often been proposed that the fertility of the soil, as far as nitrogen is concerned, should be maintained by the turning under of leguminous green-manuring crops. There is no doubt that such a practice would aid in solving the so-called nitrogen problem, but it cannot be considered a general solution of the entire problem. It is not always profitable to turn under a legume crop, either for the nitrogen or for the organic matter, or both, which it may contain. Where intensive farming is employed land is too valuable to be used in the production of the common field crops. Furthermore, there are large areas in this country, such as the wheat-growing areas, where rainfall is such a limiting factor that legumes cannot easily be grown, and if grown and turned under they would be objectionable in that decay processes would take place at such a slow rate. Again, legumes in general require for best growth a soil that is free of excessive acidity, and one that is supplied with an optimum of plant nutritents, and as many soils are acid and infertile, lime and fertilizers must be added if profitable leguminous crops are to be grown. This leads to an additional expense that would have to be charged against the leguminous crop.

Factors Determining the Crop-Producing Power of the Soil

In any system of soil management there are factors other than the use of fertilizers that may exert a major influence on crop yield. Soil moisture probably has a greater influence than any other one factor. Droughts that cause plants to suffer for water may affect the availability of nutrients as well as their accessibility to plant roots. The lime content of the soil has an influence, not only because plants demand calcium as a nutrient, but because of the influence of lime on soil acidity, soil structure, and on the microbiological activities that go on within the soil. A very high pH or a very low pH of the soil solution may bring about the transfer of soil colloids from the surface soil to the subsoil and this may result in the production of a hardpan. Likewise, organic matter and tillage do much toward maintaining in the soil a suitable physical condition for plant growth and microbiological activity. Where, however, ideal conditions as to moisture supply, lime, organic matter, and good tillage are approached, a maximum crop cannot be made unless the essential nutrient elements are present in the soil in sufficient quantities.

Fertility Lost from the Soil Through the Sale of Farm Animals

Livestock farmers often overlook the fact that when they sell their animals from their farms they are depleting the fertility of their soils. It has been estimated that by the time a 1000-pound steer has reached maturity it has taken from its feed fertility equivalent to 100 pounds of super-

FIG. 142. The author illustrating how the electro-metric determination of the pH of a soil may be easily and quickly determined. Approximately 2,250,000 samples of soil representing most of the cultivated fields of the state, were tested in this way during a recent year in this laboratory of the South Carolina Experiment Station.

phosphate, 150 pounds of sodium nitrate, and 50 pounds of limestone. Animals do not utilize large amounts of potash. Most of the potash that is taken from the soil by forage crops which are consumed by livestock is voided by the animals that consume the crops.

A good cow is expected to give 10,000 pounds of milk a year. In this amount of milk there are nitrogen and phosphoric acid equivalent to that found in 200 pounds of superphosphate and 300 pounds of ammonium sulfate. It may thus be seen that when an animal, or animal products, are sold, considerable fertility is lost from the farm. The use of commercial fertilizers must be advocated therefore, even under a livestock system of farming, if this loss is to be counterbalanced.

Home-Mixing of Fertilizers

As many of the fertilizer materials that go into commercial mixed goods can be purchased on the market by anyone, it is often advantageous for farmers to mix their own fertilizers. Mixtures made in this manner are commonly referred to as "home mixtures." Frequently, home mixtures can be made with the regular labor of the farm at a saving of 10 to 20 per cent. Andrews in 1940 was able to secure 26 per cent more fertilizer for a given expenditure when home-mixing was employed. Home mixtures have been shown by some of the agricultural experiment stations to give just as satisfactory results as similar mixtures made by manufacturers. Manufacturers often make the claim that their mixtures are superior to similar home-mixed fertilizers in that their products are more intimately mixed. It is true that the manufacturer, by the use of machinery, can mix his materials more thoroughly than the farmer, but this superiority does not materially influence crop yields.

As a drier or conditioner the farmer may use Cyanamid according to the directions already given or he may use such proteid-organics as tankage or cottonseed meal. Most proteid-organics in amounts of 200 or 300 pounds usually will give a mixture with perfect drilling condition.

Some fertilizer materials should not be mixed. For example, basic slag and ammonium sulfate, large amounts of calcium cyanamide and mixtures containing superphosphate. Calcium nitrate should not be added to any mixture.

Calcium, magnesium, and sodium sulfates have the property of reducing the free moisture in fertilizer mixtures by combining it as water of crystallization.

In comparing the cost of home-mixed fertilizers with that of factory goods the farmer often overlooks the extra labor, trouble, and time consumed on the farm in the making of home mixtures. It is probable that on many farms home-mixing should not be attempted unless the work can be done with the regular labor and during periods when other farm work is not pressing. It should not be overlooked that mixing of fertilizers is not always necessary and if the single nutrient materials can be applied separately with satisfaction a considerable saving in time and labor may be made.

The farmer should keep in mind that whatever method of obtaining his mixed goods is used, he should strive to obtain the kind and form of nutrients best suited to his soils and crops at the lowest price per unit. This may mean that during some years he may find that he should home-mix, whereas during other years he should buy ready-mixed goods. The present economic condition of the fertilizer market indicates that the quantity of synthetic fertilizers on the market will soon be such that prices will be reduced to a point where home-mixing will seldom be considered. In most cases where farmers use large quantities of fertilizers, it is best to get bids from local fertilizer dealers. Savings of amounts sufficient to prohibit home-mixing have been made in this way. In buying fertilizers farmers should take advantage of discounts for cash and order in carload lots when possible.

How to Mix Fertilizers

Although there have been many years when individual farmers have found it to be profitable to home-mix their fertilizers, the practice has not become popular in this country. Farmers hesitate to assume the responsibility of buying and mixing the separate materials. Many times they have had unsatisfactory results in securing a uniform mixture of drillable quality. Probably the greatest argument that can be put forth in favor of home-mixing of fertilizer is that it forces the farmer to study fertilizers and their influence on his crops.

After a farmer decides to buy the separate fertilizer materials, he must secure quotations and information concerning their composition and determine the cost of freight and hauling from the seller to his farm. From this information he must select his carriers and calculate the quantities desired either for the purpose of applying separately, or for the purpose of home-mixing.

After the materials have been secured, mixing may easily be accomplished. All that is needed is: (1) a coarse sand screen; (2) a grinder, or in many cases a wooden maul will prove satisfactory; (3) a tight floor; and

(4) hoes, preferably long handled mortar hoes, a rake, and shovels. If the naterial is bought in bags, as is most customary, the weight usually is iven on the bag, and scales for weighing are not required. For all practical urposes the bags of fertilizer may be halved or quartered when necessary.

The correct weights of materials should be spread out on the floor in ayers, the bulkier materials being put near or at the bottom. The materials hould then be shoveled over, leveled and shoveled again a sufficient numer of times to insure thorough mixing. If the mixture is made from very igh-analysis materials, it may be of such a salt concentration that a poor hysical condition may result or when applied to the soil it may injure seed ermination or seedlings. This danger can be reduced if some dry filler added to the mixture. Sand, soil, sawdust, or dolomitic limestone will erve the purpose admirably.

Table 127 gives the amounts of fertilizer of a given composition reuired to furnish a given percentage of plant nutrient in a ton of mixed oods.

Table 127

TABLE FOR COMPUTING FERTILIZER FORMULAS FROM MATERIALS OF KNOWN COMPOSITION

ercentage of Plant Food Desired	*Per Cent Composition of Material*											
	6	*8*	*10*	*12*	*14*	*16*	*18*	*20*	*22*	*24*	*48*	*50*
1	333	250	200	166	143	125	111	100	90	83	41	40
2	666	500	400	333	286	250	222	200	180	166	83	80
3	1000	750	600	500	429	375	333	300	272	250	124	120
4	1333	1000	800	666	571	500	444	400	364	333	166	160
5	1666	1250	1000	833	714	625	555	500	454	417	208	200
6	2000	1500	1200	1000	857	750	666	600	545	500	250	240
7	..	1750	1400	1166	1000	875	777	700	635	583	292	280
8	..	2000	1600	1333	1144	1000	888	800	729	666	333	320
9	..	..	1800	1500	1285	1125	999	900	818	750	375	360
10	..	..	2000	1666	1428	1250	1111	1000	909	833	416	400
11	..	..	..	1833	1571	1375	1222	1100	1000	916	458	440
12	..	..	..	2000	1714	1500	1333	1200	1090	1000	500	480

lculation of Fertilizer Formulas

There are four types of fertilizer formula problems that may need to solved in connection with the home-mixing of fertilizers. In the folving paragraphs an example of each of these problems will be given and solution explained.

I. The determination of the number of pounds of fertilizer materials required to make a complete fertilizer of a given analysis when each nutrient represented in the analysis is derived from a separate carrier.

PROBLEM

Calculate the ingredients for a ton of 4–10–4. Use the following materials and add filler if necessary.

Ammonium sulfate..........	20% nitrogen
Superphosphate..............	18% phosphoric acid
Manure salts................	20% potash

SOLUTION

The fertilizer mixture is to contain 4 per cent of nitrogen. One ton of this mixture would carry .04 × 2000 or 80 pounds of nitrogen. If ammonium sulfate carrying 20 per cent nitrogen is used, 80 divided by .20, or 400, would represent the number of pounds of ammonium sulfate required. In a like manner, it will be found that the required amounts of superphosphate and manure salts are 1112 and 400 pounds, respectively. The total amount of materials used in this case is 0912 pounds. It will thus be necessary to add 88 pounds of filler to make a ton of the required mixture

II. The determination of the number of pounds of fertilizer materials required to make a complete fertilizer of a given analysis when a part o one or more of the nutrients is derived from two or more materials.

PROBLEM

Calculate the ingredients for a ton of 4–10–4. Use the following ma terials but derive ½ of the nitrogen from tankage and ½ from sodiun nitrate.

Tankage.................	8% nitrogen and 3% phosphoric acid
Sodium nitrate............	16% nitrogen
Superphosphate...........	18% phosphoric acid
Muriate of potash.........	50% potash

SOLUTION

The fertilizer mixture is to contain 10 per cent phosphoric acid, but a tankage is to supply a part of the nitrogen it will also introduce into th mixture some phosphoric acid, and this quantity of phosphoric aci must be known before the required amount of superphosphate can b determined.

One ton of the mixture would carry .04 × 2000 or 80 pounds of nitro gen. One-half, or 40 pounds, must be furnished by tankage and one-half by sodium nitrate. If the tankage carries 8 per cent of nitrogen, 40 divided by .08 or 500 pounds, is the amount of tankage required to furnish the nitrogen. In a similar manner the sodium nitrate required will be found to be 250 pounds, and the muriate of potash 160 pounds.

If superphosphate carrying 18 per cent of phosphoric acid is used to supply the phosphoric acid then 200, the number of pounds of phosphoric acid required, minus 15, the number of pounds of phosphoric acid supplied in 500 pounds of tankage, or 185 pounds, divided by .18 will give 1028 pounds, which is the number of pounds of superphosphate necessary.

Totaling the ingredients required we have 1938 pounds, leaving only 62 pounds of filler necessary to make a ton.

III. The determination of the number of pounds of fertilizer materials required to make a complete fertilizer of a given analysis when two carriers can be so apportioned that no filler is required.

Problem

Calculate the ingredients for a ton of 4–10–4. Use the following ingredients and so apportion the kainite and muriate of potash that a filler will be unnecessary.

Ammonium sulfate..........	20% nitrogen
Superphosphate.............	16% phosphoric acid
Muriate of potash...........	50% potash
Kainite.....................	12% potash

Solution

Using the method employed in problem 1 and problem 2, we find that 1250 pounds of superphosphate and 400 pounds of ammonium sulfate are required to furnish the phosphoric acid and nitrogen, respectively. This is a total of 1650 pounds, which leaves 350 pounds to be supplied by the muriate of potash and kainite if no filler is to be used.

Our problem now is to make such a mixture that 350 pounds will supply 80 pounds of potash. A simple algebraic method of solving this problem is as follows:

Let X = lbs. of kainite
and
$350 - X$ = lbs. of muriate of potash
Then: $.12X$ = lbs. of K_2O in kainite.
$.50 \times (350 - X)$ or $(175 - .50X)$ = lbs. of K_2O in muriate of potash
If: 80 = the lbs. of K_2O to be supplied

Then: $.12X + 175 - .50X = 80$
or
$$.38X = 95$$
or
$X = 250$ lbs., or the weight of the kainite required
and
$350 - X$ or $350 - 250 = 100$ lbs., or the weight of muriate of potash required.

IV. The determination of the number of pounds of materials, which carry two or more nutrient elements, required to make a mixture which will contain an equivalent quantity of nutrients as that found in some standard application.

PROBLEM

Calculate the quantities of Ammo-Phos A and Ammo-Phos B required to make a mixture which will contain the same amounts of nitrogen and phosphoric acid as is found in a 250-pound application of a 4–10–4.

Ammo-Phos A......... 11% nitrogen and 46% phosphoric acid
Ammo-Phos B......... 16.5% nitrogen and 20% phosphoric acid

SOLUTION

A 250-pound application of a 4–10–4 fertilizer contains 10 pounds of nitrogen and 25 pounds of phosphoric acid. The amounts of Ammo-Phos A and Ammo-Phos B required to make a mixture containing exactly 10 pounds of nitrogen and 25 pounds of phosphoric acid can be determined algebraically as follows:

Let X = total lbs. of Ammo-Phos A to be furnished
and Y = total lbs. of Ammo-Phos B to be furnished
Then: $.11X + .165Y = 10$ (lbs. of nitrogen) (1)
and $.46X + .20Y = 25$ (lbs. of phosphoric acid) (2)
Multiplying (1) × 40, $4.40X + 6.60Y = 400$ (3)
Multiplying (2) × 33, $15.18X + 6.60Y = 825$ (4)
Subtracting (3) from (4), $10.78X = 425$
Whence: $X = 39.4$ or the amount of Ammo-Phos A required
Substituting $X = 39.4$ in (2), $18.124 - .20Y = 25$
Whence: $.20Y = 6.876$
$Y = 34.4$ or the amount of Ammo-Phos B required.

PROOF

	Nitrogen	*Phosphoric Acid*
39.4 lbs. 11/46 Ammo-Phos contains................	4.33	18.12
34.4 lbs. 16.5/20 Ammo-Phos contains..............	5.67	6.88
Totals..	10.00 lbs.	25.00 lbs.

18

Application of Fertilizers, and Their Influence on Germination and Seedling Growth

The increased use in recent years of high-analysis fertilizers has led both the fertilizer manufacturer and the farmer to give greater attention to the problems arising from varying the time, the method, and the rate of fertilizer application.

Physical Characteristics of Mixed Fertilizers

Manufacturers of mixed fertilizers make an effort to produce mixtures that will have good physical characteristics. Fertilizer mixtures must not set into a hard cake nor should they become mushy on damp days. They must be in such a physical condition that they will pass readily through the fertilizer distributor. In former years the manufacturer of complete fertilizers employed calcium cyanamide largely for the purpose of putting his mixture in a good physical condition, and the organic nitrogenous fertilizers of plant or animal origin for keeping his product that way. A few hundred pounds of cottonseed meal, tankage, tobacco stems, or fish scrap will serve the latter purpose well.

Today, many manufacturers are ammoniating their superphosphates and superphosphate mixtures, and this procedure yields a product of good physical characteristics. This is especially true if ammonium nitrate is added with the ammonia solution or liquor. When this method is employed manufacturers find that it is not necessary to add calcium cyanamide for the purpose of producing a good physical condition in their products.

Time of Applying Fertilizers

Inasmuch as plants absorb the greater part of their nutrients during the first half of their growing period, fertilizers should be applied either before planting or during the seedling growth period, or at both times, depending upon the fertilizer, the soil, the crop, and the season. Plants characterized by a determinate growth, such as wheat, may absorb sufficient phosphoric acid within a few weeks to take care of their needs to maturity, whereas

Table 128

PERCENTAGE OF NUTRIENTS ABSORBED BY COTTON AT VARIOUS STAGES OF GROWTH. THE COMPOSITION OF THE MATURE PLANT IS TAKEN AS 100

Composition	*First Square*	*First Bloom*	*First Open Boll*	*Mature Plant*
Dry matter	12.2	28.8	48.5	100
Total mineral (ash)	28.6	63.1	82.0	100
Nitrogen	34.0	66.0	84.2	100
Phosphorus	36.6	76.8	95.3	100
Sulfur	43.1	75.0	84.7	100
Potassium	35.5	62.9	76.0	100
Calcium	32.8	74.0	84.0	100
Magnesium	38.0	69.3	89.4	100

Table 129

PERCENTAGE OF NUTRIENTS ABSORBED PER ACRE BY POTATO CROP DURING WEEKLY INTERVALS OF GROWTH

Period of Growth (wks.)	*Plant Food Absorbed per Acre by Plants and Tubers During Each Period*					
	Nitrogen (N)	*Phosphorus (P_2O_5)*	*Potassium (K_2O)*	*Magnesium (MgO)*	*Calcium (CaO)*	*Totals*
0–7	11	6	12	4	3	9
8	6	7	6	6	7	6
9	10	11	12	10	8	10
10	24	22	24	20	17	23
11	13	14	16	15	18	16
12	32	34	24	29	17	28
13	1	4	5	8	17	5
14	3	2	1	8	13	3
Totals	100	100	100	100	100	100

plants with an indeterminate growth, such as tomatoes, present an entirely different situation. In Table 128 are given the percentages of various nutrients found in the cotton plant at various stages of growth as determined by White (1914), of the Georgia Experiment Station, and in Table 129 is given similar information with reference to Irish potatoes as determined by Carolus (1937).

In those sections where cost of farm labor is high and the cost of machinery low farmers usually prefer to apply complete fertilizers rather than single ingredients, and to apply at one time most of the fertilizer which the crop is to receive, usually just before or at time of seeding. In European countries, where the opposite economic conditions prevail, the custom is to apply single ingredients by hand at times other than at seeding time.

Because crops do not absorb much nitrogen, that is, on a basis of pounds per acre, during the first few weeks of growth, and because some forms of nitrogen may be lost by leaching during this period if not utilized by plants, it may sometimes be desirable to apply the nitrogen just before planting in an available but leach-resistant form or forms.

All phosphatic and potassic fertilizers usually are applied at or before planting. Side applications of potash may be made to some crops at intervals during the first few weeks of growth, especially when symptoms of potash starvation are apparent early in the growth of the plants. This practice is employed sometimes as a means of reducing the injury to transplants when such crops as tomatoes are grown. All proteid-organic nitrogenous fertilizers should be applied before seeding. Water-soluble nitrogenous fertilizers which are subject to leaching may be applied before seeding or at intervals during the growing season. Split applications of nitrates are generally recommended. All ammoniacal fertilizers are leach-resistant and may be applied before planting. This is true also of urea. With most crops it would seem best to apply nitrogen under the crop in some available but leach-resistant form, such as urea, and side dress only in those rare cases where extra nitrogen may be needed just prior to the time of maximum absorption by the crop.

Top and side dressings of carriers of readily available nitrogen may be made at any time during the first half of the growing season of the crop but both the quantity and form of the fertilizer carrier should be given consideration. Side dressings of fertilizer are especially adapted for crops having a long growing season or for crops grown in sections where the soil is sandy and the rainfall high. The first application of soluble nitrogen may be put under the crop, or it may be applied just after securing a stand. Usually the second application of soluble and easily leachable nitrogen is made to cotton just after chopping, to corn when the plants are about knee high, and to winter small grain when they are starting growth early in the spring. Top dressings of soluble nitrogen may be made to pastures wherever there is evidence that nitrogen is needed, provided the grass is dry. As a rule, pastures should be fertilized very early in the spring in the South, and as soon as the snow has gone in the North. Some grasses will grow even when it is still quite cold, if they can obtain available plant nutrients.

In general, fertilizers should not be applied late in the growing season for if absorbed they either accumulate in old organs or give rise to secondary growth of shoots and leaves.

Methods Used in Applying Fertilizers

Only during the last 25 years have agronomists come to appreciate the value of proper placement of fertilizer with reference to seed or plant and even to-day the value of correct placement is not generally recognized by farmers. Many farmers use complete fertilizers with approximately the correct ratio of nutrient elements and apply the correct amounts of plant nutrients per acre, but fail to secure the results that should be expected because the method of application employed does not place the fertilizer in the soil where best absorption by plants can take place.

FIG. 143. Starter solutions pay. The corn in the best looking row (*left of center*) received 450 gallons per acre of a fertilizer solution made up by dissolving 3 pounds of 13–26–13 in 50 gallons of water in addition to 500 pounds per acre of dry 5–10–5 fertilizer. The center row received only water and the dry fertilizer at the same rates. The other rows in the field received only the dry fertilizer. (Courtesy, Dr. V. A. Tiedjens and Better Crops with Plant Food.)

Fertilizers may be distributed by means of machinery or by hand, but the application with machinery is much to be preferred as it saves labor and gives a more uniform distribution. There are two general methods commonly employed in applying fertilizers: (1) broadcasting over the surface of the field; and (2) localizing, as is done when fertilizer is placed in a drill or hill. In addition, fertilizers may be dissolved in irrigation water and applied as such, as is sometimes done in the American Southwest, or fertilizers may be put in holes around the plant to be fertilized, as is done in fertilizing trees. When the fertilizer is broadcast it may be left on the surface of the soil or worked into and incorporated with the soil. In those sections where the broadcasting of fertilizers is generally advocated, broadcasting machines should be used. These machines distribute the fertilizer fairly uniformly over the surface thus permitting it to be harrowed uniformly into the soil.

In many sections of the country one of the easiest and most frequently employed methods of applying fertilizer, although it is not now generally recommended, has been through the distributor attachment of the seed drill or seed planter. In the East and South large quantities are applied in this way, often very successfully. In the West, however, particularly on soils receiving a limited rainfall, often as little as 200 pounds of a complete fertilizer applied in this way is sufficient to injure the crop. It must therefore be recognized that the best method of applying fertilizers in any section must depend upon a number of factors such as the fertilizer, the soil, the climatic conditions, and the crop grown.

In many localities where fertilizers are put in the hill or drill prior to seeding, it is still the common practice with most farmers to attempt to mix the fertilizer with the soil so that the fertilizer will not come in direct contact with the germinating seed. Recent experimental results have shown that this practice may give better results than that of allowing the fertilizer to come in direct contact with the seed but it is not the best placement in the soil for mixed fertilizers for most row crops. As put in the drill by the distributor fertilizers may be placed under, over, or at one or both sides of the seeds, depending upon the kind of distributor used. Most cotton farmers are not equipped to plant cotton seed with planters that have a fertilizer attachment. Separate fertilizer distributors generally are employed to distribute the fertilizer and to mix it with the soil before the cotton seed are planted. In the North, farmers often have found that 100 to 300 pounds of a phosphate-potash fertilizer applied in the row to corn gave good results, but the same rate of application applied broadcast and worked into the surface soil has not yielded satisfactory results. The best results have been secured from broadcasting and plowing under heavy applications or by placing the fertilizer in the bottom of the furrow.

Theoretically, it would appear that the best placement of fertilizer would be that which would allow the crop to secure the nutrients in optimum quantities at the times required by the crop. Certainly, plants should not be stunted for lack of nutrients at any stage of growth.

FIG. 144. Fertilizer placement at the sides of the seed. View, one of the control experiments of the Agricultural Research Division of the American Cyanamid Company. (Courtesy, *American Hortigraphs and Agronomic Review.*)

Experimental evidence indicating the best practice of applying fertilizer is still somewhat unsatisfactory. That variations in the placement of fertilizers do have a marked influence on crop yields is shown by the work reported by Smalley (1931), in which yields of seed-cotton varying from 604 to 1648 pounds were secured from various placements of 800 pounds per acre of a 4–8–4 fertilizer. Similar results with many other crops have more recently been reported by many other investigators.

For many years there were those who argued that hill or row application of fertilizer would cause a bunching of roots around the fertilizer applied, to an extent that would discourage the development of a larger root system. It is now generally accepted that hill or row application may promote a heavier growth of the root systems of plants within the fertilizer zone, but apparently this does not greatly restrict the extension of the root system in all directions outside the fertilized zone. It appears that rainfall and soil conditions influence the extent of root bunching around the fertilizer applied, for Collings and Warner (1927) were unable to find any great development of those portions of the root systems of cotton that were in the immediate vicinity of the fertilizer. The cotton was grown during a dry year on Cecil sandy loam, and 600 pounds per acre of a 4–10–2 was applied.

The depth in the soil at which fertilizer should be placed appears to vary with the soil, the crop and the composition of the fertilizer. In Table 130 are given the results secured at the North Louisiana Experiment Station by Stubbs and Lee (1890), from an experiment which was planned to determine the best depth to apply fertilizer for cotton. The highest yields were made when the fertilizer was placed 2 to 3 inches deep. Tiedjens (1940) concluded that for vegetables the application of lime and superphosphate under the row, in the bottom of the furrow or even in the subsoil for several inches, with nitrogen and potash applied as side dress-

Table 130

YIELDS OF SEED-COTTON FOLLOWING THE APPLICATION OF FERTILIZER AT VARIOUS DEPTHS

Depth Applied	*Pounds of Seed Cotton per Acre*	
	560 lbs. of Complete Fertilizer Used	*1120 lbs. of Complete Fertilizer Used*
6 to 8 in.	920	740
4 to 5 in.	920	680
2 to 3 in.	1030	1400
Top dressed	870	1140

ings, would probably give better results than any other method until the surface soil and subsoil have been properly adjusted for pH with liming materials.

It is the consensus of opinion that drilling fertilizers for field crops gives better results than broadcasting, particularly on soils of low fertility. The advantages of localized application are less marked on soils of high natural fertility.

Localized placement of fertilizers results in a more complete utilization of nutrients by the crop fertilized and in a smaller residual influence than broadcast application. Furthermore, it has been amply demonstrated that increased early growth and hastened maturity result from localized application. During the World War II period solutions of mixed fertilizers were satisfactorily used in New Jersey for starting and side dressing of transplanted vegetable plants. Prince and Tiedjens (1945) found ordinary mixed fertilizers satisfactory for this purpose.

Truog and Harper (1923) found that with the small grains the application of fertilizers by the drill row method gave better results than when applied broadcast. This was also the conclusion reached with reference to cotton and corn by Williams, Kilgore, and Russell (1914, 1915), of the North Carolina Experiment Station. Their results are shown in Table 131. These results represent the averages of 8 tests conducted at various times from 1902 to 1909. It has been pointed out by Haskell (1923) that

FIG. 145. Proper placement of fertilizer and potato seed piece. (Courtesy, Deere and Co.)

Table 131

INFLUENCE OF METHOD AND TIME OF APPLYING FERTILIZER ON THE YIELD OF CORN AND COTTON AT THE NORTH CAROLINA EXPERIMENT STATION

Method of application	Yield		Increase over checks	
	Corn (bu.)	Seed Cotton (lbs.)	Corn (bu.)	Seed Cotton (lbs.)
All applied in drill before planting........	18.10	1028.3	6.95	524.6
½ in drills, ½ side dressing, July 1........	18.70	914.8	9.10	477.3
Broadcast before planting..............	16.60	866.3	8.10	465.3
4 to 5 in. beneath the seed.............	18.50	921.4	10.25	532.7

because of its labor-saving benefits drill application of fertilizers will be the most common practice for most crops, until it is proved to be lower in efficiency than broadcast distribution. If large quantities of fertilizers are to be applied at any one time on properly limed soils it would be better for farmers to divide their applications and use both the broadcasting and placement-in-the-row methods. Certainly combinations of broadcasting and side dressing have their place in the production of many vegetable and field crops.

The Amount of Fertilizer to Apply

The amount of fertilizer that should be applied to the various field and horticultural crops is dependent upon many variables, such as crop, soil, and season, as well as the analysis of the fertilizer, the intensity of the agriculture, the cost of the fertilizer, and the selling price of a unit of the crop grown. The per acre applications of complete fertilizer commonly recommended for the general field crops of this country usually vary from 150 to 800 pounds, and for truck and root crops from 800 to 2000 pounds. Skinner and Ruprecht (1930) have reported results showing that the greatest increase in yields of celery was secured from an application of 10,000

Table 132

EFFECT OF VARYING QUANTITIES OF A COMPLETE FERTILIZER CONTAINING 6 PER CENT AMMONIA, 6 PER CENT PHOSPHORIC ACID, AND 6 PER CENT POTASH, ON YIELD OF CELERY*

Fertilizer Applied per Acre (lbs.)	*Yield per acre of marketable celery*			
	Meisch's Special Planted 1/5/24, Cut 5/12/24 (Crates)	*Golden Heart Planted 9/15/24, Cut 2/23/25 (Crates)*	*Meisch's Special Planted 1/15/25, Cut 5/11/25 (Crates)*	*Average Crates*
None	0	0	0	0
1,000	118	345	306	256
2,000	122	444	582	383
3,000	174	763	632	523
4,000	310	606	710	542
5,000	372	781	684	612
6,000	458	839	701	664
8,000	569	826	588	661
10,000	634	829	719	727

* Sources of fertilizer ingredients: Phosphoric acid from superphosphate; potash from sulfate of potash; nitrogen from ⅙ each ammonium sulfate, sodium nitrate, cottonseed meal, fish scrap, tankage and dried blood.

Table 133

MOST COMMONLY USED GRADES FOR PRINCIPAL FERTILIZED CROPS IN STATES CONSUMING 75 PER CENT OF THE TOTAL FERTILIZER TONNAGE

State	Cotton		Corn		Potatoes		Wheat		Tobacco	
	Grade	Application per Acre (lbs.)	Grade	Application per Acre (lbs.)	Grade	Application per Acre (lbs.)	Grade	Application per Acre (lbs.)	Grade	Application per Acre (lbs.)
Alabama	6–8–4	300	3–8–5	100	4–10–7	1,000	..	..	..	..
Florida	3–8–5	300	4–8–4	200	4–8–8	2,000	..	..	3–8–5	1,000
Georgia	4–8–4	350	3–9–3	200	..	..	2–10–4	200	3–8–5	1,000
Indiana	..	..	2–12–6	125	0–10–10	500	2–12–6	125	2–12–6	200
Maryland	..	..	2–12–4	100	6–6–5	2,000	2–8–5	250	4–8–12	800
Mississippi	4–8–4	300	4–8–4	150	6–10–7	1,000	..	..	..	..
North Carolina	3–8–3	400	3–8–3	200	6–6–5	2,000	3–8–3	200	3–8–5	1,000
New York	..	..	4–12–4	300	4–8–7	2,000	2–8–10	250	..	..
Ohio	..	..	2–12–6	125	4–8–8	1,000	2–12–6	200	4–10–6	400
Pennsylvania	..	..	2–8–5	200	2–8–10	1,000	3–12–6	250	4–8–7	750
South Carolina	3–8–3	400	3–8–3	200	6–6–5	2,000	3–8–3	200	3–8–5	600
Virginia	0–10–4	400	3–8–3	200	6–6–5	2,000	0–16–0	300	3–10–6	1,000
Louisiana	4–8–4	200	..	..	4–12–4	800	..	..	..	..
Maine	..	..	3–10–6	600	4–8–10	2,000	..	..	..	..

FIG. 146. "Juno spots" commonly seen in Kansas wheat fields are easily visible half a mile away. They are most frequently seen in fields that have been grazed, and are commonly attributed to deposits of feces or urine. They show a general lack of fertility in the rest of the field. (Courtesy, Kansas Agricultural Experiment Station.)

pounds of a 6–6–6 fertilizer. The most profitable application was found to be 4 to 5 tons per acre. Their results are given in Table 132. Meadows and pastures are usually top dressed with soluble fertilizers at the rate of 150 to 500 pounds per acre. Orchards are fertilized at the rate of 1 to 15 pounds per tree, depending upon the fertilizer materials used and the kind and size of the tree.

It is now generally agreed among agronomists and horticulturists that many of our truck, tobacco, and potato growers are using as much, if not more, fertilizer than they can economically justify.

The quantity of fertilizer to apply to a crop is not necessarily the amount that would make the highest yield. The rate of application in most cases should be determined by the relation existing between the commercial value of the fertilizer and the value of the increase in the crop produced. The law of diminishing returns is an important factor to be considered in fertilizer practice, and that rate of application should be used which is the most profitable. *The choice of a complete fertilizer should be made with the idea of supplying a given number of pounds of nitrogen, phosphoric acid, and potash rather than because the fertilizer is of a given analysis in which nutrients are present in a given ratio.*

In Table 133 is given, according to Moss and Mehring (1938) the principal grades applied to 5 of the most frequently fertilized crops in the states consuming 75 per cent of the total fertilizer tonnage.

Influence of Size of Particle and Density on Drillability

As a rule, if a fertilizer contains much material finer than 200-mesh or coarser than 5-mesh uniform distribution will be difficult and sometimes impossible. According to data secured by Mehring *et al.* (1935), fertilizers give best results when the particles are about 80- to 150-mesh in size, although they found that any particle sizes between 5- and 200-mesh gave satisfactory drillability if other conditions were reasonably satisfactory.

The density of a fertilizer upon application also influences its uniformity of distribution. In Table 134 is given the approximate density of some commonly used fertilizer materials.

Table 134

THE APPROXIMATE DENSITY OF SOME COMMONLY USED FERTILIZER MATERIALS

Materials	*Pounds per Cubic Foot*
Ammonium sulfate	62.0
Castor pomace	44.5
Cottonseed meal	38.0
Muriate of potash	68.0
Sodium nitrate	72.5
Superphosphate	56.0
Uramon	46.5

Efficiency of a Fertilizer Application

Soils are dynamic and should not be thought of merely as a mass of inert material. Physical, chemical, and biological changes are continually taking place in soils, and these changes, acting in unison with climatic forces and the plant itself, influence the efficiency of any fertilizer application. As a general rule, much larger amounts of nitrogen, phosphoric acid, and potash should be applied to the soil than will be removed during the growing season by the crop, for it is impossible for the crop to recover all or even a greater part of the nutrients applied.

Salter (1931) has reported that at the Ohio Experiment Station only about 40 per cent of the entire effect of a single application of 500 pounds of a 4–16–4 fertilizer has been found to be realized during the year the fertilizer is applied. Thirty per cent of the fertilizer was utilized by crops during the second year, 17 per cent during the third year, and 12 per cent in the fourth year. The residual influence of muriate of potash, as reported by Cooper and Wallace (1936), is shown in Table 135.

Six factors largely control the efficiency of a fertilizer application. They are (1) the crop, (2) the method of application, (3) the chemical composi-

FIG. 147. Plants may be grown equally well in solution, washed sand, or soil. (Courtesy, Dr. D. R. Hoagland and the California Agricultural Experiment Station.)

tion of the soil and its reaction, (4) the organic matter content of the soil, (5) the physical condition and drainage of the soil, and (6) weather conditions.

Table 135

AVERAGE YIELDS OF SEED COTTON IN POUNDS PER ACRE FROM RESIDUAL POTASH, 1932–1935. (MURIATE OF POTASH APPLIED BROADCAST IN 1932 AT THE RATES INDICATED.) SANDHILL EXPERIMENT STATION

Rate of Muriate Applied	*Total Yield of Seed Cotton in Pounds per Acre*					*Increase over Preceding Increment*	*Increase over No Potash*	
	1932	*1933*	*1934*	*1935*	*Average*		*Lbs.*	*Per Cent*
None	1029	858	727	195	702	..	..	..
100 lbs.	1453	1281	1032	699	1116	414	414	59.0
200 lbs.	1835	1479	1265	1045	1406	290	704	100.3
400 lbs.	1889	1713	1308	1202	1528	122	826	117.7

The influence of crop is illustrated by the apple. Because bearing apples appear to be able to store comparatively large quantities of nutrients and to absorb them over a longer period than that of their rapid period of vegetative growth, the trees' response to nutrient application is slow and many contradictory recommendations are often made.

Fertilizer Distributing Machinery

Many types of fertilizer distributing machinery may be found on the market. They vary as to type, depending upon the rate and time of application and placement desired and upon the nature of the crop. They also vary widely as to their ability to prevent the contact of the fertilizer and seed. Some are designed for broadcasting while others are designed for placement in spots or broken bands, or in continuous rows. Coe (1931) at the New Jersey Experiment Station, made a survey of 124 planter sets for drilling, and found that 71, or 56 per cent of the total number, allowed direct contact of the fertilizer and seed, either in the delivery tube or in the row before the seed and the fertilizer were covered with soil. This situation still exists although much improvement has been made recently in the design of fertilizer-distributing equipment by a few companies. Much of the equipment now on farms is obsolete in design. Many distributors cannot be adjusted to apply smaller amounts than 200 pounds of fertilizer per acre with any degree of uniformity. A good machine should be able to apply as little as 50 pounds per acre. Formerly it was difficult to secure distributing machinery which would drill less than 125 pounds. Today machines are on the market which will drill as little as 25 pounds per acre and do it efficiently.

Probably the most important consideration in choosing a fertilizer distributor is its ability to place the fertilizer in accord with sound agro-

FIG. 148. Applying fertilizer in the bottom of the furrow with a home-made distributor. (Courtesy, American Potash Institute.)

nomic principles. In addition, attention should be given to the uniformity of delivery, ease of adjustment, range in rates, size of hopper, visibility of feed, adaptation to different types of fertilizers, hopper agitators, and ease of emptying and cleaning.

In general, fertilizer distributing machinery may be classed in three groups: (1) implements which are designed principally for the application of fertilizer; (2) implements which have been designed as attachments to apply fertilizers at the same time as performing such field operations as planting or cultivating; and (3) machines for applying fertilizer with irrigation water.

The first group would include (1) broadcasting distributors of the lime sower type, (2) broadcasting distributors of the drill type, (3) row distributors, (4) hand operated distributors of the wheelbarrow or planter type and (5) the fertilizer horn. The second group would include (1) fertilizer grain drills, (2) cultivators with fertilizer attachments, (3) disk harrow fertilizer attachments, (4) planter attachments, and (5) plant setters with fertilizer attachments.

Many different types of feeds are used in fertilizer distributors. The simplest feed is that of the lime-sower type in which the rate of distribution is governed by opening or closing holes in the bottom of the hopper. The most common type is the star or forced feed. In this type of distribution a star-shaped wheel revolves carrying definite volumes of fertilizer under a gate, the height of which governs the quantity of fertilizer distributed. Among other types of feeds employed are the revolving plate, the knocker bottom, the screw conveyor, the endless belt, and the paddle wheel.

Irregular distribution of fertilizer by fertilizer distributors is the rule, and is attributed by Cumings *et al.* (1930) to such variables as variable

FIG. 149. Equipment for applying dry limestone or fertilizer, or a solution of fertilizer, in the subsoil. (Courtesy, V. A. Tiedjens.)

FIG. 150. Photograph showing residual fertilizer leaf near corn hill after corn plan has grown to maturity. (Courtesy, Iowa Agricultural Experiment Station.)

wheel slippage, cycles of delivery, lack of refinement of the distributing machine, tilting of the machine, changes in the depth of fertilizer in the hopper and to the inability of the fertilizer to flow uniformly. Cumings, Mehring and Serviss (1931) have shown also that with a majority of the fertilizer distributing machinery it is either difficult or impossible to adjust them closely to the rate of distribution desired.

Effects Arising from Irregular Distribution of Fertilizers

It appears now to be a well-established fact that on some soils the placing of fertilizers that carry phosphorus and potassium—elements that may be readily fixed in soils—in restricted areas results in increased effectiveness. But the residual effect of applying fertilizer in the drill or row is often objectionable in sections where crops that are seeded broadcast are planted later in the rotation. The fields become spotty in fertility and this results not only in a very uneven appearance of the crops that have been seeded broadcast, but it interferes very seriously with the orderly maturing of the crops. Often this is noted in fields of small grain following corn or cotton that have been fertilized in the drill. In such cases the evidence would strongly indicate that these fields should receive a heavy broadcast application of fertilizer, and that it would be a profitable practice.

Mehring and Cumings (1930) have made a statistical study of the effect of irregular distribution of fertilizers on the growth of cotton plants. They concluded that a more rapid and uniform growth, earlier flowering, earlier maturity, and larger yields of cotton were produced by uniform applications of fertilizers than by irregular applications. Muntz and Gaudeschon (1909) have shown that diffusion of some nutrients in the soil is extremely slow and that areas of varying fertility may exist in the soil for long periods of time.

Sayre and Clark (1935) found considerable movement of fertilizer salts in the soil in both upward and downward directions, but the lateral movement of the salts tested was very limited.

In Table 136 is shown, according to Nelson and Wheeting (1941), the concentration of fertilizer constituents at certain distances from fertilizer band after irrigations. The complete fertilizer applied was a mixture of sodium nitrate, treble superphosphate, and muriate of potash.

Table 136

CONCENTRATION OF FERTILIZER CONSTITUENTS AT CERTAIN DISTANCES FROM FERTILIZER BAND AFTER IRRIGATION

Sample No.	Distance from Band		After 36-hr. Irrigation				
	Below (in.)	To Side (in.)*	NO_3, p.p.m.†	P, p.p.m.‡	K, p.p.m.§	Cl, p.p.m.†	K, p.p.m.†
1	4	0	0.7	699.3	449.7	0	73.3
2	8	0	0.4	621.0	151.2	0	40.4
3	12	0	0	679.1	107.5	0	
9	0	4R	37.2	629.1	2,946.1	31.37	1,606.7
10	0	8R	60.8	650.7	242.5	14.36	64.8
11	4	8R	4.6	645.3	174.0	0	39.1
12	8	8R	0.5	669.6	183.8	0	35.8
33	0	4L	29.1	664.2	231.4	2.13	55.4
34	4	8L	0.5	629.1	224.9	0	43.7
35	8	8L	0.2	677.7	340.2	0	82.8
56	0	14R	1.7	639.9	117.3	0	43.7
57	4	14R	0.5	621.0	143.4	0	16.3
58	8	14R	0.2	661.5	138.2	0	32.6

* R signifies to the right and L to the left.
† Water extract.
‡ 0.2 *N* nitric acid extract.
§ 1.0 *N* ammonium acetate extract.

Influence of Fertilizer on Germination of Seed

Coe (1926) showed that very small amounts of soluble fertilizers in contact with the seed are capable of delaying the time when a stand of a crop is secured. This is found to be true of quantities of fertilizer that are less than the minimum for which the fertilizer attachments upon commonly used seeding machines are capable of adjustment. For this reason planters which are so designed as to cause, or permit, a direct contact of fertilizer with seed should not be used. This is borne out also by the work reported by Giles and Collins (1942) from North Carolina. Average results for a 9-year period show that 95 per cent of the cottonseed germinated when the fertilizer was placed in bands to one side of seed, 75 per cent when the fertilizer was mixed with soil under the seed, and 62 per cent when the fertilizer was placed in a band under the seed.

The manner in which fertilizers act upon the seed to prevent or to retard germination has not been explained satisfactorily. Shive (1916) suggests that retarded germination is correlated with the amount of water absorbed by the seed, which in turn is dependent upon the concentration of the soil solution. It is probably the influence of fertilizer salts on the concentration of the soil solution that produces the delayed germination of field crop seeds so often observed following an application of commercial fertilizer. Haskell (1923) has pointed out that if this is true, the problem of the method of application is of more importance in the relatively dry West than in the humid East. Unpublished data secured by Collings show that high rates of application of soluble salts can be employed without injury to germination of soybeans when the moisture con-

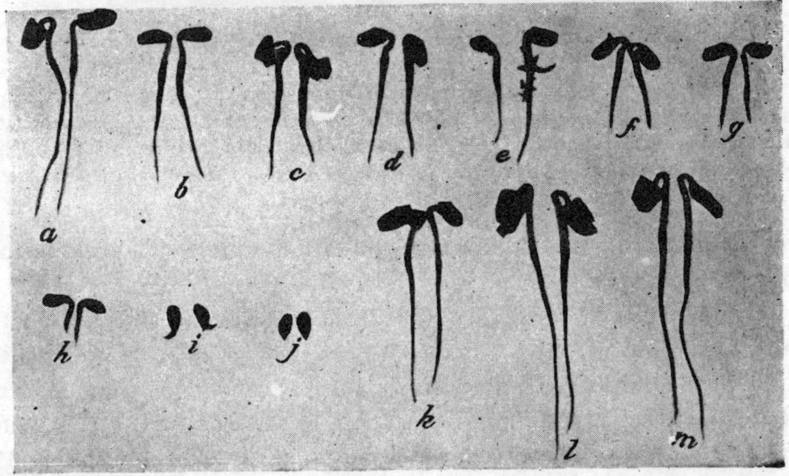

FIG. 151. Typical cotton seedlings 6 days after planting, for various placement of 800 pounds per acre of 4–8–4 fertilizer in Norfolk very fine sandy loam. (*a*) No fertilizer; (*b*) 1.75-inch band 4 inches below seed; (*c*) 1.75-inch band 3 inches below seed; (*d*) 1.75-inch band 2 inches below (seed pressed into soil and covered loosely); (*e*) 1.75-inch band 2 inches below; (*f*) 1.75-inch band 1 inch below; (*g*) 3.5-inch band 1 inch below; (*h*) one-eighth contact, seven-eighths at sides; (*i*) one-fourth contact, three-fourths at sides; (*j*) all contact; (*k*) mixed with soil below seed; (*l*) bands 1.5 inches to each side 2 inches below level of seed; (*m*) bands 3.5 inches to sides 2 inches below level of seed. (Courtesy, U.S. Department of Agrieulture.)

FIGS 152 AND 153. Cotton planted on Ruston fine sandy loam, with an 8–16–8 fertilizer applied as follows: (*Above*) mixed with the soil below the seed at rates varying from 200 pounds to 600 pounds per acre. (*Below*) In bands 2¼ inches to each side and 2 inches below the level of the seed at varying ratios. (Courtesy, U.S. Department of Agriculture.)

tent of the soil is high, but the same applications to soils of a low moisture content result in injury.

Both Hicks (1900) and Ewart (1912) showed that the superphosphates are less injurious when in contact with seed than are the common carriers of nitrogen and potash. Injury from phosphates, sulfates, chlorides, and nitrates increases in the order named. Millar and Mitchell (1927) found that applications of 16 per cent superphosphate, up to and including 350 pounds per acre, in direct contact with seeds of white beans did not retard germination. On the other hand, Morse (1935) reported that the germination of corn may be retarded by the toxic influence of the fluorine in some superphosphates if the seed are allowed to come in contact with the superphosphate. Rusche (1912) determined the relative injury of heavy applications of soluble fertilizers on the germination of wheat, oats, peas, beans, lupines, white clover, alsike clover, and seradella. He found oats to be resistant to all chemicals used, but with all other seeds germination was retarded by all chemicals. He found the injury to vary, however, from a short delay in germination to serious injury. It is thus apparent that crop seeds vary in their susceptibility to germination injury by the same fertilizer salt.

Cumings, Mehring, and Serviss (1931) observed that a poor stand of cotton caused by germination injury from fertilizer may often produce a good yield and ascribe this to inhibition of germination of weak seeds and of the growth of weak seedlings.

Where Fertilizers Should Be Placed with Reference to Seed

The best placement of fertilizers with reference to seed or plant depends upon the kind of crop, the nature of the soil, the type of fertilizer salt, and the climatic conditions that obtain. Theoretically, it would appear inadvisable to put fertilizer in the soil either above or below the seed, even though it is separated from the seed by a layer of soil. It seems improbable that a developing radicle or plumule would be benefited by passing through a layer of soluble salts. Also, fertilizer placed above the seed may be so near the surface that it will not be effective during dry years. In Table 137 are given 4-year results of a fertilizer placement study made at the South

FIG. 154. Applying ammonium sulfate in plow furrow. (Courtesy, W. R. Paden and the South Carolina Agricultural Experiment Station.)

Carolina Experiment Station, as reported by Hall and Harrell (1936), and in Table 138 the results secured by Collins (1939) at the North Carolina Experiment Station. Tests similar to these have recently been made in all the fertilizer consuming sections of the United States. They show that the placing of complete fertilizer to one side of the seed for crops grown in rows and cultivated like cotton and corn, seems to be preferable to placing

Table 137

YIELD OF SEED COTTON AS INFLUENCED BY FERTILIZER PLACEMENTS, 1930–1933. PEE DEE EXPERIMENT STATION

Postion of Fertilizer	*Pounds of Seed Cotton per Acre*				*4-yr. Average Yield*
	1930	*1931*	*1932*	*1933*	
1¾″ band—2 in. below seed	1376	371	1214	1335	1074
1¾″ band—3 in. below seed	1406	735	1453	1574	1292
3½″ band—2 in. below seed	1400	772	1338	1774	1321
3½″ band—3 in. below seed	1328	1017	1609	1974	1482
Band 1½″ to each side—1″ below seed level	1410	1057	1668	2011	1536
Band 1½″ to each side—2″ below seed level	1502	1147	1539	1951	1535
Band 1½″ to each side—3″ below seed level	1648	1143	1684	2037	1628

it either above or below. In this way injury to germinating seed and to seedlings may be avoided while at the same time an abundance of nutrients is so placed that rapid and early growth is favored. For small grain the fertilizer can be put in the drill with the seed, for the rows of drilled crops are close together and the amount of fertilizer applied under each plant is necessarily small.

Table 138

YIELD OF SEED COTTON AS INFLUENCED BY FERTILIZER PLACEMENT, 1931–1938

	1931	*1932*	*1933*	*1934*	*1935*	*1936*	*1937*	*1938*	*Average*	*Average Value at 3.5¢*
1. Fertilizer in a band under seed	1631	1632	756	758	224	769	1357	1330	1057	$37.00
2. Fertilizer placed to the sides and below seed level	1593	1661	1579	1036	946	1259	1471	1425	1371	47.99
3. Fertilizer mixed with the soil under seed	1541	1640	1454	1003	125	1025	1431	1290	1189	41.62

Salter (1931) concluded as a result of placement studies with corn in Ohio that the fertilizer should be placed either in a circular band approximately 2 inches wide and 3 inches inside diameter, or in two parallel bands each about 2 inches by 8 inches and separated 3 inches, giving in either case a minimum horizontal distance of about ¾ inch between the seed and fertilizer. The fertilizer band should probably be within a zone from ¾

FIG. 155. Applying soluble fertilizer in irrigation water. (Courtesy, Walter Fitts and the Nebraska Agricultural Experiment Station.)

inch above to 1 inch below the seed, the evidence being slightly in favor of the deeper placement.

Zimmerley (1936) found that for snap beans, peas, kale, spinach, onions, tomatoes, early cabbage and lima beans fertilizer applied in bands from 2 to 3 inches on each side of the seed and 1 to 2 inches below the level of the seed was less injurious to germination and gave higher yields than other methods of application.

One advantage of applying superphosphate in bands or in the granular form is that plant roots may be able to absorb the phosphate before it becomes reverted. It seems reasonable to suppose then that the band methods of applying complete fertilizers would have even greater use on acid soils.

Skinner (1931) believes that the best placement of fertilizer will depend in part upon the weather conditions. He reports that with cotton, during a dry year, the most advantageous position was about 4 inches to the side of the seed, whereas during normal or wet seasons the best results have been secured when the fertilizer was applied 2 to 3 inches under the seed and mixed with the soil.

The caustic and hydrated forms of lime, whether used as such or applied in materials like calcium cyanamide or basic slag, should not be placed in the soil in direct contact with seeds. When used in considerable amounts such materials should either be applied in advance of planting or by the use of machinery designed to separate laterally the seed and the material containing the lime.

It is to be expected that the band methods of applying fertilizers would have their greatest use on acid soils but because this is true it does not necessarily follow that the band methods are best for properly limed soils. Only a general rule can be given as to the best width of fertilizer bands, the best depth to place the bands, or the best distance between seed and band. Bands are usually made ¾ inch to 2 or more inches in width. They

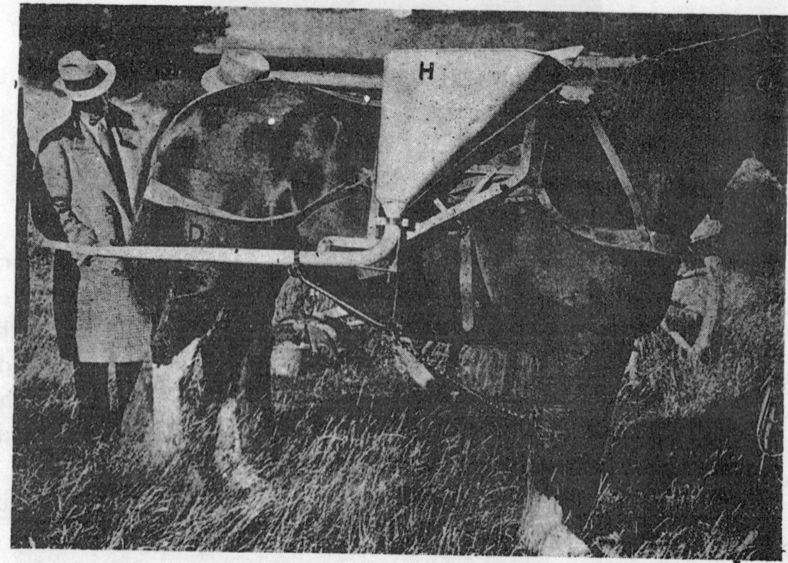

FIG. 156. Apparatus used for the broadcasting of fertilizer in the mountainous areas of New Zealand. (Courtesy, Societe Commerciale des Potasses d'Alsace.)

are usually placed from seed level to 1 to 2 inches below the level of the seed and 3/4 inch to 2½ inches from the seed. Likewise, only a general rule can be given as to whether continuous bands or broken bands should be used with hilled crops. It would seem that broken bands may be best with light applications of fertilizer, and continuous bands with relatively heavy applications.

During the last decade many experiments have been carried out to test different fertilizer placement "patterns" with our most important crops under a wide variety of soil and climatic conditions. Certain generalizations can now be made but much additional work is needed.

It is not to be expected that a single pattern of fertilizer placement would be found superior with all crops and under all conditions. Nevertheless, recent investigation has shown that under many conditions and with many crops the placing of the fertilizer in bands at the sides of the seed or plant has been the most efficient placement. In many investigations the most damaging placement has been direct contact of fertilizer and seed. The common practice of mixing the fertilizer with the seed in the row, or placing it under the seed, has shown good results sometimes, but often the injurious effect of the fertilizer is seen, especially if the soil remains fairly dry after planting.

Heavy applications of fertilizer, in addition to that put in the row, may be profitable, especially on soils of low ionic exchange capacity. Such applications may be made on the surface of the soil and turned under or, preferably, they may be put directly in the bottom of the furrow. Tiedjens (1945) repotted not only good results in New Jersey from plowing under fertilizers, but he secured even better results when the fertilizer was applied in solution in the plow furrow.

Recommendations of the Joint Committee on Fertilizer Application

Following six years of study the National Joint Committee on Fertilizer Application published (1938) its recommendations on methods of applying fertilizers. The recommendations for some field and horticultural crops are given below.

"*Cotton:* Fertilizers when applied simultaneously with planting of the seed should be placed in narrow bands approximately 2½ inches to one or both sides of the seed row and 1½ to 2 inches below the level of the seed.

"*Corn:* When corn is check-planted, the fertilizer should be dropped at the hill and not drilled. Hill applications are best placed in bands 6 to 8 inches long on each side of the seed and separated from it by ½ to 1 inch of fertilizer-free soil

FIG. 157. Effect of row fertilization of the early growth of corn. (*Center*) Complete fertilizer applied in the row. (*Right and left*) Complete fertilizer broadcast. (Courtesy, A. G. Weidman and the Michigan Agricultural Experiment Station.)

and in a depth zone from 1 inch below to slightly above the seed level. Narrow bands, 1 inch or less in width, are satisfactory for quantities up to 200 pounds per acre. For heavier rates the width of band or the lateral separation of seed and fertilizer should be increased.

"For drilled corn the fertilizer should be drilled uniformly in continuous bands in the same relative position as suggested for hill applications. Placement close to the seed is less damaging with heavy applications by the drilling method than with hill application.

"*Tobacco:* Fertilizer should be applied in bands about 3 inches to each side of the row at a depth of 1 inch below the root crown of the plant. In cases where the ordinary simple distributors are employed in existing practices, precaution should be exercised to avoid too great a concentration of the fertilizer immediately under or around the seedling roots.

"*Potatoes:* Fertilizer should be applied in bands on each side of the seed piece with 2 inches of fertilizer-free soil interposed. The fertilizer bands should be on a level with or slightly below the seed piece. On sloping land, in order to avoid fertilizer shifting too close to the seed, it is recommended that the depth of fertilizer placement be about an inch below the seed level.

"*Snap beans:* Fertilizer should be applied in bands 1½ to 2 inches to each side of the seed and 1 to 1½ inches below the seed level.

"*Lima beans:* Fertilizer should be applied in bands, 2 inches to each side of the seed and 1 inch below seed level.

"*Peas:* Fertilizer should be applied in bands, 2 inches to each side and 1 inch below seed level for peas planted in wide rows, using a bean planter and not a grain drill. For canning peas, planted in narrow rows, in the eastern part of the United States, placement 2½ inches to one side and 1 inch below the seed level with a grain drill is recommended. No definite recommendations are being made for the Middle West.

"*Kale and salad greens:* Fertilizer should be placed in bands 2 to 4 inches to each side of the seed and 2 inches below the seed level on soils where fertilizer is needed at time of planting.

"*Spinach:* Fertilizer should be applied in bands, midway between the rows and 2 inches deep, if rows are not more than 9 inches apart, where as much as

1,000 pounds of fertilizer to the acre is used. It should be applied in bands 2 to 3 inches to the side of the row and 2 inches deep for wider rows; or where light applications of fertilizer are made for narrow-row plantings.

"*Field beans:* The fertilizer, even at the comparatively low rates of application ordinarily used, should be separated from the seed and in the light of present results should be placed in a band at a depth 1½ to 2 inches below the seed level either directly under or not more than 1½ inches to the side of the seed.

"*Sugar beets:* Observations indicate that sugar beet roots extend rapidly downward but slowly horizontally. It has been definitely indicated that early benefit of fertilizer to the young seedlings is advantageous. Thus the placements showing greatest promise to date for average conditions are those in which the fertilizer does not adversely affect seed germination yet is in a position to stimulate the young seedling, such as application in a band 1 to 2 inches below the seed level, either directly under the seed or ½ to 1 inch to the side of the row. This is a tentative recommendation and subject to change when additional experimental results are obtained."

Influence of Fertilizer on the Cold Resistance of Plants

Truog and Harper (1923), at the Wisconsin Experiment Station, have demonstrated that the use of fertilizer may lower the freezing point of the soil solution as much as 3° to 4° F. In climates similar to that of Wisconsin this difference often means the difference between a stand of corn and a "freeze out" during a late spring cold spell.

Influence of Salt Concentration on Crop Development

In certain unpublished and published data secured at the New Jersey Experiment Station, Collings (1927) was able to demonstrate that although good germination of soybeans may be secured when high rates of application of fertilizer salts are applied, the roots of seedlings soon become stunted and injured, probably as a result of plasmolysis induced by the high concentration of the soil solution. Other investigators have made

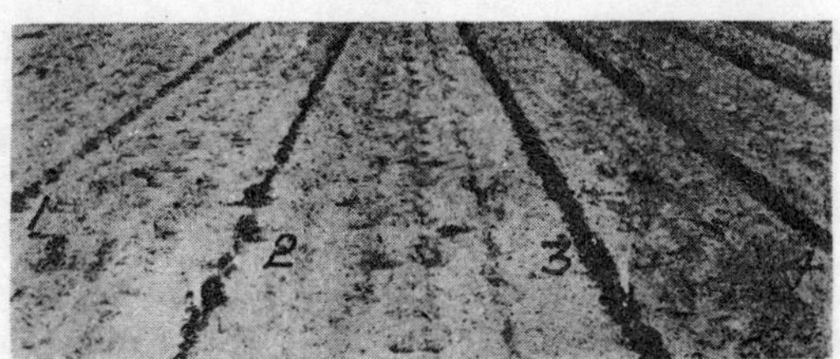

FIG. 158. Cotton photographed May 14, 1932. Fertilizer placement test. Fertilizer placed in (1) 3.5-inch band—3.0 inches below seed; (2) 3.5-inch band 4 inches below seed; (3) bands 1.5 inches to each side—3 inches below level of seed; (4) bands 1.5 inches to each side—2 inches below level of seed. (Courtesy, South Carolina Agricultural Experiment Station.)

FIG. 159. Showing the difference in size of Bluegill Bream due to food supply. Both of these fish are 18 months old. The one above grew in a fertilized pond. The one below grew in an unfertilized pond. (Courtesy, Alabama Agricultural Experiment Station.)

similar observations. Root burn may result from lower fertilizer concentrations of the soil solution than is capable of producing serious injury to germination. Some salts, when present in low concentrations, appear to cause injury by interfering with the metabolism of the plant while others do not seem to injure the plant until an osmotic pressure in the soil solution is produced that is greater than that of the cell sap.

Ross and White (1939) found that the order in which some commonly used fertilizer materials fall when arranged according to their effect per unit of total plant food in changing the concentration of the soil solution was that given in the following table.

Table 139

ASCENDING ORDER OF SOME COMMON FERTILIZER MATERIALS WHEN ARRANGED ACCORDING TO THEIR EFFECT PER UNIT OF TOTAL PLANT FOOD IN CHANGING THE CONCENTRATION OF THE SOIL SOLUTION

Order	*Fertilizer*	*Order*	*Fertilizer*
1	Double superphosphate	8	Urea
2	Monocalcium phosphate	9	Potassium chloride
3	Ordinary superphosphate	10	Ammonium nitrate
4	Free ammonia	11	Ammonium sulfate
5	Cottonseed meal	12	Manure salt
6	Potassium sulfate	13	Sodium nitrate
7	Potassium nitrate	14	Kainite

In all sections of the country where commercial fertilizer salts are used the firing of crops has been observed. Firing under natural field conditions appears to be an injury resulting from drouth. During dry periods, as water is removed from the soil by evaporation from the surface of the soil and by transpiration by plants, the concentration of the soil solution becomes increasingly greater until a point is reached when the crop is unable to secure sufficient water and the symptoms of firing result. As conditions producing firing are reached more quickly in soils containing soluble salts, it is natural to expect that crops will "fire" more quickly on soils which have received fertilizers, especially those with a high soluble salt content. Also, it is probable that such injury is greater on the drier lands of the West than on the humid soils of the East. Kansas farmers, for instance, have not been able so far to apply soluble nitrogen and potash fertilizers successfully, although soluble phosphates are applied advantageously to their wheat crops. This may be due in part to the limited rainfall of Kansas.

Fertilization of Fish Ponds

During the last few years considerable interest has been manifested in the fertilization of farm fish ponds for the production of fish. Many ponds, varying in size from an acre to two or three acres, have been built according to the specifications of various governmental agencies. These specifications usually provide that the lowest point in the pond be at least 5 feet deep, so that during hot weather the fish can seek cool depths, but that much of the area under water be relatively shallow so as to provide the best feeding grounds for the fish.

As fish feed largely upon microscopic life, the abundance of fish in a pond is determined largely by the quantity of the microscopic plant and animal life present and the quantity of microscopic plant life present in a pond is determined largely by the amount of the plant nutrients, nitrogen, phosphoric acid, and potash that may be present in the water of the pond. Fish culturists have shown that economically it is a sound practice to apply commercial fertilizers to the surface of the water for the purpose of supplying these nutrients.

The fertilization of fish ponds should necessarily vary depending upon the location of the pond, the depth of the pond, and the composition of the water entering the pond. About 400 to 500 pounds per acre, per year, of a complete fertilizer is generally recommended. The fertilizer should be applied to the water surface in split applications during the spring or

FIG. 160. Fertilizer being applied to a small fish pond. Fertilized fish ponds yield pleasure and profit. (Courtesy, B. S. Swingle and the Alabama Agricultural Experiment Station.)

early summer. It is sometimes best to scatter the fertilizer about the edges of the pond so that rains will carry it into the pond gradually.

Tests conducted at the Alabama Experiment Station, as reported by Robertson (1939), have shown that as much as 580 pounds of fish per acre of pond can be produced annually by proper fertilization. The Alabama Experiment Station recommends using 100 pounds per acre of a commercial mixture analyzing 6–8–4, together with 10 pounds of sodium nitrate, at weekly intervals during the spring and at four- to eight-week intervals during the remainder of the summer. The aim of fertilization should be to keep the microscopic plant life of the pond at such a concentration that the pond water presents a slightly greenish cast.

Bibliography

A

Adams, J. R., and A. R. Merz: Hygroscopicity of fertilizer materials and mixtures, *Ind. Eng. Chem.*, **21,** 305 (1929).

Alben, A. O. ,and H. M. Boggs: Zinc content of soils in relation to pecan rosette, *Soil Sci.*, **41,** 329 (1936).

———, J. R. Cole, and R. D. Lewis: New development in treating pecan rosette with chemicals, *Phytopathology*, **22,** 979 (1932).

Albrecht, W. A.: Artificial manure production, *Missouri Agr. Exp. Sta. Bull.*, 258, 1927.

———: Plant nutrition and the hydrogen ion. V. Relative effectiveness of coarsely ground and finely pulverized limestone. *Soil Sci.*, **61,** 265 (1946).

———: Potassium in the colloid complex and plant nutrition, *Ibid.*, **55,** 13 (1943).

———, and R. A. Schroeder: Plant nutrition and the hydrogen ion. I. Plant nutrients used most effectively in the presence of a significant concentration of hydrogen ions, *Ibid.*, **53,** 313 (1942).

———, and N. C. Smith: Saturation degree of soil and nutrient delivery to the crop. *J. Am. Soc. Agron.*, **32,** 148 (1940).

Aldrich, D. G., E. R. Parker, and H. D. Chapman: Effects of several nitrogenous fertilizers and soil amendments on the physical and chemical properties of an irrigated soil, *Soil Sci.*, **59,** 299 (1945).

Allison, F. E.: Cyanamid, its uses as a fertilizer material, *U.S. Dep. Agr. Cir.* 64, 1929.

———: Forms of nitrogen assimilated by plants, *Quart. Rev. Biol.*, **6,** 313 (1931).

———, and R. C. Cook: The effect of ammonium sulphate on soil acidity, *Soil Sci.*, **3,** 507 (1917).

———, L. A. Pink, and M. S. Sherman: Comparative availability of organic and inorganic phosphates as shown by the Neubauer method, *J. Am. Soc. Agron.*, **33,** 919 (1941).

Allison, R. V.: Leaching studies from Florida soils with various sources of nitrogen, *Com. Fertilizer*, July, **1939,** 12.

———, O. C. Bryan, and J. H. Hunter: The stimulation of plant response on the raw peat soils of the Florida Everglades through the use of copper sulphate and other chemicals. *Florida Agr. Exp. Sta. Bull.* 190, 1927.

Alway, F. J.: A nutrient element slighted in agricultural research, *J. Am. Soc. Agron.*, **32,** 913 (1940).

———, A. W. Marsh, and W. J. Methley: Sufficiency of atmospheric sulphur for maximum crop yields, *Soil Sci. Soc. Am. Proc.*, **2,** 229 (1937).

———, and J. H. Nesom: Effectiveness of calcium metaphosphate and fused rock phosphate on alfalfa. *J. Am. Soc. Agron.*, **36,** 73 (1944).

Ames, J. W., and G. E. Boltz: Sulphur in relation to soils and crops, *Ohio Agr. Exp. Sta. Bull.*, 292, 1916.

Anderson, M. S., and W. M. Noble: Comparison of various chemical quick tests on different soils, *U.S. Dep. Agr. Misc. Pub.*, 259, 1937.

Anderson, P. G., M. F. Morgan, and N. T. Nelson: The phosphorus requirements of old tobacco soils, *Connecticut Agr. Exp. Sta. Bull.*, 7, 1927.

Andre, G.: Déplacement de la potasse contenue dans certaines roches feldspathiques par quelques substances employees comme engrais. *Compt. rend. acad. sci.* (Paris), **157,** 856 (1913).

Andrews, W. B.: The elimination of differences in the evaluation of fertilizer analyses, *J. Am. Soc. Agron.*, **32,** 495 (1940).

Archibald, J. G., and P. R. Nelson: The chemical composition of grass from plots fertilized and grazed intensively, *Ibid.*, **21,** 686 (1929).

Arnon, D. I.: The essential nature of molybdenum for the growth of higher plants, *Chron. Bot.*, **6** (1940), No. 3, p. 56.

———, and P. R. Stout: Molybdenum as an essential element for high plants, *Plant Physiol.*, **14,** 599 (1939).

Ashley, G. H.: Notes on the greensand deposits of the Eastern United States, *U.S. Geol. Survey Bull.*, 660, 27 (1918).

Askey, H. O.: Successful use of cobalt salts for pasture top dressing in the treatment of stock ailments at Glenhope, Nelson, New Zealand, *J. Sci. Tech.*, **20,** 315a (1939).

Aston, B. C.: Mineral content of pasture research, *New Zealand J. Agr.*, **39,** 18 (1929).

Atwater, W. O.: On the acquisition of atmospheric nitrogen by plants, *Am. Chem. J.*, **6,** 365 (1884).

B

Bailey, E. H.: Hydrogen-ion concentration of the important soils of the United States in relation to other profile characteristics. II. Pedalfers and soils transitional between Pedocals and Pedalfers, *Soil Sci.*, **59,** 239 (1945).

Bain, H. F., and H. S. Mulliken: Nitrogen survey: U.S. Dept. Commerce, Supplement to commerce reports, *Trade Information Bull.*, 170, Jan., 1924.

Baker, O. E.: The outlook for land utilization in the United States, *U.S. Extension Service Cir.* 168, 1931.

Baker, W. G.: Residual effects of acid phosphate and rock phosphate, *J. Am. Soc. Agron.*, **17,** 172 (1925).

Barnette, R. M.: Pot and field experiments with common salt, *Ibid.*, **17,** 125 (1925).

———, J. P. Camp, J. D. Warner, and O. E. Gall: The use of zinc sulphate under corn and other field crops, *Univ. Florida Agr. Exp. Sta. Bull.*, 292 1936.

———, and J. B. Hester: Influence of inorganic nitrogen compounds on reaction and replaceable bases of Norfolk sand, *Soil Sci.*, **30,** 431 (1930).

———, and J. D. Warner: A response of chlorotic corn plants to the application of zinc sulfate to the soil, *Ibid.*, **39,** 145 (1935).

Bartholomew, R. P.: The availability of phosphatic fertilizers, *Arkansas Agr. Exp. Sta. Bull.* 289, 1933.

———: Fluorine, its effect on plant growth and its relation to the availability to plants of phosphorus in phosphate rock, *Soil Sci.*, **40,** 203 (1935).

———: Availability of phosphate rock in soils of varying degrees of acidity, *J. Am. Soc. Agron.*, **29,** 293 (1937).

———, and G. Janssen: The rate of absorption of potassium by plants and its possible effect upon the amount of potassium remaining in soils from applications of potassium fertilizers, *Arkansas Agr. Exp. Sta. Bull.* 265, 1931.

Bauer, F. C.: The relation of organic matter and the feeding power of plants to the utilization of rock phosphates, *Soil Sci.*, **12,** 21 (1921).

———: The foraging power of plants for phosphate rock, *J. Am. Soc. Agron.*, **15,** 99 (1923).

Bear, F. E.: "Soil Management," New York, John Wiley & Sons, 1927.

———: "Theory and Practice in the Use of Fertilizers," New York, John Wiley & Sons, 1929.

———, and R. M. Salter: The residual effect of fertilizers, *West Va. Agr. Exp. Sta. Bull.* 160, 1916.

Beaumont, A. B., G. J. Larsinos, P. Piekenbrock, and P. R. Nelson: The assimilation of nitrogen by tobacco, *J. Agr. Research*, **43,** 559 (1931).

Beeson, K. C.: The occurrence of mineral nutritional diseases of plants and animals in the United States, *Soil Sci.*, **60,** 9 (1945).

———, and W. H. Ross: Chemical reactions in fertilizer mixtures, *Ind. Eng. Chem.*, **29,** 1176 (1937).

Bennett, H. H., and W. C. Lowdermilk: General aspects of the soil erosion problem, *U.S. Dep. Agr. Yearbook*, **1938,** 581.

Bennett, J. P.: The treatment of lime-induced chlorosis with iron salts, *California Agr. Exp. Sta. Cir.* 321, 1931.

———: Iron in leaves, *Soil Sci.*, **60,** 91 (1945).

Bizzell, J. A.: The comparative effect of ammonium sulphate and sodium nitrate on the composition of certain vegetable crops, *Soil Sci. Soc. Am., Proc.*, **2,** 343 (1937).

———, and T. L. Lyon: Composition of drainage waters from lysimeters at Cornell University, *Proc. and Papers, First Internat. Cong. Soil Sci.*, **2,** 342 (1927).

Blackwell, C. P., and T. S. Buie: Fertilizer experiments, *South Carolina Agr. Exp. Sta. Bull.* 209, 1921.

———, and ———: Cotton production, *Ibid.*, 219, 1924.

Blair, A. W.: Reducing losses on the farm, *Com. Fertilizer*, **41,** 40 (October, 1930).

———, and A. L. Prince: Some changes brought about in cylinder soils by long continued crop and fertilizer treatment, *Soil Service*, **18,** 31 (1924).

———, and ———: The influence of lime on the reaction of subsoils, *J. Agr. Research*, **48,** 469 (1934).

Bledsoe, R. P.: Sources of nitrate of soda, *Georgia Agr. Exp. Sta. Press Bull.* 440, 1937.

———, and L. C. Olson: Experiments with liquid fertilizer, *Com. Fertilizer*, October, **1944,** 19.

Bortner, C. E.: Toxicity of manganese to turkish tobacco in acid Kentucky soils. *Soil Sci.*, **39,** 15 (1935).

Boynton, D., and O. C. Compton: Leaf analysis in estimating the potassium, magnesium, and nitrogen needs of fruit trees, *Ibid.*, **59,** 339 (1945).

Brand, C. J.: American fertilizer practices, Part XI, *Nat. Fertilizer Assoc.*, Washington, 1929.

———: An air trip to South America, *Fertilizer Rev.*, **18,** 2 (1943).

———: Post-war relations between government and industry, *Proc. Nat. Fertilizer Assoc.*, **1944,** 40.

Bray, R. H.: Phosphorus availability and crop response. *Soil Sci. Soc. Am., Proc.*, **1937,** 215.

———: Nitrate tests for soils and plant tissues, *Soil Sci.*, **60,** 219 (1945).

Breazeale, J. E.: Nutrition of plants considered as an electrical phenomenon. *J. Agr. Research*, **24,** 41 (1923).

Brenchley, Winifred E.: The influence of copper sulphate and manganese sulphate upon the growth of barley, *Ann. Bot.*, **24,** 571 (1910).

———: "Inorganic Plant Poisons and Stimulants," Cambridge University Press, 2d ed., 1927.

———, and K. Warrington: The role of boron in the growth of plants, *Ann. Bot.*, **41,** 167 (1927).

Briggs, L. J., and J. F. Breazeale: Availability of potash in certain orthoclase-bearing soils as affected by lime or gypsum, *J. Agr. Research*, **8,** 21 (1927).

Brooks, A. N.: Control of celery pink rot, *Florida Agr. Exp. Sta. Press Bull.* 567, 1942.

Brown, B. E.: Concerning some effects of long continued use of sodium nitrate and ammonium sulphate on the soil, *Pennsylvania Agr. Exp. Sta. Ann. Rpt.* 1908, **1909,** 84.

Brown, P. E., and F. B. Smith: The effects of artificial farm manures on soils and crops, *Iowa Agr. Exp. Sta. Res. Bull.* 127, 1930.

———, and H. W. Warner: Production of available phosphorus from rock phosphate by composting with sulphur and manure, *Soil Sci.*, **4,** 269 (1917).

Browne, C. A.: Potash—its production and use, *Am. Fertilizer*, July 27, 1935.

Burgess, P. S.: The reaction of soils in the field as influenced by the long continued use of fertilizer chemicals, *Rhode Island Agr. Exp. Sta. Bull.* 189, 1922.

Burns, J. H., *et al.:* Report on the fixation and utilization of nitrogen, *War Dep. Ordnance Office, Nitrate Division Bull.* 2041, 1922.

Burstrom, H.: Uber die Schwermetallkatalyse der Nitrat-Assimilation Planta, *Archiv fur Wissenschatftliche Botanik*, **29,** 292–305 (1938–39).

Bushnell, J.: The phosphorus content of a sandy loam containing sufficient available phosphorus for vegetable crops, *Soil Sci.*, **51,** 153 (1941).

Butler, B. S., and H. S. Gale: Alunite, a newly discovered deposit near Marysvale, Utah, *U.S. Geol. Survey Bull.*, 511, 1912.

C

Caldwell, W. E., and N. C. Brady: The effect of calcium on certain characteristics of peanut fruit, *J. Am. Soc. Agron.*, **37,** 696 (1945).

Cameron, F. K., and J. M. Bell: Calcium sulphate in aqueous solutions. *U.S. Dep. Agr. Bur. of Soils, Bull.* 33, 1906.

Camp, A. F.: Zinc as a nutrient in plant growth, *Soil Sci.*, **60,** 157 (1945).

Carolus, R. L.: Chemical estimations of the weekly nutrient level of a potato crop, *Am. Potato J.*, **14,,** No. 5 (1937).

Carpenter, C. C.: Riboflavin-Vitamin B_2 in soil, *Science*, **28,** 109 (1943).

Chandler, W. H.: Zinc as a nutrient for plants, *Botan. Gaz.*, **98,** 625 (1937).

———, D. R. Hoagland, and P. L. Hibbard: Little leaf or rosette in fruit trees, *Proc. Am. Soc. Hort. Sci.*, **28,** 556 (1931); **29,** 255 (1932).

Chandler, W. V.: Phosphorus absorption by five Alabama soils as influenced by reaction base saturation, and free sesquioxides, *J. Am. Soc. Agron.*, **33,** 1 (1941).

Chapman, H. D., and S. M. Brown: The effects of sulphur deficiency on citrus, *Hilgardia*, **14,** 185 (1941).

Chucka, J. A.: Magnesium deficiency in Aroostook potato soils, *Am. Potato J.*, **11,** 29 (1934).

Clark, K. G., and M. S. Sherman: World production and consumption in pre-war years, *Fertilizer Rev.*, **20,** 3 (1945).

———, and ———: Prewar world production and consumption of Plant Foods in Fertilizers, *U.S. Dep. Agr., Mis. Bull.*, 593, 1946.

Clarke, F. W.: The data of geochemistry, *U.S. Geol. Survey, Bull.* 770, 1924.

Clemson, T. G.: Year Book of Agricultural Progress and Discovery for 1855 and 1856, *Am. Farmer*, **11,** 236 (1856).

———: Fertilizers, Rpt. Commissioner of Patents, United States Government, 1859.

Clevenger, C. B., and L. G. Willis: Immediate effect of fertilization upon soil reaction, *J. Am. Soc. Agron.*, **27,** 833 (1935).

Coe, D. G.: Unpublished data. Fellowship established by the American Cyanamid Company at N. J. Agr. Exp. Sta., 1931.

———: The effects of various methods of applying fertilizers on crop yields, *Soil Sci.*, **21,** 127 (1926).

———: Effects of various methods of applying fertilizers on crops and on certain soil conditions, *Soil Sci.*, **21,** 7 (1926).

Coleman, R.: Phosphorus fixation by the coarse and fine clay fractions of kaolinitic and montmorillonitic clays, *Soil Sci.*, **58,** 71 (1944).

Collander, R.: Selective absorption of cations by higher plants, *Plant Physiol.*, **16,** 691 (1941).

Collings, G. H.: "The Production of Cotton," New York, John Wiley & Sons, 1926.

———: The influence of boron on the growth of the soybean plant, *Soil Sci.*, **23,** 83 (1927).

———: Observed methods of mining and refining French and German potash salts, *Proc. Assoc. Southern Agr. Workers*, **1929,** 146.

———, and J. D. Warner: Root development of cotton on Cecil sandy loam during 1926, *J. Am. Soc. Agron.*, **19,** 839 (1927).

Collins, E. R.: The application of fertilizers to southern crops, *Com. Fertilizer*, July, **1939,** 20.

Collison, R. C., H. G. Beattie, and J. D. Harlan: Lysimeter investigations. III. Mineral and water relations and final nitrogen balance in legume and nonlegume crop rotations for a period of 16 years, *New York State Agr. Exp. Sta. Tech. Bull.* 212, 1933.

———, and H. J. Conn: Artificial manure from straw, *New York (Geneva) State Agr. Exp. Sta. Bull.* 573, 1929.

———, and J. E. Mensching: Lysimeter investigation, *New York Agr. Exp. Sta. Tech. Bull.* 193, 1932.

Collison, S. E., and S. S. Walker: Loss of fertilizer by leaching, *Florida Agr. Exp. Sta. Bull.* 132, 1916.

Connor, S. D.: The availability of phosphates, *Am. Fertilizer*, **72,** 17 (1930).

———: Finely ground vs. coarse fertilizer, *Fertilizer Rev.*, **6,** 9 (1931).

———, and J. E. Adams: Availability of Tennessee raw rock phosphate in relation to fineness and other factors, *J. Am. Soc. Agron.*, **18,** 1103 (1926).

Constable, E. W., and D. E. Miles: Soil testing methods and apparatus designed for economy in time and labor, *Ibid.*, **33,** 623 (1941).

Cooper, H. P.: Ash constituents of pasture grasses, their standard electrode potentials and ecological significance, *Plant Physiol.*, **2,** 193 (1930).

———: The relation of standard electrode potentials and ionization potentials of atoms to cation exchange in soils, and to the absorption and utilization of nutrients by plants, *Proc. Internat. Cong. Soil Sci.*, IV, Commission IV, **1930,** 164.

———: Certain factors affecting the availability, absorption, and utilization of magnesium by plants, *Soil Sci.*, **60,** 107 (1945).

———, and W. H. Garman: Effect of applications of sodium on the composition and yield of cotton at different levels of potash fertilization, *Soil Sci.*, **7,** 331 (1942).

———, and R. W. Wallace: Comparative yields of seed cotton from the use of calcium nitrate and sodium nitrate, *South Carolina Agr. Exp. Sta. Ann. Rep.*, **1936,** p. 111.

———, and ———: Residual effect of potash on cotton, *Ibid.*, **1936,** p. 111.

———, and ———: Results from an experiment on rate and time of applying potash fertilizers to cotton, *Ibid.*, **1937,** p. 127.

———, and J. K. Wilson: Relation to ash constituents of pasture plants to the oxidation-reduction potentials of nutrients, *Soil Sci.*, **30,** 421 (1930).

———, *et al.*: Relative value of different brands of sodium nitrate in cotton production, *South Carolina, Agr. Exp. Sta. Cir.* 56, 1937.

Cooper, J. F.: Florida program meets all present needs, *Farm For Victory*, No. 16, **1945,** 11.

Corrie, F. E.: Phosphoric acid in agriculture, *Am. Fertilizer*, **72,** 44 (1930).

Crane, F. H.: "Interrelation of Calcium, Phosphorus and Hydrogen-ion Concentration in Relation to the Growth of Plants," *Master's Thesis*, Univ. Illinois, 1926.

Crawley, J. T.: Fixation of phosphorus in soil, *J. Am. Chem. Soc.*, **24,** 1114 (1902).

Crocker, W.: The necessity of sulphur carriers in artificial fertilizers, *J. Am. Soc. Agron.*, **15,** 129 (1923).

———: Sulphur deficiency in soils, *Soil Sci.*, **60,** 149 (1945).

Cumings, G. A., A. L. Mehring, and W. H. Sachs: Field and laboratory studies of fertilizer distributors for cotton, *Agr. Engin.*, **11,** 149 (1930).

———, ———, G. H. Serviss, and W. H. Sachs: Progress report on mechanical application of fertilizers to cotton in South Carolina, 1930, *U.S. Dep. Agr. Cir.* 192, 1931.

Curtis, H. A.: "Fixed Nitrogen," New York, Chemical Catalog Co., 1932.

D

Davidson, O. W.: Nutrient deficiency symptoms in deciduous fruits, *Hunger Signs In Crops, Nat. Fert. Assoc.*, 1941.

Davis, F. L., and C. A. Brewer: The effect of liming on the absorption of phosphorus and nitrogen by winter legumes, *J. Am. Soc. Agron.*, **32,** 419 (1940).

Davis, L. E.: Adsorption of phosphates by non-calcareous Hawaiian soils, *Soil Sci.*, **40,** 129 (1935).

Davis, R. O. E., and W. Scholl: Ammoniated peat—effect of varying the conditions of ammonia treatment on nitrogen quality, *Ind. Eng. Chem.*, **31,** 185 (1939).

De France, J. A.: The killing of weed seed in compost by the use of certain fertilizers and chemicals, *Proc. Am. Soc. Hort. Sci.*, **43,** 336 (1943).

Dennis, R. W. G.: A review of developments in the applications of boron in agriculture and horticulture, *Fertilizer, Feeding Stuffs and Farm Supplies J.*, Sept. 8, 12, Oct. 6, 20, 1937.

———, and D. G. O'Brien: Boron in agriculture, *West Scotland Agr. College Research Bull.* 5, 1937.

De Turk, E. E., L. K. Wood, and R. H. Bray: Potash fixation in corn belt soils, *Soil Sci.*, **55,** 1 (1943).

Dickey, R. D., and G. H. Blackmon: A preliminary report on little-leaf of the peach in Florida, *Florida Agr. Exp. Sta. Bull.* 344, 1940.

Dickman, S. R., and E. E. De Turk: A method for the determination of the organic phosphorus of soils, *Soil Sci.*, **43,** 29 (1938).

Digby, K.: "A Discourse Concerning the Vegetation of Plants," London, 1669.

Dolbear, S. H.: "Potash Reserves of the United States," Samuel H. Dolbear, New York, 1946.

Dowell, C. T.: Crops and soils information, *Louisiana Agr. Exp. Sta. Bull.* 283, 1937.

E

Earley, E. B.: Minor element studies with soybeans. I. Varietal reaction to concentrations of zinc in excess of nutritional requirement, *J. Am. Soc. Agron.*, **35,** 1012 (1943).

Eaton, F. M.: Deficiency, toxicity, and accumulation of boron in plants, *J. Agr. Research*, **69,** 237 (1944).

Elmore, K. L., O. E. Huffman, and W. W. Wolf: Defluorination of phosphate rock in the Molten State, *Ind. Eng. Chem.*, **34,** 40 (1942).

Erdman, L. W., and W. B. Bollen: Field experiments with gypsum in Iowa, *Iowa Agr. Exp. Sta. Bull.* 232, 1925.

Ernst, F. A.: "Fixation of Atmospheric Nitrogen," New York, D. Van Nostrand Co., 1928.

———, and M. S. Sherman: The world's inorganic nitrogen industry, *Ind. Eng. Chem.*, **19,** 196 (1927).

Ewart, A. J.: The influence of superphosphates on the germination of wheat, *J. Dep. Agr. Victoria*, **10**, 256 (1912).

F

Felix, E. R.: Correction of unproductive muck by the addition of copper, *Phytopathology*, **17,** 49 (1927).

Ferguson, W., and L. E. Wright: Micro-element studies with special reference to the element boron, *Canadian Dom. Dep. Agr., Exp. Farms Service Contribution* No. 523, 1940.

Fireman, M., and L. V. Wilcox: Effect of sodium nitrate on permeability of western soils, *Calif. Citrograph*, **29,** 196 (1944).

Finch, A. H., and A. F. Kinnison: Zinc treatment of pecan rosette, *Arizona Agr. Exp. Sta. Ext. Cir.* 82, 1934.

Fippin, E. O.: Plant nutrient losses in silt and water in the Tennessee River system, *Soil Sci.*, **60,** 233 (1945).

Fitts, J. W.: Nebraska soil studies, *Com. Fertilizer*, May, **1945,** 24.

Flor, H. H.: Chlorotic dieback of flax grown on calcareous soils, *J. Am. Soc. Agron.*, **35,** 259 (1943).

Forbes, E. B., F. M. Beegle, and J. E. Mensching: Mineral and organic analysis of foods. *Ohio Agr. Exp. Sta. Bull.* 255, 1913.

Fraps, G. S., and J. F. Fudge: The effect of sulphur and sulphuric acid upon the development of soil acidity at different depths. *J. Am. Soc. Agron.*, **28,** 1012 (1936).

———, and ———: The nature of the phosphates dissolved by various soil extractants, *J. Am. Soc. Agron.*, **37,** 532 (1945).

Fudge, B. R.: Magnesium deficiency in relation to yield and chemical composition of seedy and commercially seedless varieties of grapefruit, *Proc. Florida State Hort. Soc.*, **51,** 34 (1938).

G

Gaddum, L. W., and L. H. Rogers: A study of some trace elements in fertilizer materials, *Florida Agr. Exp. Sta. Bull.* 290, 1936.

Gale, H. S., and W. B. Hicks: Mineral resources of the United States for 1917, *U.S. Geol. Survey*, 1920.

———, and R. W. Richards: Preliminary report on the phosphate deposits in southeastern Idaho and adjacent parts of Wyoming and Utah, *U.S. Geol. Survey, Bull.* 430, **1910,** 457.

Gall, O. E., and R. M. Barnette: Toxic limits of replaceable zinc to corn and cowpeas grown on three Florida soils, *J. Am. Soc. Agron.*, **32,** 23 (1940).

Gardner, F. D., C. F. Noll, and P. S. Baker: Thirty-five years' results with fertilizers, *Penn. Agr. Exp. Sta. Bull.* 146, 1917.

Gardner, V. R.: Report of the Michigan Agricultural Experiment Station, 1930–1931, June 30, 1931.

Garman, W. H.: Unpublished data, *South Carolina Agr. Exp. Sta.*, 1946.

Garner, W. W., and D. E. Brown: Experiments with tobacco, *Maryland Agr. Exp. Sta. Bull.* No. 225, 1919.

———, and J. E. McMurtrey: A complete fertilizer for the tobacco crop, *American Fertilizer*, **72,** 36 (1930).

———, ———, J. D. Bowling, and E. G. Moss: Role of chlorine in nutrition and growth of the tobacco plant and its effect on the quality of the cured leaf. *J. Agr. Research*, **40,** 145 (1930).

Gaskill, E. F.: Comparison of different potash fertilizers, *Mass. Agr. Exp. Sta. Rpt.* 1917, **1918,** 23a.

Gericke, W. F.: On the physiological balance in nutrient solutions for plant cultures, *Am. J. Bot.*, **9,** 180 (1922).

Gilbert, B. E., F. T. McLean, and L. J. Hardin: The relation of manganese and iron to lime induced chlorosis, *Soil Sci.*, **22,** 437 (1926).

———, and F. R. Pember: A study of the availability of ammoniated superphosphate and various unusual phosphatic carriers by means of vegetative pot tests, *Rhode Island Agr. Exp. Sta. Bull. 256*, 1936.

Giles, G. W., and E. R. Collins: The one-mule farmer needs a new machine, *Better Crops with Plant Food*, Aug.–Sept., **1942,** 19.

Gilligan, G. M.: The effect of degree of base saturation of soils upon the fixation of phosphorus and potassium and the availability of phosphorus, *Delaware Agr. Exp. Sta. Bull.* 215, 1938.

Gourley, J. H., and J. W. Wander: The lateral distribution of potassium in an orchard soil, *J. Am. Soc. Agron.*, **31,** No. 7, 1939.

Graham, E. R.: Primary minerals in the silt fraction as contributors to the exchangeable-base level of acid soils, *Soil Sci.*, **49,** 277 (1940).

———: Calcium transfer from mineral to plant through colloidal clay, *Ibid.*, **51,** 65 (1941).

Gray, D. T.: Using nitrogen only on delta soil, *Arkansas Agr. Exp. Sta.*, *Data Sheet* No. 4, 1938.

Gregory, F. G.: The differential effect of the ions of three-salt solutions on the growth of potato plants in sand culture, *Proc. Roy. Soc. London*, [B], **102,** 311 (1928).

H

Haddon, C. B.: Biennial report of the northeast Louisiana Experiment Station, St. Joseph, Louisiana, 1935–1936.

Haenseler, C. M., and T. R. Moyer: Effect of calcium cyanamide on the soil microflora with special reference to certain plant parasites, *Soil Sci.*, **43,** 133 (1937).

Hall, A. D.: The effect of the long continued use of sodium nitrate on the constitution of the soil, *Trans. Chem. Soc. (London)*, **85,** 950 (1904).

———: "The Soil, An Introduction of the Scientific Study of the Growth of Crops," New York, E. P. Dutton & Co., 1915.

———: "Fertilizers and Manures," New York, E. P. Dutton & Co., 1921.

———, and C. T. Gimmingham: The interaction of ammonium salt and the constitution of the soil, *J. Chem. Soc. (London)*, **91,** 677 (1907).

———, and C. G. T. Morison: On the function of silica in the nutrition of cereals, *Proc. Roy. Soc., London*, [B], **77,** 455 (1906).

Hall, E. E., and F. M. Harrell: Cotton production studies, Annual report, *South Carolina Agr. Exp. Sta. Rpt.*, **1936,** 78.

Halversen, W. V., and W. G. Hoge: The Azotobacter plaque test as applied to the determination of phosphate deficiency in Idaho soils, *J. Am. Soc. Agron.*, **34,** 503 (1942).

Hamner, Karl C.: Unpublished data, *U.S. Dep. Agr., Plant Soil and Nutrition Lab.*, Ithaca, N. Y., 1946.

Hardestry, J. O., and F. W. Parker: Use of ammonium nitrate in mixed fertilizers, *Com. Fertilizer*, December, **1943,** 32.

Harmer, P. M., and E. J. Benne: Sodium as a crop nutrient, *Soil Sci.*, **60,** 137 (1945).

Harper, H. J.: A study of the secondary effects of hill fertilization, *Iowa Agr. Exp. Sta. Research Bull.* 87, 1925.

———: Sulphur content of Oklahoma rainfall, *Proc. Oklahoma Acad. Sci.*, **23**, 73 (1943).

Harris, H. C., R. W. Bledsoe, and P. W. Calhoun: Responses of cotton to sulphur fertilization, *J. Am. Soc. Agron.*, **37**, 323 (1945).

Harrison, W. H., and S. Das: The retention of soluble phosphates in calcareous and non-calcareous soils, *India Dep. Agr. Mem. Chem. Ser.* 5, p. 195 (Abstract) *Exp. Sta. Record*, **46**, 121 (1921).

Hart, E. B., and W. H. Peterson: Sulphur requirements of farm crops in relation to the soil and air supply, *Wisconsin Agr. Exp. Sta. Res. Bull.* 14, 1911.

———, and W. E. Tottingham: Relation of sulphur compounds to plant nutrition, *J. Agr. Research*, **5**, 233 (1915).

Hartwell, B. L.: The percentage of total phosphorus in flat turnips as influenced by the amount available in the soil, *Rhode Island Agr. Exp. Sta. Bull.* 154, 1913.

———: The manurial value of a modification of orthoclase-bearing rock where only potassium was deficient, *J. Am. Soc. Agron.*, **11**, 327 (1919).

———: Relative growth response of crops to each fertilizer ingredient and the use of this response in adapting a fertilizer analysis to a crop, *Ibid.*, **13**, 353 (1922).

———: The influence of fertilizers on crop quality, *Proc. Nat. Fertilizer Assoc.*, Sec. II, 1932.

———, and F. K. Crandall: The substitution of stable manure by fertilizers, green manures and peat, *Rhode Island Agr. Exp. Sta. Bull.* 188, 1922.

———, and S. C. Damon: The value of sodium when potassium is sufficient, *Ibid.*, 177, 1919.

———, and ———: Relative lime needs of sulphate of ammonia and nitrate of soda and of different crops, *J. Am. Soc. Agron.*, **19**, 843 (1927).

———, and F. R. Pember: The effect of the "wet process" on the availability of low-grade nitrogenous materials, *Ind. Eng. Chem.*, **4**, 441 (1912).

———, and ———: The presence of aluminum as a reason for the difference in the effect of so-called acid soil on barley and rye, *Soil Sci.*, **6**, 259 (1918).

———, and P. H. Wessels: The effect of sodium manuring on the composition of plants, *Rhode Island Agr. Exp. Sta. Bull.* 153, 1913.

Harvey, E. W.: Factors affecting returns in comparative experiments with sulphate of ammonia and nitrate of soda, *Am. Fertilizer*, **63**, 13, Dec. 26, 1925.

Haskell, S. B.: Methods of distribution of phosphorus fertilizers, *J. Am. Soc. Agron.*, **15**, 141 (1923).

Haskins, H. D., H. R. De Rose, M. W. Goodwin, and J. W. Kuzmeski: Inspection of commercial fertilizers, *Mass. Agr. Exp. Sta. Bull.* 51, 1929.

Hass, A. R. C.: Deficiency chlorosis in citrus, *Soil Sci.*, **42**, 435 (1936).

———: Influence of chlorine on plants, *Ibid.*, **60**, 53 (1945).

Heck, A. F.: Phosphate fixation and penetration in soils, *Ibid.*, **37**, 343 (1934).

———: Availability and fixation of phosphorus in Hawaiian soils, *J. Am. Soc. Agron.*, **27**, 874 (1935).

Hellriegel, H.: Welche Stickstoffquellen Stehen der Pflanze zu Gebote? *Tagesblatt der 59 Versamml. deut. Naturf. u. Aerzte*, Berlin, **1886**, 290.

Hendricks, S. B., and W. L. Hill: The inorganic constitution of bone, *Science*, **96**, 255 (1942).

Hepler, J. R., and H. R. Kraybill: Effect of phosphorus upon the yield and time of maturity of the tomato, *New Hampshire Agr. Exp. Sta. Tech. Bull.* 28, 1925.

Hester, J. B.: The influence of phosphate fertilization upon the amphoteric properties of Coastal Plain soils, *J. Am. Soc. Agron.*, **29**, 10 (1937).

———: The relationship of liming to soil fertilization, *Am. Fertilizer*, June 8, 1940.

———: Boron and manganese in tomato production, *Ibid.*, Sept. 27, 1941.

Hester, J. B.: The importance of phosphorus in the profitable production of high-value crops, in "Manual on Phosphates," Baltimore, Davidson Chemical Corp., 1942.

———, M. M. Parker, and H. H. Zimmerley: Liming Coastal Plain Soils, *Virginia Truck. Exp. Sta. Bull.* 91, 1936.

———, and Shelton, F. A.: The influence of acid-neutral fertilizers on vegetable crop production in eastern Virginia, *Ibid.*, 90, 1936.

———, and ———: Seasonal variation of pH in field soils, a factor in making lime recommendations, *J. Am. Soc. Agron.*, **25,** 299 (1933).

———, and ———: Soil organic matter investigations upon Coastal plain soils, *Virginia Truck. Exp. Sta. Bull.* 94, 1937.

Hibbard, P. L.: Factors influencing phosphate fixation in soils, *Soil Sci.*, **39,** 337 (1935).

———: A soil zinc survey in California. *Ibid.*, **49,** 63 (1940).

Hibbard, R. P.: Negative results on physiological balance in soil cultures, *Plant Physiol.*, **2,** 1 (1927).

Hicks, G. H.: The germination of seeds as affected by certain chemical fertilizers, *U.S. Dep. Agr. Div. Bot. Bull.* 24, 1900.

Hicks, W. B.: Potash, *U.S. Geol. Survey, Mineral Resources U.S.*, Part II, **1918,** 385.

Hilbert, J. E., L. A. Pinck, M. S. Sherman, and T. H. Tremearne: Organic Phosphates: I. Fixation studies with three different soil types, *Soil Sci.*, **46,** 409 (1938).

Hill, M. F., and O. C. Bryan: The nutritive relation of copper on different soil types in Florida, *J. Am. Soc. Agron.*, **29,** 809 (1937).

Hoagland, D. R.: The effect of hydrogen- and hydroxyl-ion concentration on the growth of barley seedlings, *Soil Sci.*, **3,** 547 (1917).

———: "Lectures on the Inorganic Nutrition of Plants," Waltham, Mass., Chronica Botanica Co., 1944.

———, and D. I. Arnon: Physiological aspects of availability of nutrients for plant growth, *Soil Sci.*, **51,** 431 (1941).

Hoffer, G. N.: Testing corn stalks chemically to aid in determining their plant food needs, *Indiana Agr. Exp. Sta. Bull.* 298, 1926.

———: Deficiency symptoms in corn and small grains, "Hunger Signs in Crops," *Nat. Fert. Assoc.*, **1941,** 55.

———, and R. H. Carr: Accumulation of iron and aluminum compounds on corn plants and its probable relation to root rots, *J. Agr. Research*, **23,** 801 (1923).

Holmes, R. S.: Copper and zinc content of certain United States soils, *Soil Sci.*, **56,** 359 (1943).

———, W. E. Hearn, and H. G. Byers: The chemical composition of soils and colloids of the Norfolk and related soil series, *U.S. Dep. Agr. Tech. Bull.* 594, 1938.

Holt, M. E., and N. J. Volk: Sodium as a soil nutrient and substitute for potassium, *J. Am. Soc. Agron.*, **37,** 821 (1945).

Hopkins, C. G.: Soil fertility and permanent agriculture, New York, Ginn & Co., 1910.

———, and W. H. Sachs: Radium as a fertilizer, *Illinois Agr. Exp. Sta. Bull.* 177, 1915.

Howard, P. E.: Plant food consumption by countries, 1928. *Fertilizer Rev.* **5,** 5 (1930).

———: Survey of the fertilizer industry, *U.S. Dep. Agr. Cir.* 129, 1931.

———: American fertilizer, *U.S. Dep. Agr. Misc. Pub.* 143, 1932.

Hutchinson, G. E.: Aluminum in soils, plants and animals, *Soil Sci.*, **60,** 29 (1945).

Hutchinson, H. B., and N. H. J. Miller: The direct assimilation of inorganic and organic forms of nitrogen by higher plants, *Centr. Bak.*, **30,** 513 (1910).

J

Jacks, G. V., and H. Scherbatoff: Soil deficiencies and plant diseases, *Imp. Bur. Soil Sci., Tech. Bull.* 31, 1934.

Jacob, K. D., *et al.*: The composition and distribution of phosphate rock with special reference to the United States, *U.S. Dep. Agr., Tech. Bull.* 364, **1933,** 90.

———, and W. H. Arminger: Field experiments with alkylation-acid superphosphate, *J. Am. Soc. Agron.*, **36,** 281 (1944).

Jacobson, H. G. M., and T. R. Swanback: Manganese content of certain Connecticut soils and its relation to the growth of tobacco, *Ibid.*, **24,** 237 (1932).

Jamison, V. C.: Adsorption and fixation of copper in some sandy soils of central Florida, *Soil Sci.*, **53,** 287 (1942).

Janssen, G., and R. P. Bartholomew: The translocation of potassium in tomato plants and its relation to their carbohydrate and nitrogen distribution. *J. Agr. Research*, **38,** 447 (1929).

Jenkins, E. H.: Studies on the tobacco crop of Connecticut, *Connecticut Agr. Exp. Sta. Bull.* 180, 1914.

———: Domestic supplies of potash, *Ibid.*, 198, 1917.

———, and J. P. Street: Manure from the sea, *Ibid.*, 194, 1917.

Jensen, C. A.: Effect of decomposing organic matter on the solubility of certain inorganic constituents of the soil, *J. Agr. Research*, **9,** 253 (1917).

Jewett, T. N.: Loss of ammonia from ammonium sulphate applied to alkaline soils, *Soil Sci.*, **53,** 401 (1942).

Joffe, J. S., and A. K. Levine: The relation of potassium fixation to the exchange capacity of soils, *Soil Sci. Soc. Am., Proc.*, **1939,** 157.

Johnson, B. L.: Potash, *Dep. Com., Bur. Mines, Economic Paper* 16, 1933.

———, and E. W. Tucker: Phosphate rock. United States Bur. Mines Minerals Yearbook, 194, 1944.

Johnson, E. S., and W. H. Dore: The relation of boron to the growth of the tomato plant, *Science*, N.S., **67,** 324 (1928).

———, and ———: The influence of boron on the chemical composition and growth of the tomato plant, *Plant Physiol.*, **4,** 31 (1929).

Johnson, M. O.: The spraying of yellow pineapple plants on manganese soils with iron sulphate solutions, *Hawaii Agr. Exp. Sta. Press Bull.* 51, 1916.

Johnson, T. C.: Effects of organic matter in maintaining soil fertility for truck crop production, *Proc. Am. Soc. Hort. Sci.*, **1924,** 277.

———, and H. H. Zimmerly: Effect of certain sodium and potassium salts on sweet-potato production in eastern Virginia, *Ibid.*, **20,** 155 (1923).

Jones, H. A., and B. E. Brown: Plant-nutrient deficiency symptoms in the potato, "Hunger Signs In Crops," Nat. Fert. Assoc., **1941,** 99.

Jones, L. R., and W. A. Orton: The orange hawkweed or "paint brush," *Vermont Agr. Exp. Sta. Bull.* 56, 1897.

K

Kapp, L. C.: Preliminary report on the effect of certain chemicals on rice production and their effect on the rice soil, *Arkansas Agr. Exp. Sta. Bull.* 277, 1932.

Karraker, P. E.: The comparative effect of muriate and sulphate of potash on the composition and quality of white Burley tobacco, *Kentucky Agr. Sta. Bull.* 334, 1932.

———, H. F. Miller, C. E. Bortner, and J. R. Todd: Greenhouse tests of the availability of phosphorus in certain phosphate fertilizers, *Ibid., Bull.* 413, 1941.

Kelly, J. B., and A. R. Midgley: Phosphate fixation: an exchange of phosphate and hydroxyl ions, *Soil Sci.*, **55,** 167, 1943.

Kelley, W. P.: The assimilation of nitrogen by rice, *Hawaii Agr. Exp. Sta. Bull.* 24, 1911.

Kennen, F. G.: Reactions occurring during the ammoniation of superphosphate, *Ind. Eng. Chem.*, **22,** 1378 (1930).
Killinger, G. B., R. E. Blaser, E. M. Hodges, and W. E. Stokes: Minor elements stimulate pasture plants, *Florida Agr. Exp. Sta. Bull.* 384, 1943.
Klemet, R.: *Naturwissenschaften*, **26,** 145 (1938).
Knight, H. C.: Fertilizer research, *Com. Fertilizer*, **55,** 8 (1937).
Knot, J. E.: The effect of certain mineral elements on the color and thickness of onion scales, *Cornell Univ. Agr. Exp. Sta. Bull.* 552, 1933.
Kolodny, L., and W. R. Robbins: Availability of fixed potassium in plants, *Soil Sci.*, **49,** 303 (1940).
Kraybill, H. R.: Fertilizer in relation to the quality of crops, *Am. Fertilizer*, **72,** 40 (July, 1930).
Kurtz, T., E. E. De Turk, and R. H. Bray: Phosphate adsorption by Illinois soils, *Soil Sci.*, **61,** 111 (1946).
Kuykendall, R.: *Miss. Agr. Exp. Sta. Service Sheets*, 218, 223. 224, 1938.

L

Lagatu, H., and L. Maume: Antagonisme du calcaire, a l'egard de l'absorption deal potasse sur la vigue, Comm. Congres Int. du vin et du pin maritime, Bordeaux, 1982.
Laurie, A., and A. Wagner: Deficiency symptoms of greenhouse flowering crops, *Ohio Agr. Exp. Sta. Bull.* 611, 1940.
Lawes, J. B., and J. H. Gilbert: On the sources of the nitrogen of vegetation; with special reference to the question whether plants assimilate free or uncombined nitrogen, *Abstract Proc. Roy. Soc. Lond.*, **10,** 544 (1860). (Rothamsted Memoirs, **1,** No. 18.)
Leclerc, J. A., and J. F. Breazeale: Plant food removed from growing plants by rain and dew, *U.S. Dep. Agr. Yearbook*, **1908,** 389.
Lehr, J. J.: The importance of sodium for plant nutrition, *Soil Sci.*, **52,** 237 (1941).
Lipman, C. B.: A critique of the hypothesis of the lime-magnesium ratio, *Plant World*, **19,** 83, (1916).
———: Importance of silicon, aluminum, and chlorine for higher plants, *Soil Sci.*, **45,** 189 (1938).
Lipman, J. G.: Suggestions concerning the terminology of soil bacterias, *Bot. Gaz.*, **51,** 454 (1911).
———: "The Menace of Overproduction," New York, John Wiley & Sons, 1930.
———: The nitrogen outlook, *J. Am. Soc. Agron.*, **24,** 227 (1932).
———, and A. W. Blair: Twenty years' work on the availability of nitrogen in nitrate of soda, ammonium sulphate, dried blood and farm manures, *Soil Sci.*, **5,** 291 (1918).
———, and ———: Use of concentrated fertilizers for crops grown on soils varying in mechanical composition, *New Jersey Agr. Exp. Sta. Ann. Rept.*, **1929,** 325.
———, ———, and A. L. Prince: Investigations relative to the use of nitrogenous plant-foods, 1913–1927, *New Jersey Agr. Exp. Sta. Bull.* 519, 1931.
———, and A. B. Conybeare: Preliminary notes on the inventory and balance sheet of plant nutrients in the United States, *Ibid.*, *Bull.* 607, 1936.
———, and G. MacKinney: Proof of the essential nature of copper for higher green plants, *Plant Physiol.*, **6,** 593 (1931).
———, and H. C. McLean: Experiments with sulphur-phosphate composts conducted under field conditions, *Soil Sci.*, **5,** 243 (1918).
Loew, O: The relation of lime and magnesia to plant growth, *U.S. Dep. Agr., Bur. Plant Ind. Bull.* 1, 1901.

Lyon, T. L., and J. A. Bizzell: Lysimeter experiments, *Cornell Univ. Agr. Exp. Sta. Mem.* 12, 1918.

———, and H. O. Buckman: "The Nature and Properties of Soils," New York, The Macmillan Co., 1922.

M

MacIntire, W. H.: A new explanation of what happens to superphosphate in limed soils, *Tenn. Agr. Exp. Sta. Bull.* 176, 1941.

———: The effect of liming materials upon the solubility of potassium compounds in the soil, *Better Crops with Plant Food*, p. 9, Nov., 1943.

———, and L. J. Hardin: Compatibility of fused rock phosphate with superphosphates, *Ind. Eng. Chem.*, **32**, 574 (1940).

———, ———, and F. D. Oldham: Behavior of a new source of available magnesium for phosphatic fertilizers, *Ibid.*, **30**, 61 (1938).

———, ———, ———, and J. W. Hammond: Development of P_2O_5 insolubility in phosphatic mixtures, *Ibid.*, **29**, 758 (1937).

———, ———, S. H. Winterberg, and J. W. Hammond: Nature and liming value of quenched calcium, silicate slag, *Soil Sci.*, **50**, 219 (1940).

———, and B. W. Hatcher: Evidence in support of a concept as to the end product of superphosphate in soils, *Ibid.*, **53**, 43 (1942).

———, and K. B. Sanders: The fixation of the potash of a green manure by liming materials, *Ibid.*, **29**, 109 (1930).

———, ———, and W. M. Shaw: The availability of hydrated lime, limestone, and dolomite of two degrees of fineness, with supplements of red clover hay, as measured by lysimeter leachings, *J. Am. Soc. Agron.*, **25**, 285 (1933).

———, W. M. Shaw, and B. Robinson: Influence of limestone and dolomite upon sulfate retention from annual additions of potassium sulfate. *Soil Sci.*, **51**, 73 (1941).

———, ———, ———, and J. B. Young: Potassium retention from annual additions of chloride, sulphate, and nitrate, as influenced by limestone and by dolomite. *Ibid.*, **55**, 321 (1943).

———, ———, J. B. Young, and B. Robinson: The effect of 12-year residues of lime and magnesia upon the outgo of subsequent additions of potash, *J. Am. Soc. Agron.*, **28**, 202 (1936).

———, ———, ———, and ———: Behavior of calcium, magnesium and potassium sulphates as influenced by limestone and by dolomite—a lysimeter study, *Soil Sci.*, **46**, 229 (1938).

———, and P. M. Shuey: Chemical changes in mixtures of superphosphate with dolomite and with limestones, *Ind. Eng. Chem.*, **24**, 933 (1932).

———, and S. H. Winterberg: Quenched calcium silicate slag *Tenn. Agr. Exp. Sta. Bull.* 184, 1943.

———, ———, H. W. Dunham, and L. B. Clements: Response of Sudan grass to ammonium hydroxide in pot cultures, *Soil Sci. Soc. Am., Proc.*, **8**, 1944.

———, ———, B. W. Hatcher, and G. Palmer: Fused tricalcium phosphate: relation of degree of defluorination to fertilizer value of quenched fusions of rock phosphate, *Soil Sci.*, **57**, 425 (1944).

MacLachlan, J. D.: Control of water core of turnips by spraying with borax, *Sci. Agr.*, **24**, 7 (1944).

Mansfield, G. R.: "The Economic Geology of Fertilizer Minerals. World Conditions as to Mineral Raw Materials for the Fertilizer Industry," Nat. Fertilizer Assoc., **1926**, 13.

———: "The Phosphate Deposits of the World. World Conditions as to Mineral Raw Materials for the Fertilizer Industry," Nat. Fertilizer Assoc., **1926**, 94.

Mansfield, G. R.: Potash in the United States, *J. Chem. Education*, **7**, 737 (1930).

———: Recent studies of reserves of domestic phosphate, *Am. Inst. Mining Met. Eng., Tech. Pub.* No. 1208, 1940.

———, and L. Boardman: Nitrate deposits of the United States, *U.S. Dep. Int., Geol. Survey Bull.* 838, 1932.

Marais, J. S.: The comparative agricultural value of insoluble mineral phosphates of aluminum, iron, and calcium, *Soil Sci.*, **13**, 355 (1922).

Marsh, A. W., and W. L. Powers: Responses of plants to additions of manganese on some Oregon soils, *J. Am. Soc. Agron.*, **37**, 1 (1945).

Martin, J. P.: Symptoms of malnutrition manifested by the sugar cane plant when grown in culture solutions from which certain essential elements are omitted, *Hawaiian Planters' Rec.*, **38**, 3 (1934).

———, and Y. Wang: Utilization of plant residues for the production of artificial manures, *J. Am. Soc. Agron.*, **36**, 373 (1944).

Martin, W. H.: The relation of sulphur to soil acidity and to the control of the potato scab, *Soil Sci.*, **9**, 393 (1920).

Matson, G. C.: The phosphate deposits of Florida, *U.S. Geol. Survey Bull.* 604, 1915.

Mayer, P.: "The Technology of Chilean Nitrate. World Conditions as to Mineral Raw Materials for the Fertilizer Industry," Nat. Fertilizer Assoc., **1926**, 154.

Maynard, T. P.: The potash-bearing slates of Georgia, *Mfrs'. Record*, Sept. 19, **1918**, 73.

McCool, M. M.: Comparison of agronomic value of the insoluble nitrogen derived from urea-ammonia liquor 37 and other sources, *Contrib. Boyce Thompson Inst.*, **11**, No. 6 (1941).

McGeorge, W. T.: The relation of potential alkalinity to the availability of phosphate in calcareous soils, *Soil Sci.*, **39**, 443 (1935).

———: Acidulated fertilizers for alkaline soil, *California Citro-grower*, **21**, 368 (1936).

———: Some modifications in the Neubauer method, *Soil Sci.*, **58**, 389 (1944).

———, and J. F. Breazeale: Phosphate solubility studies on some unproductive calcareous soils, *Arizona Agr. Exp. Sta. Bull.* 35, 1931.

———, and W. A. Frazier: Sulphur placement in the fertilization of alkaline soils, *Soil Sci. Soc. Am., Proc.*, **1939**, 288.

McHargue, J. S.: Manganese and plant growth, *Ind. Eng. Chem.*, **18**, 172 (1926).

———: The role of manganese in agriculture, *Soil Sci.*, **60**, 115 (1945).

———, W. S. Hodgkiss, and E. B. Offutt: The boron content of some important forage crops, vegetables, fruits and nuts, *J. Am. Soc. Agron.*, **32**, 622 (1940).

———, E. B. Offutt, and W. S. Hodgkiss: Spectroscopic methods for the determination of boron in plant and animal materials, *Soil Sci. Soc. Am., Proc.*, **4**, 308 (1939).

———, W. R. Roy, and J. G. Pelphrey: Iodine in some fertilizer materials, *Am. Fertilizer*, **73**, 40 (Nov., 1930).

———, D. W. Young, and R. K. Calfee: The effect of certain fertilizer materials on iodine content of important foods, *J. Am. Soc. Agron.*, **27**, 559 (1935).

McKaig, N., W. A. Carns, and A. B. Bowen: Fertilizer ratio experiment with soybeans, *South Carolina Agr. Exp. Sta. Rept.*, **1937**, 136.

McLarty, H. R.: British Columbia uses boron for fruit, *Canadian Dep. Agr. Sci. Service Contrib.*, No. 611, 1940.

McMurtrey, J. E.: Boron deficiency in tobacco under field conditions, *J. Am. Soc. Agron.* **27**, 271 (1935).

———: Distinctive plant symptoms caused by deficiency of any one of the chemical elements essential for normal development, *Botan. Rev.*, **4**, 183 (1938).

———: Plant nutrient deficiency in tobacco, "Hunger Signs In Crops," *Nat. Fertilizer Assoc.*, **1941**, 15.

McMurtrey, J. E., Jr., and W. O. Robinson: Neglected soil constituents that effect plant and animal development, *U.S. Dep. Agr. Yearbook*, **1938,** 807.

McVickar, M. H.: Manganese status of some important Ohio soil types and up-take of manganese by Kentucky bluegrass, *J. Am. Soc. Agron.*, **34,** 123 (1942).

Mehlich, A., E. B. Fred, and E. Truog: The Cunninghamella plaque method of measuring available phosphorus in soil, *Soil Sci.*, **38,** 445 (1934).

Mehring, A. L.: What the southern farmer pays for the filler in his fertilizer, *Com. Fertilizer Yearbook*, **1936,** 21.

———: Physiological acidity of commercial mixed fertilizers, 1929–1936, *Com. Fertilizer*, July, **1938,** 13.

———: High analysis fertilizers as related to The Victory Program, *Better Crops with Plant Food*, Jan., **1942,** 11.

———, E. Bailey, and H. M. Wallace: A survey of fertilizers grades and plant food consumption in the United States for the year ended June 30, 1943, *Proc. Nat. Fertilizer Assoc.*, **1944,** 31.

———, and G. A. Cumings: Factors affecting the mechanical application of fertilizers to the soil, *U.S. Dep. Agr. Tech. Bull.* 182, 1930.

———, L. S. Deming, and H. A. Willett: A survey of plant food consumption in the United States in the year ended June 30, 1939, *Bull. Nat. Fertilizer Assoc.*, 1940.

———, and A. J. Peterson: The equivalent physiological acidity or basicity of American fertilizers, *J. Assoc. Agr. Chemists*, **27,** 95 (1934).

———, and ———: Changes in composition of American fertilizers, 1880 to 1932, *U.S. Dep. Agr., Cir.* 315, 1934.

———, H. M. Wallace, and M. Drain: Nitrogen, phosphoric acid, and potash consumption in the United States, by years and by states, with preliminary figures for 1944, *J. Am. Soc. Agron.*, **37,** 595 (1945).

———, L. M. White, W. H. Ross, and J. E. Adams: Effects of particle size on the properties and efficiency of fertilizer, *U.S. Dep. Agr., Tech. Bull.* 485, 1935.

Merkle, F. G.: Soil testing, *Penn. Agr. Exp. Sta. Bull.* 398, 1940.

Merz, A. R.: Production and agricultural use of sodium nitrate, *U.S. Dep. Agr., Cir.* 436, 1937.

———: Selecting fertilizers, *U.S. Dep. Agr., Cir.* 487, 1940.

———, and C. C. Fletcher. Production and Agricultural use of sodium nitrate, *U.S. Dep. Agr., Cir.* 436, 1940.

Metzger, W. H.: The nature, extent, and distribution of fertilizer residues in the soil of some old fertility plats, *Soil Sci.*, **47,** 15 (1939).

Midgley, A. R.: Overliming acid soils, *J. Am. Soc. Agron.*, **24,** 822 (1932).

———, and D. E. Dunklee: The effect of lime on the fixation of borates in soil, *Soil Sci. Soc. Am., Proc.*, **1939,** 302.

Millar, C. E.: Hill fertilization of corn, *Mich. Agr. Exp. Sta. Quarterly Bull.*, **5,** No. 2, p. 69, Nov. 1922.

Millar, C. E., and J. F. Mitchell: Effect of rate and method of application of fertilizer on the germination of white beans, *J. Am. Soc. Agron.*, **19,** 270 (1927).

Miller H. G.: Relation of sulphates to plant growth and composition, *J. Agr. Research*, **17,** 87 (1919).

Mooers, C. A.: New fertilizers on the market for 1928. Suggestions for their use, *Tenn. Agr. Exp. Sta. Circ.* 18, 1928.

———: The comparative values of different phosphates, *Tenn. Agr. Exp. Sta. Bull.* 141, 1929.

———: An evolution of the Neubauer and the Cunninghamella and *Aspergillus niger* methods for the determination of the fertilizer needs of the soil, *Soil Sci.*, **46,** 211 (1938).

Morse, H. H.: The toxic influence of fluorine in phosphatic fertilizers on the germination of corn, *Ibid.*, **39,** 177 (1935).

Moser, F.: The adaptability of rapid chemical tests for use in determining the nutrient needs of South Carolina soils, *J. Am. Soc. Agron.*, **31,** 188 (1939).

Moss, E. G., J. E. McMurtrey, Jr., and W. M. Lunn: Fertilizer experiments with flue-cured tobacco, *North Carolina Dep. Agr., Bull.*, June, **1927,** 406.

Moyer, T. R., and A. W. Blair: Availability studies on calcium cyanamide and other nitrogenous materials, *New Jersey Agr. Exp. Sta. Ann. Rep.*, **1930,** 318.

Munsell, R. I., and B. A. Brown: The boron content of certain forage and vegetable crops, *J. Am. Soc. Agron.*, **35,** 401 (1943).

Muntz, A., and H. Gaudechon: Sur la diffusion des engrais salins dans la terre, *Compt. rend. acad. sci.* (*Paris*), **148,** 253 (1909).

Murphy, R. C.: "Bird Islands of Peru," New York, G. P. Putman's Sons, 1925.

Muscle Shoals Commission: Muscle Shoals, U.S. Printing Office, Washington, 1931.

N

Nagaoka, M.: On the action of various insoluble phosphates upon rice plants, *Bull. Coll. Agr. Imp. Univ. Tokyo*, **6,** 215 (1904).

Neller, J. R.: Phosphorus content and buffer capacity of plant sap as related to the physiological effect of phosphorus fertilizers in fibrous low-moor peat, *J. Agr. Research*, **51,** 287 (1935).

Nelson, C. E., and L. C. Wheeting: Fertilizer placement under irrigation in Washington, *J. Am. Soc. Agron.*, **33,** 105 (1941).

Neubauer, H., and W. Schneider: Die Nahrstoffaufnahme der Keimpflanzen und ihre Anwendung auf die Bestimmung des Nahrstoffgehalts der Boden, *Z. f. Pflanzenernähr.*, **1923,** 329.

Nightingale, G. T.: Physiological-chemical functions of potassium in crop growth, *Soil Sci.*, **55,** 73 (1943).

Nolan, T. B.: Potash brines in the Great Salt Lake Desert, Utah, *Geol. Survey Bull.* 795, 1928.

Noll, C. F.: The effects of phosphate on early growth and maturity, *J. Am. Soc. Agron.*, **15,** 87 (1923).

———: The comparative effects of different phosphates on yield and maturity. Fiftieth anniversary of the Jordan soil fertility plats, *Penn. Agr. Exp. Sta. Tech. Bull.* 273, 1932.

———, F. D. Gardner, and C. J. Irvin: Fiftieth anniversary of the general fertilizer tests, *Penn. Agr. Exp. Sta. Bull.* 264, 1931.

———, C. J. Irvin, and F. D. Gardner: Field experiments with phosphates, *Ibid.*, *Bull.* 315, 1935.

Nordengren, Sven, and H. Lehrecke: Possible development of the superphosphate industry, *Com. Fertilizer*, Aug., **1940,** 15.

O

O'Brien, R. L., *et al.:* "Chemical Nitrogen." U.S. Tariff Commission. Rep. 114, Ser. 2, 1937.

Olsen, S. R., and B. T. Shaw: Chemical Mitscherlich and Neubauer methods for determining available potassium in relation to crop response to potash fertilization, *J. Am. Soc. Agron.*, **35,** 1 (1943).

P

Paden, W. R.: Effect of nitrogen from various sources upon the yields of seed cotton on unlimed and limed soils, *S. Carolina Agr. Exp. Sta. Rep.*, **1939,** 34.

———: Relation of lime to efficiency of fertilizers, *Com. Fertilizer*, Nov., **1942,** 22.

Pardee, J. G.: The Garrison and Philipsburg phosphate fields, Montana, *Geol. Survey Bull.* 640, 1917.

Parker, F. W.: The availability of phosphoric acid in precipitated phosphates, *Com. Fertilizer*, **42,** 28 (May, 1931).

———: The importance of calcium and magnesium to the nitrogen and fertilizer industry, *Am. Fertilizer*, Jan., 1932.

———, and J. G. Fudge: The concentration of organic and inorganic phosphorus in the soil solution and the availability of organic phosphorus to plants, *Soil Sci.*, **24,** 119 (1927).

———, and W. H. Ross: Problems in formulating fertilizers for 1943–44, *Proc. Nat. Fertilizer Assoc.*, 1943.

———, and J. W. Tidmore: The influence of lime and phosphatic fertilizers on the phosphorus content of the soil solution and of soil extracts, *Soil Sci.*, **21,** 425 (1926).

Parks, R. A., C. B. Lyon, and S. L. Hood: Some effects of boron supply on the chemical composition of tomato leaflets, *Plant Physiol.*, **19,** 404 (1944).

Patterson, H. J.: Experiments upon the use of potash as a fertilizer, *Maryland Agr. Exp. Sta. Bull.* 89, 1903.

Peech, M., and Leah English: Rapid microchemical soil tests, *Soil Sci.*, **57,** 167 (1944).

Pember, F. R.: Studies of both pot and solution culture of the phosphorus and potassium requirements of the barley plant during its different periods of growth, *Rhode Island Agr. Exp. Sta. Bull.* 169, 1917.

Perkins, A. T., and H. H. King: Phosphatic fixation by soil minerals. II. Fixation by iron, silicon, and titanium oxides, *Soil Sci.*, **58,** 243 (1944).

Pettinger, N. A.: Effect of fertilizers on the chlorine content of the sap of corn plants, *J. Agr. Research*, **44,** 919 (1932).

Pierre, W. H.: Effect of nitrogenous fertilizers on soil acidity, *Ind. Eng. Chem.*, **23,** 1440 (1931).

———: Hydrogen-ion concentration, aluminum in the soil solution, and percentage base saturation as factors effecting plant growth on acid soils, *Soil Sci.*, **31,** 183 (1931).

———: Determination of equivalent acidity and basicity of fertilizers, *Ind. Eng. Chem.*, **5,** No. 4, 1933.

———: The effect of mono-, di-, and tricalcium phosphates on the reaction of soils of different degrees of acidity, *J. Am. Soc. Agron.*, **26,** 278 (1934).

———, and C. A. Bower: Potassium absorption by plants as effected by cationic relationships, *Soil Sci.*, **55,** 23 (1943).

———, and G. M. Browning: The temporary injurious effect of excessive acid soils and its relation to the phosphate nutrition of plants, *J. Am. Soc. Agron.*, **27,** 742 (1935).

———, and A. D. Stuart: Soluble aluminum studies. IV. The effects of phosphorus in reducing the detrimental effects of soil acidity on plant growth, *Soil Sci.*, **36,** 211 (1933).

Piland, J. R., C. F. Ireland, and H. M. Reisenauer: The importance of borax in legume seed production in the south, *Ibid.*, **57,** 75 (1944).

Pitz, W.: Effect of elemental sulphur and of calcium sulphate on certain of the higher and lower forms of plant life, *J. Agr. Research*, **5,** 771 (1915).

Piper, C. S.: The availability of manganese in the soil, *J. Agr. Sci. (England)*, **21,** 762 (1931).

Powers, W. L.: Boron as a fertilizer for western Oregon, *Soil Sci.*, **90,** 36 (1939).

Powers, W. L.: The minor elements in Oregon soil fertility and plant nutrition, *Oregon Agr. Exp. Sta. Cir.* 223, 1940.

Pridmore, J. C.: The goal of fertilizer consumption in the South, *Proc. Nat. Fertilizer Assoc.*, 1926.

Prince, A. L., and V. A. Tiedjens: Water solutions of mixed fertilizers, *Com. Fertilizer*, Jan., **1945**, 19.

———, and H. W. Winsor: The availability of nitrogen in garbage tankage and in urea in comparison with standard materials, *Soil Sci.*, **21,** 59 (1926).

Proul, E. C., *et al.*: Commercial fertilizers, *Indiana Agr. Exp. Sta. Bull.* 253, 11 (1921).

Purvis, E. R.: The present status of boron in American agriculture, *Soil Sci. Soc. Am., Proc.*, **1939,** 316.

———, and W. J. Hanna: Vegetable crops affected by boron deficiency in eastern Virginia, *Virginia Truck. Exp. Sta. Bull.* 105, 1940.

R

Raleigh, G. J.: Silicon in plant growth, *Soil Sci.*, **60,** 133 (1945).

Reed, H. J., and H. G. Hall: Report of Moses Fell Annex Farm, *Indiana Agr. Exp. Sta. Circ.* 172, 1930.

Reed, H. S.: Effects of zinc deficiency on cells of vegetative buds, *Am. J. Botan.*, **28,** 10 (1941).

Reeve, E., A. L. Prince, and F. E. Baer: The boron needs of New Jersey soils, *New Jersey Agr. Exp. Sta. Bull.* 709, 1944.

———, and J. W. Shive: Potassium-boron, and calcium-boron relationships in plant nutrition, *Soil Sci.*, **57,** 1 (1944).

Rehling, C. J., and E. Truog: Activated sludge, *Ind. Eng. Chem.*, **11,** 281 (1939).

Reimer, F. C., and H. V. Tartar: Sulphur as a fertilizer for alfalfa in southern Oregon, *Oregon Agr. Exp. Sta. Bull.* 163, 1919.

Richardson, H. L.: Studies on calcium cyanamide, III. *J. Agr. Sci.*, (*England*), **22,** 348 (1932).

Rigg, T., and H. O. Askew: Mineral content of some typical pastures in Waimea County, *New Zealand J. Agr.*, **38,** 304 (1929).

Robertson, D. L.: Fertilized ponds produce more fish, *Better Crops with Plant Food*, February, **1939,** 20.

Robinson, W. O.: The occurrence of rare earths in plants and soils, *Soil Sci.*, **56,** 1 (1943).

———, and G. Edgington: Minor elements in plants and some accumulator plants, *Soil Sci.*, **60,** 15 (1945).

Rogers, C. H., and J. W. Shive: Factors affecting the distribution of iron in plants, *Plant Physiol.*, **7,** 227 (1932).

Rogers, G. S.: The phosphate deposits of South Carolina, *Geol. Survey Bull.* 580, 1915.

Rogers, L. H., O. E. Gall, and R. M. Barnette: The zinc content of weeds and volunteer grasses and planted land covers, *Soil Sci.*, **47,** 237 (1939).

Romaine, J. D.: Consider plant food content of crop, *Better Crops with Plant Food*, March, **1940,** 6.

Ross, W. H., J. R. Adams, J. Y. Yee, and V. L. Goddy: Preparation of ammonium nitrate for use as a fertilizer, *Com. Fertilizer*, Nov. **1943,** 16.

———, ———, ———, C. W. Whittaker, and K. S. Love: Preparation of Ammonium Nitrate for use as a fertilizer, *U.S. Dep. Agr., Tech. Bull.* 912, 1946.

———, and A. L. Mehring: Influence of concentration as a factor in the retail price of fertilizers, *Ind. Eng. Chem.*, **12,** 430 (1934).

———, and ———: Plant food concentration as a factor in the retail prices of fertilizers, *Ind. Eng. Chem.*, **12**, 430 (1934).

———, and ———: Methods for reducing the retail cost of cotton fertilizers, *Com. Fertilizer Yearbook*, 1935, 1936.

———, and ———: Mixed fertilizers, *U.S. Dep. Agr. Yearbook*, **1938**, 522.

———, A. R. Merz, and C. R. Wagner: The recovery of potash as a by-product in the cement industry, *U.S. Dep. Agr. Bull.* 572, 1917.

———, and L. M. White: How to prevent fertilizer burn, *Better Crops with Plant Food*, **23**, 21 (1939).

Royster, P. H., *et al.:* Blast furnace process for the production of phosphatic and potassic fertilizer materials, *U.S. Dep. Agr. Tech. Bull.* 543, 1937.

Rubins, E. J., and F. E. Baer: Carbon-nitrogen ratios in organic fertilizer materials in relation to the availability of their nitrogen, *Soil Sci.*, **54**, 411 (1942).

Rudolfs, W.: Sulphur oxidation in inoculated and uninoculated green-sand mixtures and its relation to the availability of potassium, *Ibid.*, **14**, 307 (1922).

Rusche, A.: Beeinflussung der Keimfähigkeit verschiedener Kulturpflanzen durch Salzdüngung, *J. Landw.*, **60**, 305 (1912).

Russell, E. J.: The agricultural value of organic manures, *J. Ed. Agr.*, London, **26**, 228 (1919).

———: Plant nutrition and crop production, *Univ. Cal. Press.*, Berkeley, 1926.

———: Soil conditions and plant growth, New York, Longmans, Green & Co., 1915 and 1932.

S

Sackett, W. G.: A bacteriological method for determining soil deficiencies, *Proc. Nat. Fertilizer Assoc.*, **1930**, 47.

Salter, R. M.: About 40 per cent of plant food used by crops first year, *Fertilizer Rev.*, **6**, 10, (1931).

———, *et al.:* Improved placement of fertilizers in the hill for corn, *Ohio Agr. Exp. Sta., Bimonthly Bull.* 156, May-June, 1932.

———, and E. E. Barnes: The efficiency of soil and fertilizer phosphorus as affected by soil reaction, *Ohio Agr. Exp. Sta. Bull.* 553, 1935.

———, and C. J. Schollenberger: Farm manure, *Ibid., Bull.* 605, 1939.

Samuels, G., and C. S. Piper: Gray speck (manganese deficiency) disease of oats, *J. Agr. S. Australia*, **31**, 696 (1928).

Saussure, Theodore de: "Recherches chimiques sur la vegetation," Paris, 1804.

Sayre, C. B., and A. W. Clark: Changes in composition of granular and powdered fertilizers in the soil, *J. Am. Soc. Agron.*, **30**, 30 (1938).

———, and ———: Rates of solution and movement of different fertilizers in the soil and the effects of the fertilizers on the germination and root development of beans, *New York Agr. Exp. Sta. Tech. Bull.* 231, 1935.

Scarseth, G. D.: The mechanism of phosphate retention by natural alumino-silicate colloids, *J. Am. Soc. Agron.*, **27**, 596 (1935).

———: Methods of diagnosing plant nutrient needs, *Better Crops with Plant Food*, May, **1943**, 11.

———, and J. W. Tidmore: The fixation of phosphates by clay soils, *J. Am. Soc. Agron.*, **26**, 152 (1934).

Schollenberger, C. J., and F. R. Dreibelbis: Effect of cropping with various fertilizers, manure, and lime treatments upon the exchangeable bases of plot soils, *Soil Sci.*, **29**, 371 (1930).

Schreiner, O.: Organic phosphorus in soils, *J. Am. Soc. Agron.*, **15,** 117 (1923).

———: New factors in fertilizer practice, *Com. Fertilizer*, **41,** 24 (October, 1930).

———: Less common fertilizer elements, *Am. Fertilizer*, **74,** 48 (1931).

———, *et al.*: A beneficial organic constituent of soils; creatinine, *U.S. Dep. Agr. Bur. Soils, Bull.* 83, p. 44, 1911.

———, and M. S. Anderson: Determining the fertilizer requirements of soils, *U.S. Dep. Agr. Yearbook*, **1938,** 469.

———, and P. F. Dawson: Manganese deficiency in soils and fertilizers, *Ind. Eng. Chem.*, **19,** 400 (1927).

———, A. R. Merz, and B. E. Brown: Fertilizer materials, *U.S. Dep. Agr. Yearbook*, **1938,** 487.

———, and J. J. Skinner: Nitrogenous soil constituents and their bearing on soil fertility, *U.S. Dep. Agr., Bur. Soils, Bull.* 87, 1912.

———, and ———: Adaptation of fertilizers for cotton soils, *Am. Fertilizer*, **81,** 5 (1934).

Schultz, A. R., and W. Cross: Potash-bearing rocks of the Leucite Hills, *U.S. Geol. Survey Bull.* 512, 1912.

Schuster, C. E., and R. E. Stephenson: Sunflower as an indicator plant of boron deficiency in soils, *J. Am. Soc. Agron.*, **32,** 607 (1940).

Scott, L. E.: Boron nutrition of the grape, *Soil Sci.*, **59,** 55 (1944).

Scott, P. R., and W. C. Robertson: Gypsum or copi, *J. Dept. Agr. Victoria*, **24,** 65 (1926).

Scripture, P. N., and G. S. McHargue: Boron supply in relation to carbohydrate metabolism and distribution in the radish, *J. Am. Soc. Agron.*, **37,** 360 (1945).

Searralles, J. J.: Comparative efficiency of calcined phosphates, *Soil Sci.*, **44,** 175 (1937).

Shedd, O. M.: The relation of sulphur to soil fertility, *Kentucky Agr. Exp. Sta. Bull* 188, 1914.

Shelton, E. M.: Experiments with wheat, *Kansas Annual Rep.* 1888, p. 54, 1889.

Sherman, G. D.: The role of manganese in crop production. II. *Com. Fertilizer*, Dec., **1943,** 24.

———, and P. M. Harmer: Manganese deficiency of oats on alkaline organic soils, *J. Am. Soc. Agron.*, **33,** 1080 (1941).

———, J. S. McHargue, and W. S. Hodgkiss: The production of a lime-induced manganese deficiency on an eroded Kentucky soil, *Ibid.*, **34,** 1076 (1942).

Shive, J. W.: The effect of salt concentration on the germination of seeds, *New Jersey Exp. Sta. Rep.* 1916, p. 455, 1916.

———: The adequacy of the boron and manganese content of natural nitrate of soda to support plant growth in sand cultures, *New Jersey Agr. Exp. Sta. Bull.* 603, 1936.

Shutt, F. T.: The influence of grain growing on the nitrogen and organic matter content of the western prairie soils of Canada, *J. Agr. Sci.*, **15,** 162 (1925).

Singewald, J. T., and B. L. Miller: Genesis of the Chilean nitrate deposits, *Econ. Geol.*, **11,** 103 (1916).

Skeen, J. R.: Greensand as a source of potassium for green plants, *Am. J. Botan.*, **12,** 607 (1925).

Skinner, J. J.: Use of commercial fertilizers in the growing of cotton, *Com. Fertilizer*, **42,** 42 (March, 1931).

———: Results of fertilizer experiments on Norfolk fine sandy loam and on Norfolk sandy loam, *U.S. Dep. Agr. Tech. Bull.* 225, 1931.

———: Improved fertilizers for southern soils, *Am. Fertilizer*, **88,** No. 13; **89,** Nos. 1, 2 and 3, 1938.

———: Ammonium nitrate as a fertilizer for cotton *Com. Fertilizer*, Feb., **1944,** 48.

———, and J. H. Beattie: Influence of fertilizers and soil amendments on soil acidity, *J. Am. Soc. Agron.*, **9,** 25 (1917).

———, and T. S. Buie: Sources of ammonia, *S. Carolina Agr. Exp. Sta. Bull.* 227, 1926.

———, H. B. Mann, E. R. Collins, E. T. Batten, and R. P. Bledsoe: Adapting high analysis and concentrated fertilizers to cotton soils, *Soil Sci.*, **44,** 1 (1937).

———, and R. W. Ruprecht: Fertilizer experiments with truck crops, *Florida Agr. Exp. Sta. Bull.* 218, 1930.

———, C. B. Williams, and H. B. Mann: Effects of synthetic nitrogen and concentrated fertilizers on cotton and sweet potatoes, *N. Carolina Agr. Exp. Sta. Bull.* 266, 1929.

Slater, C. S., R. S. Holmes, and H. G. Byers: Trace elements in the soil from the erosion experiment stations, with supplementary data on other soils, *U.S. Dep. Agr. Tech. Bull.* 552, 1937.

Smalley, H. R.: Research and demonstrations—the foundation of progress, *Proc. Nat. Fertilizer Assoc.*, **1931,** 34.

———: Several plant foods should be considered, *Fertilizer Rev.*, **6,** 6 (Oct., 1931).

———, and R. H. Engle: Fertilizer use and its effect on crops in 1942, *Fertilizer Rev.*, **18,** 2 (1943).

———, and ———, and A. Willett: American fertilizer practices, *Nat. Fertilizer Assoc.*, 1939.

Smith, A. M.: Ammonium nitrate in mixed fertilizers, *Com. Fertilizer*, Oct., **1944,** 21.

Smith, C. L., A. O. Alben, and J. R. Cole: Progress report of pecan rosette control experiments in Texas, *Proc. Tex. Pecan Growers' Assoc. 14th Ann. Meeting*, **1934,** 41.

Smith, G. E., P. H. Heinze, and A. E. Murneek: Effect of soil moisture and rainfall on chemical transformations in Cyanamid granules, *Soil Sci.*, **55,** 313 (1943).

Smith, H. I.: Potash developments in southeastern New Mexico, Paper read at New York meeting of Am. Inst. Mining & Met. Eng., Feb., 1933.

Sommer, A. L.: Copper and plant growth, *Soil Sci.*, **60,** 71 (1945).

———, and C. B. Lipman: Evidence on the indispensable nature of zinc and boron for higher green plants, *Plant Physiol.*, **1,** 231 (1926).

Spencer, V. E., and F. M. Willhite: Phosphate studies. II. Chemical availability of phosphorus in various organic and inorganic carriers, as indicated by the Neubauer tests, *Soil Sci.*, **58,** 151 (1944).

Sprague, H. B., and E. E. Evaul: Experiments with turf grasses in New Jersey, *New Jersey Agr. Sta. Bull.* 497, 1930.

Stewart, G. R., E. C. Thomas, and J. Horner: The comparative growth of pineapple plants with ammonia and nitrate nitrogen, *Soil Sci.*, **20,** 227 (1925).

Stewart, R.: Sulphur in relation to soil fertility, *Illinois Agr. Exp. Sta. Bull.* 227, 1920.

———: Organo-phosphates as potential fertilizers, *Proc. Second Dearborn Conf. Agr. Ind., Sci.*, 1936, p. 8, Nov. 1940.

Stewart, S.: Biennial Report of the North Louisiana Experiment Station, Calhoun, Louisiana, 1935–1936.

Storey, H. H., and R. Leach: A sulphur-deficiency disease of the tea bush, *Ann. Applied Biol.* **20,** 23 (1933).

Stubbs, W. C., and J. G. Lee: Report of the North Louisiana Experiment Station, *Louisiana Agr. Exp. Sta.* (*First Series*) *Bull.* 27, 1890.

T

Thatcher, R. W., and A. C. Arny: The effect of different rotation systems and of fertilizers on the protein content of oats, *J. Am. Soc. Agron.*, **9,** 344 (1917).

Thomas, R. P.: The use of rapid soil tests in the United States, *Ibid.*, 28 **1936,** 411.

Thor, A. U.: The availability and downward movement of rock phosphate in Illinois soils when used liberally for 25 to 30 years and influence of the treatment on available potash and total nitrogen, *J. Am. Soc. Agron.*, **25,** 661 (1933).

Thorne, C. E., *et al.*: The maintenance of soil fertility, a quarter century's work with manure and fertilizers, *Ohio Agr. Exp. Sta. Bull.* 336, 1919.

Thorne, D. W., W. D. Laws, and A. Wallace: Zinc relationships of some Utah soils, *Soil Sci.*, **53,** 463 (1942).

Tidmore, J. W. and Williamson, J. T.: Experiments with commercial nitrogenous fertilizers. *Alabama Agr. Exp. Sta. Bull.* 238, 1932.

Tiedjens, V. A.: Improved quality and greater yields of vegetables through proper use of lime and fertilizer, *Com. Fertilizer*, **41,** 8 (Nov., 1940).

———: Bigger yields from fertilizer, *Better Crops with Plant Food*, Jan., **1945,** 23.

Tolman, B., and G. L. Stoker: Sulphur and nitrogen deficiency relationships in sugar beets grown for seed in Oregon, *J. Am. Soc. Agron.*, **33,** 1072 (1941).

Toth, S. J.: The stimulating effects of silicates on plant yields in relation to anion displacement, *Soil Sci.*, **47,** 123 (1939).

Tottingham, W. E.: The sulphur requirement of the red clover plant, *J. Biol. Chem.*, **36,** 429 (1918).

———, and C. Hoffman: Nature of the changes in the solubility and availability of phosphorus in fermenting mixtures, *Wisconsin Agr. Exp. Sta. Research Bull.* 29, 1913.

Trelease, S. F., and Helen M. Trelease: Selenium as a stimulating and possibly essential element for indicator plants, *Am. J. Botan.*, **25,** 372 (1938).

Truog, E.: The utilization of phosphates by agricultural crops, including a new theory regarding the feeding power of plants, *Wisconsin Agr. Exp. Sta. Research Bull.* 41, 1916.

———: Meeting agriculture's old and new problems with the aid of science, *Ibid., Bull.* 421, p. 33, 1932.

———, and H. J. Harper: Methods of applying fertilizer, *Ibid., Bull.* 352, p. 99, 1923.

———, and R. L. Jones: Fate of soluble potash applied to soils, *Ind. Eng. Chem.*, **30,** 882 (1938).

Turk, L. M.: Synthetic manure production in Michigan, *Mich. Agr. Exp. Sta. Bull.* 157, 1936.

Turrentine, J. W.: The potash resources of the United States, *Proc. Nat. Fertilizer Assoc.*, p. 78, 1926.

———: "Potash, a Review, Estimate and Forecast," New York, John Wiley & Sons, 1926.

———: "Potash in North America," A.C.S. Monograph No. 91, New York, Reinhold Pub. Corp., 1943.

———, C. W. Whittaker, and E. J. Fox: Potash from greensand, *Ind. Eng. Chem.*, **17,** 1177 (1925).

U

Ulrich, A.: Plant analysis as a diagnostic procedure, *Soil Sci.*, **5,** 101 (1943).

V

Vandecaveye, S. C.: Effects of soil type and fertilizer treatments on the chemical composition of certain forage and small grain crops, *Soil Sci. Soc. Am. Proc.*, **5,** 107 (1940).

Ville, M. G.: "Artificial Manures, 1879," Trans. by W. Crookes, Longmans Green & Co., London, 1909.

Vinson, A. E.: Bat guano, *Arizona Agr. Exp. Sta., Timely Hints for Farmers*, No. 109, 1915.

Voelcker, A.: On the composition of waters of land drainage, *J. Roy. Agr. Soc.*, Ser. 2, **10,** 132 (1874).

Volk, G. W.: The nature of potash fixation in soils, *Soil Sci.*, **45,** 263 (1938).

———: Availability of rock and other phosphate fertilizers as influenced by lime and form of nitrogen fertilizer, *J. Am. Soc. Agron.*, **36,** 46 (1944).

Volk, N. J., J. W. Tidmore, and D. T. Meadows: Supplements to high-analysis fertilizers with special reference to sulphur, calcium, magnesium, and limestone. *Soil Sci.*, **60,** 427 (1945).

W

Waggaman, W. H.: The manufacture of acid phosphate, *U.S. Dep. Agr. Bull.* 144, 1914.

———: The production and fertilizer value of citrate-soluble phosphoric acid and potash, *U.S. Dep. Agr. Misc. Pub.* 143, 1932.

Walker, J. C.: Histologic-pathologic effects of boron deficiency, *Soil Sci.*, **57,** 51 (1944).

———, J. P. Jolivette, and W. W. Hare: Varietal susceptibility in garden beets to boron deficiency, *Ibid.*, **59,** 461 (1945).

Wall, M. E.: The role of potassium in plants. III. *Ibid.*, **49,** 393 (1940).

Wander, I. W.: Horticulture at the Ohio Agricultural experiment station, *Ohio Agr. Exp. Sta. Special Cir.* 54, p. 24, 1938.

Warington, Katherine: The effect of boric acid and borax on the broad bean and certain other plants, *Ann. Botan.* (*London*), **37,** 630 (1923).

Warner, H. W.: Interpretation and popular presentation of data showing the economic results of fertilization, *J. Am. Soc. Agron.*, **23,** 266 (1931).

Watson, J. R.: Control of root-knot by calcium cyanamide and other means, *Florida Agr. Exp. Sta. Bull.* 136, 1917.

Weidemann, A. G.: Movement of lime in soils as determined by soil reaction. *Mich. Agr. Exp. Sta. Quart. Bull.* 18, p. 254, 1936.

Whathall, J. H., and G. L. Bridger: Fertilizer by fusion of rock phosphate with olivine, *Ind. Eng. Chem.*, **35,** 774 (1943).

Wheeler, H. J.: "Manures and Fertilizers," New York, The Macmillan Company, 1921.

Whetstone, R. R., W. O. Robinson, and H. G. Byers: Boron distribution in soils and related data, *U.S. Dep. Agr., Bur. Plant Ind. Tech. Bull.* 797, 1942.

White, H. C.: The feeding of cotton, *Georgia Agr. Exp. Sta. Bull.* 108, 1914.

White, J. W.: "Comparative Value of Sulphate of Ammonia and Nitrate of Soda," New York, Barrett Co., 1925.

———, and F. J. Holben: Development and value of Kentucky blue grass pastures, *Penn. Agr. Exp. Sta. Bull.* 195, 1925.

White, L. M., and W. H. Ross: Effect of the various grades of fertilizers on the salt content of the soil solution, *J. Agr. Research*, **59,** 81 (1939).

Whitson, A. R., and C. W. Stoddart: Phosphate content of soils as affected by methods of farming, *Wisconsin Agr. Exp. Sta., Ann. Rpt.* Part 2, **1909,** 41.

Wiancko, A. T.: Huntington experiment field, *Indiana Agr. Exp. Sta., Progress Report*, 1919–1928, 1929.

———, and Conner, S. D.: Acid phosphate vs. raw rock phosphate as fertilizer, *Indiana Agr. Exp. Sta. Bull.* 187, p. 1078, 1916.

Wilde, S. A., and R. Whittenkamp: Duff briquette fertilizers, *J. Am. Soc. Agron.*, **34,** 736 (1942).

Wilfarth, H.: Über Stickstoff Aufnahme der Pflansen. *Landw. Vers. Sta.*, **34,** 460 (1887).

Williams, C. B.: Composition of fertilizer affects maturity of cotton, *Phosphorus Digest*, February, **1931,** 9.

———, T. S. Buie, and S. V. Stacy: Effects of superphosphate upon the yield and earliness in maturity of cotton, *N. Carolina Agr. Exp. Sta. Bull.*, 314, 1937.

Williams, C. B., B. W. Kilgore, and A. R. Russell: Fertilizer experiments with corn on piedmont cecil sandy loam soil and varieties, culture, and fertilization of corn on piedmont cecil sandy loam, red clay soils, *Ibid.*, 229, 1915.

———, ———, and ———: Fertilizer experiments with cotton on piedmont cecil sandy loam, *Ibid.*, 227, 1914.

Williams, C. G.: The maintenance of soil fertility, *Ohio Agr. Exp. Sta. Bull.* 381, 1924.

Williamson, J. T.: Efficiency of ammoniated superphosphate for cotton, *J. Am. Soc. Agron.*, **27**, 724 (1935).

Willis, L. G., and H. B. Mann: Manganese as a fertilizer, *Am. Fertilizer*, **72**, 19 (1930).

———, and J. R. Piland: Ammonium calcium balance: A concentrated fertilizer problem, *Soil Sci.*, **31**, 5 (1931).

———, and ———: The influence of copper sulphate on iron absorption by corn plants, *Ibid.*, **37**, 79 (1934).

———, and ———: The function of copper in soil and its relation to the availability of iron and manganese, *J. Agr. Research*, **52**, 467 (1936).

———, and ———: A response of alfalfa to borax, *J. Am. Soc. Agron.*, **30**, 63 (1938).

———, ———, and R. L. Gay: The influence of magnesium deficiency on phosphate absorption by soybeans, *Ibid.*, **26**, 419 (1934).

———, and W. H. Rankin: Free ammonia injury with concentrated fertilizers, *Ind. Eng. Chem.*, **22**, 1405 (1930).

Wolf, B.: Experiments with boron on some New Jersey soils, *Soil Sci.*, **48**, 41 (1939).

———: Factors influencing availability of boron in soil and its distribution in plants, *Ibid.*, **50**, 209 (1940).

Wroth, J. S.: Commercial possibilities of the Texas-New Mexico potash deposits, *U.S. Dept. Commerce, Bur. Mines, Bull.* 316, 1930.

Y

Young, V. H., G. Janssen, and J. O. Ware: Cotton wilt studies, *Arkansas Agr. Exp. Sta. Bull.* 272, 1932.

Z

Zimmerley, H. H.: Discussion of machine fertilizer placement in Virginia, *Nat. Joint Committee Fert. Appl. Proc.*, **12**, 25 (1936).

Zimmerman, P. W.: Letter to author, Feb., 1941.

———: Present status of "plant hormones," *Ind. Eng. Chem.*, **35**, 596 (1943).

Index of Proper Names

S

T

U

V

W

Subject Index

B

D

E